# Regulation of Water Balance in Plants

# Regulation of Water Balance in Plants

Edited by:

Carlos Tello Lacal

Delve
PUBLISHING

www.delvepublishing.com

**Regulation of Water Balance in Plants**

*Carlos Tello Lacal*

**Delve Publishing**

2010 Winston Park Drive,

2nd Floor

Oakville, ON L6H 5R7

Canada

www.delvepublishing.com

Tel: 001-289-291-7705
001-905-616-2116

Fax: 001-289-291-7601

Email: orders@arclereducation.com



ISBN: 978-1-77361-047-4 (Hardcover)

# DECLARATION

Some content or chapters in this book are open access copyright free published research work, which is published under Creative Commons License and are indicated with the citation. We are thankful to the publishers and authors of the content and chapters as without them this book wouldn't have been possible.

# ABOUT THE EDITOR

## Carlos Tello Lacal

Carlos Tello (Valencia, Spain, 1981) studied Agricultural Engineering with specialization in Biotechnology at the Universidad Politécnica de Valencia, where he graduated with a BSc thesis on the characterization of an Arabidopsis mutant with increased tolerance to seed ageing and salt stress. He moved afterwards to Sevilla to complete his MSc and PhD at the Instituto de Recursos Naturales y Agrobiología, where he studied the mechanisms employed by plants in their adaptation to abiotic stress and more specifically the role of the Arabidopsis Na+/H+ antiporter SOS1 in the regulation of Na+ and K+ homeostasis. Since 2014, he is based in Zürich and develops several freelance science-related jobs. He has recently retrained himself in the field of cell culture in bioreactors by attending an advanced training course at the Zürcher Hochschule für Angewandte Wissenschaften.

# TABLE OF CONTENTS

# LIST OF CONTRIBUTORS

Carlos Tello

Instituto de Recursos Naturales y Agrobiología de Sevilla

**Xuejun Dong**

North Dakota State University, Central Grassland Research Extension Center, Streeter, USA

Texas A&M AgriLife Research, Uvalde, USA

**Supratim Basu**

Crop, Soil, and Environmental Sciences, University of Arkansas, Fayetteville, Arkansas, 72701, USA

**Venkategowda Ramegowda**

Crop, Soil, and Environmental Sciences, University of Arkansas, Fayetteville, Arkansas, 72701, USA

**Anuj Kumar**

Crop, Soil, and Environmental Sciences, University of Arkansas, Fayetteville, Arkansas, 72701, USA

**Andy Pereira**

Crop, Soil, and Environmental Sciences, University of Arkansas, Fayetteville, Arkansas, 72701, USA

**Tingting Mei**

Tropical Silviculture and Forest Ecology, Georg-August-University of Göttingen, Göttingen, Germany

**Dongming Fang**

Tropical Silviculture and Forest Ecology, Georg-August-University of Göttingen, Göttingen, Germany

**Alexander Röll**

Tropical Silviculture and Forest Ecology, Georg-August-University of Göttingen, Göttingen, Germany

**Furong Niu**

Tropical Silviculture and Forest Ecology, Georg-August-University of Göttingen, Göttingen, Germany

**Hendrayanto**

Department of Forest Management, Institut Pertanian Bogor, Bogor, Indonesia

**Dirk Hölscher**

Tropical Silviculture and Forest Ecology, Georg-August-University of Göttingen, Göttingen, Germany

**Marco Garrido**

PhD Program in Silvicultural and Veterinary Sciences, University of Chile, Santiago, Chile

**Paola Silva**

Soil-Plant-Water Relations Laboratory, Agricultural Production Department, Faculty of Agronomical Sciences, University of Chile, Santiago, Chile

**Edmundo Acevedo**

Soil-Plant-Water Relations Laboratory, Agricultural Production Department, Faculty of Agronomical Sciences, University of Chile, Santiago, Chile

**Claudia Cocozza**

Institute for Sustainable Plant Protection, National Research Council, Sesto Fiorentino, Italy

**Marina de Miguel**

BIOGECO, INRA, Univ. Bordeaux, Cestas, France

**Eva Pšidová**

Institute of Forest Ecology, Slovak Academy of Science, Zvolen, Slovak Republic

**L'ubica Ditmarová**

Institute of Forest Ecology, Slovak Academy of Science, Zvolen, Slovak Republic

**Stefano Marino**

Department of Agriculture, Environment and Food, University of Molise, Campobasso, Italy

**Lucia Maiuro**

Department of Agriculture, Environment and Food, University of Molise, Campobasso, Italy

**Arturo Alvino**

Department of Agriculture, Environment and Food, University of Molise, Campobasso, Italy

**Tomasz Czajkowski**

Johann Heinrich von Thünen Institute, Institute of Forest Ecosystems, Eberswalde, Germany

**Andreas Bolte**

Johann Heinrich von Thünen Institute, Institute of Forest Ecosystems, Eberswalde, Germany

**Roberto Tognetti**

Department of Biosciences and Territory, University of Molise, Pesche, Italy

The EFI Project Centre on Mountain Forests (MOUNTFOR), Edmund Mach Foundation, San Michele all'Adige, Italy

**Dhriti Singh**

National Institute of Plant Genome Research, New Delhi, India

**Ashverya Laxmi**

National Institute of Plant Genome Research, New Delhi, India

**Christoph Tim Krannich**

Institute of Biological Sciences, Department of Plant Genetics, University of Rostock, Albert-Einstein-Str. 3, 18059 Rostock, Germany

**Lisa Maletzki**

Institute of Biological Sciences, Department of Plant Genetics, University of Rostock, Albert-Einstein-Str. 3, 18059 Rostock, Germany

**Christina Kurowsky**

Institute of Biological Sciences, Department of Plant Genetics, University of Rostock, Albert-Einstein-Str. 3, 18059 Rostock, Germany

**Renate Horn**

Institute of Biological Sciences, Department of Plant Genetics, University of Rostock, Albert-Einstein-Str. 3, 18059 Rostock, Germany

**Agata Cieśla**

Biotechnology Department, Faculty of Biology, Adam Mickiewicz University, Poznań, Poland

**Filip Mituła**

Biotechnology Department, Faculty of Biology, Adam Mickiewicz University, Poznań, Poland

**Lucyna Misztal**

Biotechnology Department, Faculty of Biology, Adam Mickiewicz University, Poznań, Poland

**Olga Fedorowicz-Strońska**

Institute of Plant Genetics, Polish Academy of Science, Poznań, Poland

**Sabina Janicka**

Biotechnology Department, Faculty of Biology, Adam Mickiewicz University, Poznań, Poland

**Małgorzata Tajdel-Zielińska**

Biotechnology Department, Faculty of Biology, Adam Mickiewicz University, Poznań, Poland

**Małgorzata Marczak**

Biotechnology Department, Faculty of Biology, Adam Mickiewicz University, Poznań, Poland

**Maciej Janicki**

Biotechnology Department, Faculty of Biology, Adam Mickiewicz University, Poznań, Poland

**Agnieszka Ludwików**

Biotechnology Department, Faculty of Biology, Adam Mickiewicz University, Poznań, Poland

**Jan Sadowski**

Biotechnology Department, Faculty of Biology, Adam Mickiewicz University, Poznań, Poland

**Juan D. Franco-Navarro**

Institute of Natural Resources and Agrobiology, CSIC, Avda Reina Mercedes 10, 41012-Sevilla, Spain

**Javier Brumós**

Valencian Institute of Agrarian Research, Center of Genomics, Ctra Moncada-Náquera Km 4.6, 46113-Moncada, Valencia, Spain

Department of Plant and Microbial Biology, North Carolina State University, Raleigh, NC 27695-7614, USA

**Miguel A. Rosales**

Institute of Natural Resources and Agrobiology, CSIC, Avda Reina Mercedes 10, 41012-Sevilla, Spain

Biochemistry and Molecular Physiology of Plants (BPMP), UMR 5004 (CNRS/INRA/SupAgro/UM), 2 Place Viala, F-34060 Montpellier Cedex 2, France

**Paloma Cubero-Font**

Institute of Natural Resources and Agrobiology, CSIC, Avda Reina Mercedes 10, 41012-Sevilla, Spain

**Manuel Talón**

Valencian Institute of Agrarian Research, Center of Genomics, Ctra Moncada-Náquera Km 4.6, 46113-Moncada, Valencia, Spain

**José M. Colmenero-Flores**

Institute of Natural Resources and Agrobiology, CSIC, Avda Reina Mercedes 10, 41012-Sevilla, Spain

**Karolina M. Jarzyniak**

Laboratory of Plant Molecular Physiology, Department of Natural Products Biochemistry, Institute of Bioorganic Chemistry Polish Academy of Sciences, Poznań, Poland

Laboratory of Molecular Biology, Department of Biochemistry and Biotechnology, University of Life Sciences, Poznań, Poland

**Michał Jasiński**

Laboratory of Plant Molecular Physiology, Department of Natural Products Biochemistry, Institute of Bioorganic Chemistry Polish Academy of Sciences, Poznań, Poland

Laboratory of Molecular Biology, Department of Biochemistry and Biotechnology, University of Life Sciences, Poznań, Poland

**Christopher Hepworth**

Department of Animal and Plant Sciences, University of Sheffield, Sheffield, United Kingdom

**Carla Turner**

Department of Molecular Biology and Biotechnology, University of Sheffield, Sheffield, United Kingdom

**Marcela Guimaraes Landim**

Department of Molecular Biology and Biotechnology, University of Sheffield, Sheffield, United Kingdom

University of Brasília, Brasília, Federal District, Biological institute, Brazil

**Duncan Cameron**

Department of Animal and Plant Sciences, University of Sheffield, Sheffield, United Kingdom

**Julie E. Gray**

Department of Molecular Biology and Biotechnology, University of Sheffield, Sheffield, United Kingdom

**Anna M. Locke**

Department of Plant Biology, University of Illinois, Urbana, IL 61801, USA

**Donald R. Ort**

Department of Plant Biology, University of Illinois, Urbana, IL 61801, USA

Institute for Genomic Biology, University of Illinois, Urbana, IL 61801, USA

Global Change and Photosynthesis Research Unit, Agricultural Research Service, United States Department of Agriculture, Urbana, IL 61801, USA

# PREFACE

Water abundance is the major factor limiting distribution, growth and productivity of terrestrial plants. The Earth's atmosphere challenges plants with the contradictory needs of maximizing $CO_2$ uptake and minimizing water loss, which these organisms meet by tightly regulating their water uptake, long-distance transport and evapotranspiration rates. When challenged with drought stress, plants trigger a plethora of signaling mechanisms aimed at activating adaptive responses that help them cope with this adverse condition. Through the compilation of open access peer-reviewed articles, the present book provides a detailed overview of the strategies enabling plants to maintain their water homeostasis despite changing environmental conditions.

In Chapter number 1, the editor introduces the different subfamilies of plant aquaporins, explains the main mechanisms controlling their activity and describes their roles in regulating plant water homeostasis. The author of the study presented in Chapter 2 modified a mathematical model describing soil water infiltration and redistribution to include root growth and distribution, water uptake and effect of drought on this parameter, and soil evaporation. In Chapter 3, rice is employed as an example to dissect the different adaptive mechanisms that plants have developed against drought stress. The next three chapters explore the natural variation of traits conferring drought tolerance. In Chapter 4, sap flux measurement is employed to investigate the water use patterns of four different tropical bamboo species. Chapter 5 investigates the water relations and leaf composition of the desert phreatophyte Prosopis tamarugo grown at three different levels of water table depth. In Chapter 6, traits such as gas exchange, photosynthetic rate or leaf structure are compared for different beech (Fagus sylvatica) seedling populations subjected to drought stress. Chapters 7 and 8 deepen in the molecular basis of drought tolerance. While the former reviews the main transcription factor families involved in the activation of drought-responsive genes, the latter explores the molecular mechanisms leading to the onset of drought tolerance in order to identify candidate genes that can be employed in breeding programs. Through a reverse genetics approach, the authors of the study described in Chapter 9 functionally characterized a protein kinase functioning as a negative regulator of drought signaling in barley. The role of the micronutrient Cl- in regulating plant water balance is revealed by the authors of the study presented in Chapter 10. Chapter 11 reviews the contribution of ABA, ion, water, cuticular component and osmolite transporters to drought tolerance. Authors of the paper presented in Chapter 12 studied water balance by manipulating the environmental conditions controlling root and stomatal development. The study shown in Chapter 13 investigates soybean leaf hydraulic conductance at several growth conditions and leaf developmental stages. The last block of four chapters explores the mechanisms governing stomatal movements in response to changes in water status.

CHAPTER

1

# PLANT AQUAPORINS AND REGULATION OF WATER BALANCE

Carlos Tello

Instituto de Recursos Naturales y Agrobiología de Sevilla

## INTRODUCTION

The aerial part of plants absorbs $CO_2$ from the atmosphere and releases water vapor through stomata on the leaf epidermis, being the uptake of water by the roots and its long-distance transport to the shoot through vessels necessary for the latter. Water from the soil solution flows radially across the root epidermis, cortex, endodermis, pericycle, xylem parenchyma and vessels. In the radial and transverse walls of the exo- and endodermis, the deposition of lignin and/or suberin leads to the formation of the Casparian strip. While it constitutes an impediment to water flow in the first case, its presence at the endodermis limits the permeability of solutes into and out of the stele, which cannot continue to be transported across the cell wall (apoplastic path) and need to enter the symplasm through plasmodesmata (symplastic path) or across membranes (transcellular path) (Steudle and Peterson, 1998;

Zimmermann and Steudle, 1998; Tester and Leigh, 2001). In the vascular tissues, water moves to the aerial part through the xylem tracheids. The driving force for this movement is water evaporation in the leaves, which creates negative pressures in the xylem vessels. When the stomata are closed, plants cannot transpire and solutes are actively pumped into the root to create an osmotic driving force (Steudle, 2001). In addition to stomatal movement, changes in the hydraulic conductivity of the root and leaves are used by plants as responses to alterations in their water status.

Although water diffusion through the lipids of the plasma membrane was initially believed to be solely responsible for cell water exchange, the functional characterization of plant aquaporins after the identification of the Major Intrinsic Protein (MIP) family in the plasma membrane of mammalian lens fibers revealed the presence of water channels in plants and boosted the research in membrane transport systems and plant water balance (Gorin *et al.*, 1984; Preston *et al.*, 1992; Maurel *et al.*, 1993; Maurel, 1997). Nowadays, aquaporins are not only recognized to transport water, but also other molecules such as $CO_2$, B, Si or $H_2O_2$, proving their multiple functions in plant physiology (Maurel *et al.*, 2015). The present article introduces the different families of plant MIPs, describes their molecular structure and regulation mechanisms, and reviews their function in controlling the water balance of the plant.

## PHYLOGENY OF PLANT AQUAPORINS

MIPs are ubiquitous channels that transport water and solutes such as glycerol, urea, ammonia, metalloids and $CO_2$ across cell membranes and are represented in mammals, plants and microbes (Kruse *et al.*, 2006; Maurel *et al.*, 2008; Carbrey and Agre, 2009). MIPs can be divided in two subfamilies: water transporters or aquaporins (AQPs) and glycerol transporters or aquaglyceroporins (GLPs). Five conserved amino acids termed P1-P5 are responsible for this discrimination, being P2 (Asp) the most important one given its role in regulating pore size (Froger *et al.*, 1998; Hub *et al.*, 2008). While bacteria and archaea normally contain only one member of each type, gene duplication events followed by mutagenesis during the evolution of eukaryotic genomes have led to the functional diversification of this family of proteins (Zardoya, 2005).

In plants, the first member identified was GmNod26, an aquaporin present in the peribacteroid membrane of symbiotic nitrogen-fixing soybean roots (Fortin *et al.*, 1987). Genome-wide analyses have identified 35 genes coding for these proteins in Arabidopsis, 33 in rice, 36 in maize, 37 in tomato, 53 in Chinese cabbage, 47 in banana and 40 in chickpea (Chaumont *et al.*, 2001; Johanson *et al.*, 2001, Sakurai *et al.*, 2005; Sade *et al.*, 2009; Tao *et al.*, 2014; Hu *et al.*, 2015; Deokar and Tar'an, 2016). Plant MIPs, all of them belonging to the AQP subfamily, can be classified in five groups based on their sequence: Plasma membrane Intrinsic Proteins (PIPs), Tonoplast Intrinsic Proteins (TIPs), NOD26-like Intrinsic Proteins (NIPs), Small basic Intrinsic Proteins (SIPs) and X Intrinsic Proteins (XIPs). Mosses have Hybrid Intrinsic Proteins (HIPs) and GlpF-like Intrinsic Proteins (GIPs) in addition to the previously introduced groups, while algae contain PIP and GIP homologs, as well as five MIP classes A-E relatively unrelated to plant aquaporins (Danielson and Johanson, 2008; Anderberg *et al.*, 2011, 2012; Abascal *et al.*, 2014).

The most abundant groups, PIPs and TIPs are targeted to the plasma membrane and tonoplast, respectively. The conservation of two major PIP subclasses, which are shared by plants, mosses and algae, and five different TIP groups throughout all higher plants suggests the early divergence of these families during the plant genome evolution (Johanson *et al.*, 2001; Quigley *et al.*, 2001; Abascal *et al.*, 2014). While the function of the PIP2 group in mediating water flow across the plasma membrane is clear, several members of the PIP1 class showed lack of water transport and expression profiles differing from those of PIP2 (Suga *et al.*, 2001, 2002; Tournaire-Roux *et al.*, 2003; Suga and Maeshima, 2004; Santoni *et al.*, 2006; Törnroth-Horsefield *et al.*, 2006). Although its physiological significance remain unclear, PIPs and TIPs have also been studied with respect to their glycerol transport capacity (Biela *et al.*, 1999; Gerbeau *et al.*, 1999). TIP groups 1, 2 and 4 have been shown to be located at the membrane of the central vacuole. Among them TIP1s and TIP2s have been shown through immuno-cytochemical assays to be preferentially associated to large lytic vacuoles and vacuoles accumulating storage proteins, respectively. One member of TIP1, AtTIP1;1, accumulates in intravacuolar double plast membrane invaginations called bulbs. TIP3s are localized to the vacuole-derivative protein body. The only TIP5 member characterized so far, AtTIP5;1, is expressed in pollen mitochondria and germinating seeds (Jauh *et al.*, 1998; Jiang *et al.*, 2001; Maeshima, 2001; Saito *et al.*, 2002; Vander-Willigen *et al.*, 2006; Maurel, 2007; Soto *et al.*, 2010). Additional substrates such as $H_2O_2$ and $CO_2$ have been identified for PIPs (Hooijmaijers *et al.*, 2012; Bienert and Chaumont, 2014), while some characterized TIPs transport $NH_3$ and urea (Gerbeau *et al.*, 1999; Jahn *et al.*, 2004; Loque *et al.*, 2005). Although all NIPs are close homologues of the soybean peribacteroid membrane-located GmNod26, members of this group are also present in plasma and intracellular membranes of non-leguminous plants (Ma *et al.*, 2006; Mituzani *et al.*, 2006; Takano *et al.*, 2006; Wallace *et al.*, 2006). Phylogenetic studies have speculated that the origin of this group lies in the horizontal gene transfer of a bacterial homolog of *AqpZ*, followed by convergence evolution towards the clade of aquaglyceroporins (Zardoya *et al.*, 2002; Maurel *et al.*, 2015). All NIPs characterized so far show reduced water transport activity and are instead more permeable to small organic solutes and mineral nutrients such as beneficial (B, Se, Si) and toxic (As, Sb) metalloids (Ma *et al.*, 2006; Takano *et al.*, 2006; Bienert et al., 2007; Zhao *et al.*, 2010). The SIP group comprises only a few isoforms in seed plants (three and two in Arabidopsis and rice, respectively), but is also found in mosses and spike mosses. SIPs are characterized by extremely long branches, short N-terminal tails and ER localization. Because phylogenetic analyses have shown that algal SIPs are basal to their land plant counterparts, this group is believed to be either a true ancient plant family with members independently lost in certain Chlorophyta lineages or the product of horizontal gene transfer from algae. Their functional characterization has only shown moderate water transport capacity so far (Johanson and Gustavsson, 2002; Danielson and Johanson, 2008; Maeshima and Ishikawa, 2008; Anderberg *et al.*, 2012). The more recently discovered XIPs have been revealed as multifunctional channels permeable to water, metalloids and reactive oxygen species (ROS). Originally identified in *Physcomitrella* and dicots, they have later been reported in *Selaginella*, Dictyosteliida and fungi. Their evolutionary loss from plant lineages such as the Brassicaceae or the monocots points to functional redundancy (Danielson and Johanson, 2008; Gupta and Sankararamakrishnan, 2009; Bienert *et al.*, 2011; Anderberg *et al.*, 2012; Lopez *et al.*, 2012).

## OVERALL STRUCTURE

The aquaporin family presents highly conserved structural features, as shown by cryo-electron microscopy on 2D crystals of PvTIP3;1 and X-ray crystallography on 3D crystals of SoPIP2;1 (Daniels *et al*., 1999; Fotiadis *et al*., 2001; Kukulski *et al*., 2005). Aquaporins are 23-31 kDa proteins assembling as tetramers, each of whose monomers forms a pore. Monomers comprise six membrane-spanning α-helix domains tilted along the plane of the membrane and linked by five loops A to E. While loops B and D, in addition to the carboxy-terminal ends are always exposed to the cytosol, extracellular loops A, C and E face the apoplasm of the plasma membrane, the vacuolar space or the lumen or the secretory system. Loops B and E, which contain the conserved Asn-Pro-Ala (NPA) motif, dip from either side of the membrane into the center of the molecule to form a 7$^{th}$ transmembrane domain. Together with the other transmembrane regions, they contribute to the delineation of the pore. Close to the extracytosolic mouth of the pore, four conserved residues form an aromatic/arginine (Ar/R) constriction, which together with stereospecific recognition of the substrate mediated by H-bonding and hydrophobic interactions within the pore, functions as a selectivity filter. Aided by electrostatic repulsion, dipole orientation and transient isolation of the water molecule during its flux through the pore, aquaporins remain impermeable to protons (Fujiyoshi *et al*., 2002; Beitz *et al*., 2006; Hedfalk *et al*., 2006; Törnroth-Horsfield *et al*., 2006).

Aquaporin subfamilies can be distinguished by their specific pore profile, which determines the selectivity of the channel. NIPs have three major subclasses. A single amino acid residue at the transmembrane spanning domain 2 determines that class I NIPs (containing Ala) are permeable to larger solutes at the expenses of water transport, while those belonging to class II (containing Trp) have opposite properties. Class III NIPs allow for the passage of large diameter solutes such as salicylic acid (SA) owing to the presence of small size residues in the Ar/R selectivity filter (Wallace and Roberts, 2005; Mitani *et al*., 2008; Rougé and Barre, 2008). In TIPs, a Val to Ile substitution at loop E regulates the range of transported solutes. Similarly, an Ile residue at loop E confers permeability to $CO_2$ through interaction with a Leu residue at loop C in barley PIPs (Azad *et al*., 2012; Mori *et al*., 2014). A central pore seems to be formed at the center of the aquaporin, between the four monomers.

The observation of this structure at the central region of spinach SoPIP2;1 suggests the potential permeation of $CO_2$ through this pathway from a steric point of view. In wheat TIP2;2, the differential inhibition by $Hg^{2+}$ of water and $NH_3$ transport was interpreted as these substrates having preferred permeation pathways through the four monomer or the central pores, respectively (Bertl and Kaldenhoff, 2007; Frick *et al*., 2013). Although the differential conservation of residues Ala103, Ser190, Ala194, Phe208 and Trp209, normally referred to as P1-P5 in *E. coli AqpZ*, suggested their role in dictating the specificity of the channel for one of these substrates, site-directed mutagenesis studies did not lead to selectivity alteration, indicating that more complex evolutionary patterns must be considered (Froger *et al*., 1998; Lagrée *et al*., 1999; Savage *et al*., 2010).

## MOLECULAR MECHANISMS OF AQUAPORIN REGULATION

### Co- and Posttranslational Modifications

During the last 15 years, the development of proteomics and mass spectrometry techniques has enabled the thorough study of cotranslational and posttranslational modifications of aquaporins in spite of the high hydrophobicity of these proteins. The wide variety of modifications thus far identified includes phosphorylation, glycosylation, methylation, deamidation, amino-terminal acetylation and ubiquitination (Santoni *et al*., 2003, 2006; Nühse *et al*., 2004; Daniels and Yeager, 2005; Prado *et al*., 2013).

A general role in aquaporin gating and trafficking has been established for phosphorylation, which can take place as a response to conditions such as drought, salinity or oxidative stress (Santoni *et al*., 2003; Prak *et al*., 2008). PIP2s have been demonstrated to carry several conserved phosphorylation sites in their cytosolic B and D loops, as well as in the carboxy-terminal tail. A model based on X-ray crystallographic studies of spinach SoPIP2;1 proposed that phosphorylation of carboxy-terminal Ser274 would prevent the transition of adjacent monomers to closed-pore conformation, while that of B loop Ser115 would release loop D thereby opening the channel by disrupting its anchoring to the amino-terminal end. Phosphorylation of both residues of this protein and its tulip homolog increased channel activity upon heterologous expression in *Xenopus* oocytes (Johanson *et al.*, 1998; Törnroth-Horsefield *et al.*, 2006; Azad *et al.*, 2008; Nyblom *et al.*, 2009). A putative regulation site at the D loop Ser230 of maize ZmPIP2;1 was identified after the Ser230Ala mutant showed reduced water transport activity in oocytes. Interestingly, its homolog in SoPIP2;1 displayed threefold increased water transport in proteoliposomes when mutated to Glu, which mimicks constitutive phosphorylation (Van Wilder *et al.*, 2008; Nyblom *et al.*, 2009). In Arabidopsis, phosphoproteomic analyses identified phosphorylatable Ser and Thr residues at the amino- and carboxy-terminal tails of PIP2;1, among which Ser280 and Ser283 have been demonstrated to require this modification to enhance leaf hydraulic conductivity in the absence of light (Niittylä *et al.*, 2007; Prado *et al.*, 2013). Phosphorylation of Ser283, but not 280, directs internalized AtPIP2;1 to fuzzy structures or spherical bodies under salt stress and is required for the targeting of newly synthesized proteins to the PM (Prak *et al.*, 2008). However, the nature of the protein kinases and phosphatases involved in aquaporin reversible phosphorylation remains elusive and is expected to be revealed through large-scale interactome studies. A large-scale analysis of sucrose-induced phosphorylation led to the identification of SIRK1, which phosphorylates PIP2s at their carboxy-terminal end when starved plants are resupplied with sucrose. In turn, the kinases phosphorylating Ser115 and Ser274 of SoPIP2;1 were identified through anion exchange and peptide affinity chromatography, respectively. Recently, receptor-like kinases (RLK) RKL1 and Feronia were shown to modulate PIP activity in Arabidopsis. Similarly, leucine rich repeat (LRR)-RLK LP2 (Leaf Panicle 2) was shown to bind to rice aquaporins OsPIP1;1, OsPIP1;3 and OsPIP2;3, although the functional role of these interactions remains unknown (Sjovall-Larsen *et al.*, 2006; Wu *et al.*, 2013; Jones *et al.*, 2014; Wu *et al.*, 2015; Bellati *et al.*, 2016).

Glycosylation has only been identified in a few members of the aquaporin family, such as soybean GmNOD26 and ice plant McTIP1;2 (Miao *et al.*, 1992; Vera-Estrella *et al.*, 2004). In AtPIP2;1, the adjacent Lys3 and Glu6 of the amino-terminal cytosolic tail are di- and monomethylated, respectively, and overlap with a diacidic residue involved in ER export of the channel, suggesting a role for methylation in protein trafficking. However, the dominant nature of any mutations in these methylation sites hinders structure-function analyses (Santoni *et al*., 2006). The mechanisms leading to deamidation in Arabidopsis remain unknown, although several stimulus-induced changes have been reported in roots (Di Pietro *et al*., 2013). N-acetylation of McTIP1;2 is involved in redistributing the protein to the endosomal compartments. Both Arabidopsis PIP subclasses undergo this modification as part of their maturation process, which may determine protein stability (Vera-Estrella *et al*., 2004; Santoni *et al*., 2006). The hot pepper E3 ligase was heterologously expressed in Arabidopsis in order to study the role of ubiquitination in aquaporin function. This protein, together with an Arabidopsis analog, were capable of mediating ubiquitination of AtPIP2;1 and favoring its retention in the ER accompanied by proteasome-dependent degradation (Lee *et al*., 2009).

## Assembly

As described in section 2, aquaporins assemble in tetramers of subunits in which loops B and E delimit the central pore, being the intermolecular interactions leading to oligomer formation as yet unknown. In maize, loop E and a highly conserved Cys residue of loop A have been demonstrated to be involved in this process (Fetter *et al*., 2004; Bienert *et al*., 2012). Through FRET/fluorescence life-time imaging microscopy (FLIM) coupled with co-immunoprecipitation assays, PIP1 and PIP2 monomers were shown to form molecular interactions, pointing to the existence of heterotetramers and/or interactions between homotetramers of both subclasses, which could alter their functional properties (Zelazny *et al*., 2007). Coexpression of maize, tobacco, barley and strawberry PIP1s and PIP2s in *Xenopus* oocytes and maize protoplasts showed that, whereas PIP1 homotetramers were retained intracellularly, the interaction of PIP2 allowed for their targeting to the PM. Alternatively, PIP1 seemed to alter the sensitivity of the heterotetramer to pH and cause PIP2 transactivation (Fetter *et al*., 2004; Zelazni *et al*., 2006; Bellati *et al*., 2010; Horie *et al*., 2011; Yaneff *et al*., 2014). In yeast, heterotetramerization was shown to affect the overall sensitivity of the tetramer to protons and alter substrate specificity, which suggests its role in the fine adjustment of PIP functions at the PM (Otto *et al*., 2010; Yaneff *et al*., 2014). In Arabidopsis, overexpression of a PIP2 mutant form deficient in trafficking to the PM retained native PIPs and consequently reduced water transport, which could be caused by molecular interactions between PIPs (Sorieul *et al.*, 2011).

## Trafficking

The regulation of aquaporin trafficking plays an essential role in their function and PIP1s have been shown to present very low water transport upon expression in *Xenopus* oocytes, possibly due to lack of targeting to the PM in the absence of coexpression with PIP2s (Fetter *et al.*, 2004; Zelazny *et al.*, 2007). In some Arabidopsis and maize PIP2s, a diacidic motif at the N-terminus acts as an ER export signal, presumably by recruiting Sec24 and inducing COPII vesicle formation, much like what was seen for the $K^+$ channel KAT1. Because ZmPIP2;1 and ZmPIP2;2 can reach the PM even though they lack the diacidic motif while ZmPIP1;2 cannot despite carrying it, additional motifs involved in ER export are hypothesized (Mikosch *et al.*, 2006; Zelazny *et al.*, 2009; Hachez *et al.*, 2013). Stimulus-induced relocalization of aquaporins was first suggested for plants when osmotic stress in ice plant and salt stress in Arabidopsis led to the targeting of TIPs to endosomal compartments or vacuolar bulbs, respectively, while PIPs were found to accumulate intracellularly upon root treatment with salt and SA. The latter could be counteracted through treatment with catalase, suggesting the role of ROS triggered by these treatments in PIP relocation. More recently, $H_2O_2$ was shown to trigger AtPIP2;1 sequestration in the late endosomal compartments, which was suggested to act as a mechanism protecting aquaporins from oxidative damage (Vera-Estrella *et al.*, 2004; Boursiac *et al.*, 2005, 2008; Wudick *et al.*, 2015). As described above, phosphorylation of Ser283 provides a signal redirecting the internalized protein to fuzzy structures or spherical bodies, while salt induces its dephosphorylation (Prak *et al.*, 2008). Alternatively, auxin prevents the intracellular accumulation of AtPIP2;1 by counteracting the action of BFA (Paciorek *et al.*, 2005). Pharmacological and genetic studies have shown that aquaporins continuously cycle between the PM and the trans-Golgi network, in a process which is favored by clathrin-mediated endocytosis and blocked by the auxin analog NAA. Under osmotic and salt stress, the cycling rates of PIPs is increased and they are partially internalized, which reduces their abundance at the root cell surface and ultimately decreases water permeability. Osmotic stress also alters TIP targeting (Vera-Estrella *et al.*, 2004; Boursiac *et al.*, 2005, 2008; Dhonukshe *et al.*, 2007; Li *et al.*, 2011a; Luu *et al.*, 2012; Martinière *et al.*, 2012; Hosy *et al.*, 2015). The effect of salt stress in increasing AtPIP2;1 lateral mobility was speculated to be caused by its enhanced recruitment into the t-SNARE endocytosis machinery. The complex formed by SNARE protein AtSYP61 and AtSYP121 regulates PM water permeability by modulating the post-Golgi trafficking of AtPIP2;7, while the maize orthologue of AtSYP121 physically interacts with ZmPIP2;5 and targets it to the PM. By employing specific inhibtors, salt stress-mediated internalization of AtPIP2;1 from the PM to the vacuolar lumen was suggested to be dependent on chlatrin and kinases PI3K and PI4K (Bar-On *et al.*, 2012; Besserer *et al.*, 2012; Hachez *et al.*, 2014; Ueda *et al.*, 2016). SYP121 can form triple complexes with aquaporins and potassium channels, which could contribute to the regulation of cell turgor (Sutter *et al.*, 2006; Grefen *et al.*, 2010). Finally, membrane shape also plays a role in transmembrane protein targeting. While the mammalian aquaporin AQP10 is equally concentrated in flat and curved membranes, the potassium channel KvAP prefers tubular and highly curved membranes, being membrane tension a critical factor determining its mobility (Aimon *et al.*, 2014; Quemeneur *et al.*, 2014)

## Gating

The opening and closing of aquaporin pores, referred to as gating, is regulated by multiple factors such as phosphorylation, $H^+$ and divalent cations (Gerbeau *et al.*, 2002; Guenther *et al.*, 2003; Verdoucq *et al.*, 2008). The model describing the regulation of SoPIP2;1 gating by phosphorylation has been presented above (see section 4.a). Additionally, functional expression in *Xenopus* oocytes, pharmacological alteration of kinases and phosphatases and stopped-flow measurements established a role for phosphorylation in regulating the channel activity of soybean GmNOD26 and bean PvTIP3;1 (Maurel *et al.*, 1995; Guenther *et al.*, 2003). The reconstitution of AtPIP2;1 in proteoliposomes showed full channel inhibition when the pH dropped from 8.0 to 6.0. Functional expression of SoPIP2;1 in *Xenopus* oocytes allowed for the dissection of the mechanism by which $H^+$ regulate aquaporin gating: at low pH, the conserved His193 at loop D is charged and interacts with Asp28 and Glu31 of this domain, as well as with Ser115 of loop B, to maintain loop D in a closed pore conformation. (Tournaire-Roux *et al.*, 2003; Törnroth-

Horsefield *et al.*, 2006; Verdoucq *et al.*, 2008; Frick *et al.*, 2013). Although the molecular mechanisms remain unknown, $H^+$-independent inhibition of TIPs through a His residue at the second extracytoplasmic loop C has been reported (Soto *et al.*, 2010; Leitao *et al.*, 2012). Divalent cations bind to the target residue and mediate H-bonding between Gly20 and Asp31 at the amino-terminal region and loop D through loop B, which stabilizes the closed pore conformation. More recently, a second binding site between loop D and the carboxy-terminal end of SoPIP2;1 was identified (Alleva *et al.*, 2006; Frick *et al.*, 2013). In AtNIP7;1, the closed channel conformation is stabilized by Tyr81 of the transmembrane spanning domain 2, which interacts with Arg220 of the Ar/R filter. In mammalian aquaporin-0 and bacterial aquaporin-z, Arg189 regulates channel gating by adopting different conformations (Hedfalk *et al.*, 2006; Li *et al.*, 2011b). Other factors that apparently regulate aquaporin gating include osmotic or hydrostatic pressures and plasma membrane lipid composition (Moshelion *et al.*, 2004; Wan *et al.*, 2004; Ye *et al.*, 2004; Tong *et al.*, 2012, 2013). The aquaporin blocker most commonly employed in pharmacological assays is $Hg^{2+}$, which is believed to inhibit water permeability of these channels by binding to the thiol groups of essential Cys residues, as well as to some target His. However, its high cellular toxicity and the fact that it does not inhibit aquaporins lacking the critical Cys residues and even activates SoPIP2;1 (Daniels *et al.*, 1994; Maggio and Joly, 1995; Zhang and Tyerman, 1999; Frick *et al.*, 2013; Katsuhara *et al.*, 2014) suggest the need to employ alternative blockers such as weak acids, respiratory inhibitors (azide, cyanide), silver or gold compounds (Niemitz and Tyerman, 2002, Sutka *et al.*, 2011).

## REGULATION OF WATER BALANCE BY AQUAPORINS

### Root Water Uptake

As described in the introduction, water uptake by roots is followed by its radial transport until the xylem vessels across the different root tissues through the apoplastic, symplastic and transcellular paths. A model was developed to describe water transport in both transpiring and root pressure conditions, as well as the changes in water permeability ($Lp_r$)

under different environmental conditions (Steudle and Peterson, 1998; Maurel *et al.*, 2015). Because the transcellular path, partly contributed by aquaporins, is difficult to distinguish experimentally from its symplastic counterpart, with which it forms the cell-to-cell pathway, aquaporin blockers such as $Hg^{2+}$ were used to determine the contribution of these channels to water uptake, which was found to represent up to 57% and 64% of $Lp_r$ in tomato and Arabidopsis, respectively. The use of other blockers such as azide or weak acids, which showed a similar inhibition profile, confirmed these estimations (Maggio and Joly, 1995; Sutka *et al.*, 2011). Expression studies have focused on PIPs and TIPS due to their high abundance and role in water transport. Gene microarrays and RT-PCR of all 35 aquaporin genes of Arabidopsis showed high level of expression of these two subclasses when compared to NIPs and determined that all PIP transcripts except for AtPIP1;4 and AtPIP2;5, which are upregulated, and AtPIP2;6, which is constitutively expressed and remains unaffected, are downregulated by drought stress. AtPIP2;2 was the first member of this family shown to mediate water uptake, given the reduced $Lp_r$ observed in two knock out mutants. Similarly, mutant analyses of AtPIP1;2 confirmed the role of this channel in water conductivity of both roots and rosettes. Subsequently, a positive correlation was found between $Lp_r$ and transcript abundance of AtPIP1;1. AtPIP1;2, AtPIP1;4, AtPIP2;1, AtPIP2;3, AtPIP2;4 and AtPIP2;5 (Javot *et al.*, 2003; Alexandersson *et al.*, 2005; Postaire *et al.*, 2010; Sutka *et al.*, 2011). The knockout mutant for the $Ca^{2+}$-dependent protein kinase CPK7 showed increased $Lp_r$. Quantitative proteomics assays revealed its negative control on the cellular abundance of PIP1s, while it had no effect on PIP2s (Li *et al.*, 2015). More recently, studies with *pip2;1* mutants in the ethylene-insensitive *ein3-1/eil1-1* background demonstrated a role for ethylene in activating water transport, which was suggested to be due to enhanced S280/S283 phosphorylation of AtPIP2;1(Qing *et al.*, 2015). In maize, *in situ* and quantitative RT-PCR, as well as immunolocalization experiments, revealed strong expression of *ZmPIP2;1* and *ZmPIP2;5* in the endo- and exodermis of mature roots, suggesting the need for a bypass of the apoplastic path in the boundary of the Casparian strip. Recently, heterologous overexpression of the key regulator of ABA biosynthesis *AtLOS5* enhanced maize root hydraulic conductivity by upregulating *ZmPIP1;1*, *ZmPIP1;5* and *ZmPIP2;4* (Hachez *et al.*, 2006; Zhang *et al.*, 2016). The genome of grapevine encodes eight PIP members, of which

VvPIP1;1 and VvPIP2;2 are the most abundantly expressed (Jaillon *et al.*, 2007; Fouquet *et al.*, 2008). Through the characterization of the anatomical, molecular and biophysical aspects of root water uptake in grapevine, Gambetta and collaborators (2013) performed a quantitative tissue-specific analysis of aquaporin expression, as well as a calculation of their contribution to $Lp_r$ along the different root zones. Although most of the studies showed decreased or enhanced $Lp_r$ upon genetic manipulation of PIPs, examples of TIPs contributing to water conductivity include ginseng PgTIP1 and tomato SlTIP2;2 (Lin *et al.*, 2007; Sade *et al.*, 2009).

## Whole-Plant Response to Water Stress

With respect to their response to drought stress, plants can be classified as isohydric or anisohydric. While the stomata of the former are very sensitive to changes in water availability in order to maintain leaf water potential at a minimum, those of the latter favor gas exchange even under water deficit, which leads to large leaf water potential fluctuations under high evaporation demands or soil water scarcity. While the strategy of isohydric plants is more conservative and facilitates survival under water deprivation, that of anisohydric plants allows for better performance under mild water stress. Plants with anisohydric behavior show enhanced aquaporin expression, which increases water conductivity and buffers water potential, thereby promoting stomatal opening (Tardieu and Simonneau, 1998; Moshelion *et al.*, 2015). As an illustrative example, the grapevine anisohydric cultivar Chardonnay maintains diurnal activation of root aquaporins to meet its high transpiration demands, as opposed to the isohydric Grenache (Vandeleur *et al.*, 2009). The level of expressions of several aquaporins is altered under water stress. Drought stress reduced the expression of *NgMIP2*, *NgMIP3* an *NgMIP4* in *Nicotiana glauca*, while transiently inducing that of *SunTIP7* and *SunTIP20* in sunflower (Sarda *et al.*, 1997; Smart *et al.*, 2001). Whereas the sunflower aquaporins were upregulated during stomatal closure, the expression of aquaporins in maize stomatal complexes followed a diurnal pattern (Heinen *et al.*, 2014). In the riparian tree *Melaleuca argentea*, PIP1 expression was slightly increased upon root zone-drying, which resulted in root conductivity enhancement (McLean *et al.*, 2011). In Arabidopsis, all PIPs are downregulated upon drought treatment, except for AtPIP1;4 and AtPIP2;5, which are upregulated, and AtPIP2;6, which is unaffected.

The expression pattern varies among accessions (Alexandersson *et al.*, 2005, 2010). Drought also induces rapid accumulation of the pea *Trg-31* transcript (Guerrero *et al.*, 1990). Arabidopsis plants with increased tolerance to drought were obtained by heterologously expressing tomato *SlPIP2;1 SlPIP2;5* and *SlPIP2;7*, banana *MaPIP1;1*, *Jathropa curcas JcPIP2;7* and *JcTIP1;3* and *Thellungiella salsuginea TsTIP1;2*, while overexpression of soybean *GsPIP2;1* rendered them sensitive to salt and dehydration (Wang *et al.*, 2013; Wang *et al.*, 2014; Xu *et al.*, 2014; Khan *et al.*, 2015; Li *et al.*, 2016). Enhanced tolerance to drought stress was also achieved by overexpressing *TRANSLUCENT GREEN* (*TG*), a member of the AP2/ERBP transcription factor family, which was shown to directly bind to the promoters of *AtTIP1;1*, *AtTIP2;3* and *AtPIP2;2*, thereby activating their expression (Zhu *et al.*, 2014). In rice, the expression of *OsPIP1;3*, which is induced upon water stress in drought-tolerant cultivars, in a drought-sensitive cultivar under control of a stress promoter increased its tolerance to water deficit (Lian *et al.*, 2004). In tomato, overexpression of *SlTIP2;2* improved yield parameters and increased transpiration under both normal and stress conditions, suggesting a circumvention of the stress-induced downregulation of endogenous TIPs, which conferred an anisohydric behavior (Sade *et al.*, 2009). While antisense inhibition of tobacco PIP1s reduced leaf water potential and $Lp_r$, resulting in increased sensitivity to drought, Arabidopsis silenced plants showed normal behavior in both control and stressed conditions, but showed delayed recovery from water stress episodes when compared to wild type (Siefritz *et al.*, 2002; Martre *et al.*, 2002)

## Regulation of Leaf Hydraulics

Members of the PIP and TIP subgroups have been found in guard cells of several species such as Arabidopsis, sunflower, broad bean, spinach, maize, *Nicotiana glauca* and *Picea abies* (Sarda *et al.*, 1997; Oliviusson *et al.*, 2001; Smart *et al.*, 2001; Sun *et al.*, 2001; Leonhardt *et al.*, 2004; Fraysse *et al.*, 2005; Heinen *et al.*, 2014; Prasch *et al.*, 2015). Stomatal aquaporins are also expressed in other tissues, but their expression in guard cells shows specific regulation. In Arabidopsis, several PIPs are induced upon ABA treatment and a mutant defective in starch degradation showed reduced expression of two PIPs and three TIPs (Leonhardt et al., 2004; Prasch et al., 2015). The role of aquaporins in leaf water transport was initially studied through pharmacological inhibition, which showed reduced water permeability in leaves of Arabidopsis, sunflower and six temperate tree species upon treatment with blockers such as $Hg^{2+}$ and azide (Kaldenhoff *et al.*, 1998; Tournaire-Roux *et al.*, 2003; Aasamaa *et al.*, 2005; Nardini and Salleo, 2005). Indirect evidence for the involvement of aquaporins in stomatal movements was obtained in broad bean, where stomatal aperture was regulated by $Ca^{2+}$ in a dose-dependent manner and water conductivity increased upon hyperosmotic treatments followed by the application of membrane trafficking inhibitors, both of which suggested the activation of aquaporins in these conditions (Shope and Mott, 2006; Yang *et al.*, 2006). Additionally, overexpression of *VfPIP1* in Arabidopsis accelerated ABA-induced stomatal closure (Cui *et al.*, 2008). The use of Arabidopsis mutants showed the involvement of AtPIP1;2. AtPIP2;1, AtPIP2;2, AtPIP2;3 and AtPIP2;6 in leaf hydraulics (Kaldenhoff *et al.*, 1998; Martre *et al.*, 2002; Siefriz *et al.*, 2004; Da Ines *et al.*, 2010; Prado *et al.*, 2013). By silencing Arabidopsis *PIP1* genes through artificial microRNAs in the whole plant or in bundle sheath cells, Sade and collaborators (2014) demonstrated the role for this subclass in leaf hydraulic conductance. More recently (2015), the same group restored water conductivity in these silenced lines by ectopically expressing tobacco *NtAQP1* in the bundle sheath. Although the stomatal conductivity remains unaffected in the *atpip2;1* mutant, its stomatal movements showed normal response to light and $CO_2$ but insensitivity to ABA. The Snf1-related protein kinase (SnRK) OST1 has been shown to phosphorylate AtPIP2;1 at the cytosolic Ser121 in the presence of ABA in order to activate this aquaporin and close the stomata (Cutler *et al.*, 2010; Grondin *et al.*, 2015).

## Plant Growth

In expanding cells, an increase of turgor pressure is built up by accumulating osmolites in the cytosol and vacuole, which drives the passive water uptake through cell membranes. Turgor pressure pushes the plasma membrane against the cell wall and originates mechanical tension that causes this structure to stretch, thereby promoting cell growth (Geitmann and Ortega, 2009; Hamant and Traas, 2010; Robinson *et al.*, 2013). Aquaporins with high water transport activity are very abundant in expanding tissues (Ludevid *et al.*, 1992; Chaumont *et al.*, 1998; Higuchi *et al*., 1998). AtTIP1;1 has been involved in the differentiation of a large vacuole in fully elongated cells, given its higher expression in elongated cells and the enhanced transcript accumulation of this channel triggered by gibberellic acid ($GA_3$) (Ludevid *et al*., 1992; Phillips and Huttly, 1994). Probably due to fuctional redundancy of other homologue aquaporins, Arabidopsis *attip1;1* and *attip1;1 attip1;2* mutants displayed no apparent phenotype (Schüssler *et al*., 2008; Beebo *et al*., 2009). In old maize leaves, which have less symplastic connections with the phloem than young ones, $Hg^{2+}$ blocks tissue growth, suggesting a role for aquaporins in water flux from the phloem into the cells. Both pharmacological inhibition of aquaporins and manipulation of the ABA biosynthesis pathway, which in turns alters aquaporin expression, reduce leaf growth in this species (Hukin *et al*., 2002; Ehlert *et al*., 2009; Parent *et al*., 2009). In castor bean (*Ricinus communis*) hypocotils, transcript abundance of *RcPIP2;1* correlates with tissue expansion (Eisenbarth *et al*., 2005). In tobacco, diurnal epinastic leaf movements are caused by differential growth rate of the upper and lower surfaces of the petiole. The expression rhythm of *NtAQP1* coincides with leaf unfolding and the antisense inhibition of this gene abolishes leaf movement (Siefritz *et al*.,2004). Ectopic overexpression of *AtPIP2;1* in this species led to increased leaf growth, transpiration rate, stomatal density and photosynthesis efficiency under optimal conditions, while the transgenic plants performed very poorly under water stress (Aharon *et al*., 2003). In protoplasts of tobacco BY-2 cells, overexpression of cauliflower *BobTIP26-1* and tobacco *NtTIP1* increased cell expansion and vacuolar regeneration (Reisen *et al*., 2003; Okubo-Kurihara *et al*., 2009). In Arabidopsis, overexpression of ginseng *PgTIP1;1* led to the obtaining of transgenic plants with increased growth of vegetative and reproductive organs, as well as larger cells (Lin *et al*., 2007).

## REFERENCES

1. Aasamaa, K., Niinemets, U., Sober, A (2005) Leaf hydraulic conductance in relation to anatomical and functional traits during *Populus tremula* leaf ontogeny. Tree Physiol. 25, 1409-1418
2. Abascal, F., Irisarri, I., Zardoya, R. (2014) Diversity and evolution of membrane intrinsic proteins. Biochim et Biophys Acta. 1840, 1468-1481

3. Aharon, R., Shahak, Y., Wininger, S., Bendov, R., Kapulnik,Y., Galili, G. (2003). Overexpression of a plasma membrane aquaporin in transgenic tobacco improves plant vigor under favorable growth conditions but not under drought or salt stress. Plant Cell. 15, 439-447

4. Aimon, S., Callan-Jones, A., Berthaud, A., Pinot, M., Toombes, G.E., Bassereau, P. (2014) Membrane shape modulates transmembrane protein distribution. Dev Cell. 28, 212-218

5. Alexandersson, E., Fraysse, L., Sjövall-Larsen, S., Gustavsson, S., Fellert, M., Karlsson, M., Johanson, U., Kjellbom, P. (2005) Whole gene family expression and drought stress regulation of aquaporins. Plant Mol Biol. 59, 469-484

6. Alexandersson, E., Danielson, J.A., Råde, J., Moparthi, V.K., Fontes, M., Kjellbom, P., Johanson, U. (2010) Transcriptional regulation of aquaporins in accessions of Arabidopsis in response to drought stress. Plant J. 61, 650-660

7. Alleva, K., Niemietz, C.M., Sutka, M., Maurel, C., Parisi, M., Tyerman, S.D., Amodeo, G. (2006) Plasma membrane of *Beta vulgaris* storage root shows high water channel activity regulated by cytoplasmic pH and a dual range of calcium concentrations. J Exp Bot. 57, 609-621

8. Anderberg, H.I., Danielson, J.Å., Johanson, U. (2011) Algal MIPs, high diversity and conserved motifs. BMC Evol. Biol. 11, 110

9. Anderberg, H.I., Kjellbom, P., Johanson, U. (2012) Annotation of *Selaginella moellendorffii* major intrinsic proteins and the evolution of the protein family in terrestrial plants. Front Plant Sci. 3

10. Azad, A.K., Katsuhara, M., Sawa, Y., Ishikawa, T., Shibata, H. (2008) Characterization of four plasma membrane aquaporins in tulip petals: a putative homolog is regulated by phosphorylation. Plant Cell Physiol. 49, 1196-1208

11. Azad, A.K., Yoshikawa, N., Ishikawa, T., Sawa, Y., Shibata, H. (2012) Substitution of a single amino acid residue in the aromatic/arginine selectivity filter alters the transport profiles of tonoplast aquaporin homologs. Biochim Biophys Acta. 1818, 1-11

12. Bar-On, D., Wolter, S., van de Linde, S., Heilemann, M., Nudelman, G., Nachliel, E., Gutman, M., Sauer, M., Ashery, U. (2012) Super-

resolution imaging reveals the internal architecture of nano-sized syntaxin clusters. J Biol Chem. 287, 27158-27167

13. Beebo, A., Thomas, D., Der, C., Sanchez, L., Leborgne-Castel, N., Marty, F., Schoefs, B., Bouhidel, K. (2009) Life with and without AtTIP1;1, an Arabidopsis aquaporin preferentially localized in the apposing tonoplasts of adjacent vacuoles. Plant Mol Biol. 70, 193-209
14. Beitz, E., Wu, B., Holm, L.M., Schultz, J.E., Zeuthen, T. (2006) Point mutations in the aromatic/arginine region in aquaporin 1 allow passage of urea, glycerol, ammonia, and protons. Proc. Natl. Acad. Sci. USA. 103, 269-74
15. Bellati, J., Alleva, K., Soto, G., Vitali, V., Jozefkowicz, C., Amodeo, G. (2010) Intracellular pH sensing is altered by plasma membrane PIP aquaporin co-expression. Plant Mol Biol. 74, 105-118
16. Bellati, J., Champeyroux, C., Hem, S., Rofidal, V., Krouk, G., Maurel, C., Santoni, V. (2016) Novel aquaporin regulatory mechanisms revealed by interactomics. Mol Cell Proteomics. 15, 3473-3487
17. Bertl, A., Kaldenhoff, R. (2007) Function of a separate $NH_3$-pore in aquaporin TIP2;2 from wheat. FEBS Lett 581, 5413-5417
18. Besserer, A., Burnotte, E., Bienert, G.P., Chevalier, A.S., Errachid, A., Grefen, C., Blatt, M.R., Chaumont, F. (2012) Selective regulation of maize plasma membrane aquaporin trafficking and activity by the SNARE SYP121. Plant Cell. 24, 3463-3481
19. Biela, A., Grote, K., Otto, B., Hoth, S., Hedrich, R., Kaldenhoff, R. (1999) The *Nicotiana tabacum* plasma membrane aquaporin NtAQP1 is mercury-insensitive and permeable for glycerol. Plant J. 18, 565–570
20. Bienert, G.P., Bienert, M.D., Jahn, T.P., Boutry, M., Chaumont, F. (2011) Solanaceae XIPs are plasma membrane aquaporins that facilitate the transport of many uncharged substrates. Plant J. 66, 306-317
21. Bienert, G.P., Cavez, D., Besserer, A., Berny, M.C., Gilis, D., Rooman, M., Chaumont, F. (2012) A conserved cysteine residue is involved in disulfide bond formation between plant plasma membrane aquaporin monomers. Biochem J. 445, 101-111
22. Bienert, G.P., Chaumont, F. (2014) Aquaporin-facilitated

transmembrane diffusion of hydrogen peroxide. Biochim Biophys Acta. 1840, 1596-1604

23. Bienert, G.P., Schüssler, M.D., Jahn, T.P. (2007) Metalloids: essential, beneficial or toxic? Major intrinsic proteins sort it out. Trends Biochem Sci 33, 20-26
24. Boursiac, Y., Boudet, J., Postaire, O., Luu, D.T., Tournaire-Roux, C.C.M. (2008) Stimulus-induced downregulation of root water transport involves reactive oxygen species-activated cell signalling and plasma membrane intrinsic protein internalization. Plant J. 56, 207-218
25. Boursiac, Y., Chen, S., Luu, D.-T., Sorieul, M., van den Dries, N., Maurel, C. (2005) Early effects of salinity on water transport in Arabidopsis roots. Molecular and cellular features of aquaporin expression. Plant Physiol. 139, 790-805
26. Carbrey, J., Agre, P. (2009) Discovery of the aquaporins and development of the field, in: E. Beitz (Ed.), Aquaporins, vol. 190, Springer, Berlin Heidelberg. pp. 3–28
27. Chaumont, F., Barrieu, F., Herman, E.M., Chrispeels, M.J. (1998) Characterization of a maize tonoplast aquaporin expressed in zones of cell division and elongation. Plant Phys. 125, 1206-1215
28. Chaumont, F., Barrieu, F., Wojcik, E., Chrispeels, M.J., Jung, R. (2001) Aquaporins constitute a large and highly divergent protein family in maize. Plant Physiology 125, 1206-1215
29. Cui, X.H., Hao, F.S., Chen, H., Chen, J., Wang, X.C. (2008) Expression of the *Vicia faba VfPIP1* gene in *Arabidopsis thaliana* plants improves their drought resistance. J. Plant Res. 121, 207-214.
30. Cutler, S.R., Rodríguez, P.L., Finkelstein, R.R., Abrams, S.R. (2010) Abscisic acid: emergence of a core signaling network. Annu Rev Plant Biol. 61, 651-679
31. Da Ines, O., Graf, W., Franck, K.I., Albert, A., Winkler, J.B., Scherb, H., Stichler, W., Schäffner, A.R. (2010) Kinetic analyses of plant water relocation using deuterium as tracer – reduced water flux of Arabidopsis *pip2* aquaporin knockout mutants. Plant Biol. (Stuttg.) 12 (Suppl.1), 129-139
32. Daniels, M.J., Chrispeels, M.J., Yeager, M. (1999) Projection

structure of a plant vacuole membrane aquaporin by electron cryo-crystallography. J. Mol. Biol. 294, 1337-1349

33. Daniels, M.J., Mirkov, T.E., Chrispeels, M.J. (1994) The plasma membrane of *Arabidopsis thaliana* contains a mercury-insensitive aquaporin that is a homolog of the tonoplast water channel protein TIP. Plant Physiol. 106, 1325-1333
34. Daniels, M.J., Yeager, M. (2005) Phosphorylation of aquaporin PvTIP3;1 defined by mass spectrometry and molecular modeling. Biochemistry. 44,14443–14454
35. Danielson, J.A., Johanson, U. (2008) Unexpected complexity of the aquaporin gene family in the moss *Physcomitrella patens*. BMC Plant Biol. 8, 45
36. Deokar, A.A., Tar'an, B. (2016) Genome-Wide Analysis of the Aquaporin Gene Family in Chickpea (*Cicer arietinum L.*). Front Plant Sci. 7, 1802
37. Dhonukshe, P., Aniento, F., Hwang, I., Robinson, D.G., Mravec, J., Stierhof, Y.D., Frim, L.J. (2007) Clathrin-mediated constitutive endocytosis of PIN auxin efflux carriers in Arabidopsis. Curr Biol. 17, 520-527
38. Ehlert, C., Maurel, C., Tardieu, F., Simonneau, T. (2009) Aquaporin-mediated reduction in maize root hydraulic conductivity impacts cell turgor and leaf elongation even without changing transpiration. Plant Physiol. 150, 1093-1104
39. Eisenbarth, D.A., Weig, A.R. (2005) Dynamics of aquaporins and water relations during hypocotyl elongation in *Ricinus communis* L. seedlings. J. Exp. Bot. 56, 1831-1842
40. Fetter, K., Van Wilder, V., Moshelion, M., Chaumont, F. (2004) Interactions between plasma membrane aquaporins modulate their water channel activity. Plant Cell. 16, 215-228
41. Fortin, M.G., Morrison, N.A., Verma, D.P.S. (1987) Nodulin-26, a peribacteroid membrane nodulin is expressed independently of the development of the peribacteroid compartment. Nucleic Acids Res. 15, 813-824
42. Fotiadis, D., Jeno, P., Mini, T., Wirtz, S., Muller, S.A., Fraysse, L., Kjellbom, P., Engel, A. (2001) Structural characterization of two

aquaporins isolated from native spinach leaf plasma membranes. J. Biol. Chem. 276, 1707-1714

43. Fouquet, R., Léon, C., Ollat, N., Barrieu, F. (2008) Identification of grapevine aquaporins and expression analysis in developing berries. Plant Cell Rep. 27, 1541-1550
44. Fraysse, L., Wells, B., McCann, M.C., Kjellbom, P. (2005) Specific plasma membrane aquaporins of the PIP1 subfamily are expressed in sieve elements and guard cells. Biol. Cell. 97, 519-534
45. Frick, A., Jarva, M., Tornroth-Horsefield, S. (2013) Structural basis for pH gating of plant aquaporins. FEBS Lett. 587, 989-993
46. Froger, A. Thomas, D., Delamarche, C., Tallur, B. (1998) Prediction of functional residues in water channels and related proteins. Protein Sci. 7, 1458-1468
47. Fujiyoshi, Y., Mitsuoka, K., de Groot, B.L., Philippsen, A., Grubmüller, H., Agre, P., Engel, A. (2002) Structure and function of water channels. Curr. Opin. Struct. Biol. 12, 509-515
48. Gambetta, G.A., Fei, J., Rost, T.L., Knipfer, T., Matthews, M.A., Shackel, K.A., Walker, M.A., McElrone, A.J. (2013) Water uptake along the length of grapevine fine roots: developmental anatomy, tissue-specific aquaporin expression, and pathways of water transport. Plant Physiol. 163, 1254-1265
49. Geitmann, A., Ortega, J. (2009) Mechanics and modeling of plant cell growth. Trends Plant Sci. 14, 467-478
50. Gerbeau, P., Amodeo, G., Henzler, T., Santoni, V., Ripoche, P., Maurel, C. (2002) The water permeability of Arabidopsis plasma membrane is regulated by divalent cations and pH. Plant J. 30, 71-81
51. Gerbeau, P., Guclu, J., Ripoche, P., Maurel, C. (1999) Aquaporin Nt-TIPa can account for the high permeability of tobacco cell vacuolar membrane to small neutral solutes. Plant J. 18, 577–587
52. Gorin, M.B., Yancey, S.B., Cline, J., Revel, J.P., Horwitz, J. (1984) The major intrinsic protein (MIP) of the bovine lens fiber membrane: characterization and structure based on cDNA cloning. Cell. 39, 49–59
53. Grefen, C., Chen, Z., Honsbein, A., Donald, N., Hills, A., Blatt,

M.R. (2010) A novel motif essential for SNARE interaction with the $K^+$ channel KC1 and channel gating in Arabidopsis. Plant Cell. 22, 3076-3092

54. Grondin, A., Rodrigues, O., Verdoucq, L., Merlot, S., Leonhardt, N., Maurel, C. (2015) Aquaporins contribute to ABA-triggered stomatal closure through OST1-mediated phosphorylation. Plant Cell. 27, 1945-1954
55. Guenther, J.F., Chanmanivone, N., Galetovic, M.P., Wallace, I.S., Cobb, J.A., Roberts, D.M. (2003) Phosphorylation of soybean Nodulin 26 on serine 262 enhances water permeability and is regulated developmentally and by osmotic signals. Plant Cell. 15, 981-991
56. Guerrero, F.D., Jones, J.T., Mullet, J.E. (1990) Turgor-responsive gene transcription and RNA levels increase rapidly when pea shoots are wilted. Sequence and expression of three inducible genes. Plant Mol. Biol. 15, 11-26
57. Gupta, A.B., Sankararamakrishnan, R. (2009) Genome-wide analysis of major intrinsic proteins in the tree plant *Populus trichocarpa*: characterization of XIP subfamily of aquaporins from evolutionary perspective. BMC Plant Biol. 9, 134
58. Hachez, C., Besserer, A., Chevalier, A.S., Chaumont, F. (2013) Insights into plant plasma membrane aquaporin trafficking. Trends Plant Sci. 18, 344-352
59. Hachez, C., Laloux, T., Reinhardt, H., Cavez, D., Degand, H., Grefen, C., De Rycke, R., Inzé, D., Blatt, M.R., Russinova, Chaumont, F. (2014) Arabidopsis SNAREs SYP61 and SYP121 coordinate the trafficking of plasma membrane aquaporin PIP2;7 to modulate the cell membrane water permeability. Plant Cell. 26, 3132-3147
60. Hachez, C., Moshelion, M., Zelazny, E., Cavez, D., Chaumont, F. (2006) Localization and quantification of plasma membrane aquaporin expression in maize primary root: a clue to understanding their role as cellular plumbers. Plant Mol. Biol. 62, 305-323
61. Hamant, O., Traas, J. (2010) The mechanics behind plant development. New Phytol. 185, 369-385
62. Hedfalk, K., Tornroth-Horsefield, S., Nyblom, M., Johanson, U.,

Kjellbom, P. and Neutze, R. (2006) Aquaporin gating. Curr. Opin. Struct. Biol. 16, 447-456

63. Heinen, R.B., Bienert, G.P., Cohen, D., Chevalier, A.S., Uehlein, N., Hachez, C., Kaldenhoff, R., Le Thiec, D., Chaumont, F. (2014) Expression and characterization of plasma membrane aquaporins in stomatal complexes of *Zea mays*. Plant Mol Biol. 86, 335-350
64. Higuchi, T., Suga, S., Tsuchiya, T., Hisada, H., Morishima, S., Okada, Y., Maeshima, M. (1998) Molecular cloning, water channel activity and tissue specific expression of two isoforms of radish vacuolar aquaporin. Plant Cell Phys. 39, 905-913
65. Hooijmaijers, C., Rhee, J.Y., Kwak, K.J., Chung, G.C., Horie, T., Katsuhara, M., Kang, H. (2012) Hydrogen peroxide permeability of plasma membrane aquaporins of *Arabidopsis thaliana*. J Plant Res. 125, 147–153
66. Horie, T., Kaneko, T., Sugimoto, G., Sasano, S., Panda, S.K., Shibasaka, M., Katsuhara, M. (2011) Mechanisms of water transport mediated by PIP aquaporins and their regulation via phosphorylation events under salinity stress in barley roots. Plant Cell Physiol. 52, 663-675
67. Hosy, E., Martiniere, A., Choquet, D., Maurel, C., Luu, D.T. (2015) Super-resolved and dynamic imaging of membrane proteins in plant cells reveal contrasting kinetic profiles and multiple confinement mechanisms. Mol Plant. 8, 339-342
68. Hu, W., Hou, X., Huang, C., Yan, Y., Tie, W., Ding, Z., Wei, Y., Liu, J., Miao, H., Lu, Z., Li, M., Xu, B., Jin, Z. (2015) Genome-Wide Identification and Expression Analyses of Aquaporin Gene Family during Development and Abiotic Stress in Banana. Int J Mol Sci. 16, 19728-19751
69. Hub, J.S., de Groot, B.L. (2008) Mechanism of selectivity in aquaporins and aquaglyceroporins. Proc. Natl. Acad. Sci. 105, 1198-1203
70. Hukin, D., Doering-Saad, C., Thomas, C.R., Pritchard, J. (2002) Sensitivity of cell hydraulic conductivity to mercury is coincident with symplasmic isolation and expression of plasmalemma aquaporin genes in growing maize roots. Planta. 215, 1047-1056
71. Jahn, T.P., Moller, A.L., Zeuthen, T., Holm, L.M., Klaerke, D.A.,

Mohsin, B., Kuhlbrandt, W., Schjoerring, J.K. (2004) Aquaporin homologues in plants and mammals transport ammonia. FEBS Lett. 574, 31-36

72. Jaillon, O., Aury, J.M., Noel, B., Policriti, A., Clepet, C., Casagrande, A., Choisne, N., Aubourg, S., Vitulo, N., Jubin, C., Vezzi, A., Legeai, F., Hugueney, P., Dasilva, C., Horner, D., Mica, E., Jublot, D., Poulain, J., Bruyère, C., Billault, A., Segurens, B., Gouyvenoux, M., Ugarte, E., Cattonaro, F., Anthouard, V., Vico, V., Del Fabbro, C., Alaux, M., Di Gaspero, G., Dumas, V., Felice, N., Paillard, S., Juman, I., Moroldo, M., Scalabrin, S., Canaguier, A., Le Clainche, I., Malacrida, G., Durand, E., Pesole, G., Laucou, V., Chatelet, P., Merdinoglu, D., Delledonne, M., Pezzotti, M., Lecharny, A., Scarpelli, C., Artiguenave, F., Pè, M.E., Valle, G., Morgante, M., Caboche, M., Adam-Blondon, A.F., Weissenbach, J., Quétier, F., Wincker, P., French-Italian Public Consortium for Grapevine Genome Characterization. (2007) The grapevine genome sequence suggests ancestral hexaploidization in major angiosperm phyla. Nature. 449, 463-467
73. Jauh, G.Y., Fischer, A.M., Grimes, H.D., Ryan, C.A. Jr, Rogers, J.C. (1998) δ-Tonoplast intrinsic protein defines unique plant vacuole functions. Proc. Natl. Acad. Sci. USA. 95, 12995-12999
74. Javot, H., Lauvergeat, V., Santoni, V., Martin-Laurent, F., Guclu, J., Vinh, J., Heyes, J., Franck, K.I., Schäffner, A.R., Bouchez, D., Maurel, C. (2003) Role of a single aquaporin isoform in root water uptake. Plant Cell. 15, 509-522
75. Jiang, L., Phillips, T.E., Hamm, C.A., Drozdowicz, Y.M., Rea, P.A., Maeshima, M., Rogers, S.W., Rogers, J.C. (2001) The protein storage vacuole: a unique compound organelle. J Biol Chem. 155, 991-1002
76. Johanson, U., Gustavsson, S. (2002) A new subfamily of major intrinsic proteins in plants. Mol Biol Evol. 19, 456-461
77. Johanson, U., Karlsson, M., Johansson, I., Gustavsson, S., Sjövall, S., Fraysse, L., Weig, A.R., Kjellbom, P. (2001) The complete set of genes encoding major intrinsic proteins in *Arabidopsis* provides a framework for a new nomenclature for major intrinsic proteins in plants. Plant Physiology 126, 1358-1369

78. Johansson, I., Karlsson, M., Shukla, V.K., Chrispeels, M.J., Larsson, C., Kjellbom, P. (1998) Water transport activity of the plasma membrane aquaporin PM28A is regulated by phosphorylation. Plant Cell. 10, 451-460

79. Jones, A.M., Xuan, Y., Xu, M., Wang, R.S., Ho, C.H., Lalonde, S., You, C.H., Sardi, M.I., Parsa, S.A., Smith-Valle, E., Su, T., Frazer, K.A., Pilot, G., Pratelli, R., Grossmann, G., Acharya, B.R., Hu, H.C., Engineer, C., Villiers, F., Ju, C., Takeda, K., Su, Z., Dong, Q., Assmann, S.M., Chen, J., Kwak, J.M., Schroeder, J.I., Albert, R., Rhee, S.Y., Frommer, W.B. (2014) Border control – a membrane-linked interactome of Arabidopsis. Science. 344, 711-716

80. Kaldenhoff, R.,Grote, K., Zhu, J.J., Zimmermann, U. (1998) Significance of plasmalemma aquaporins for water-transport in Arabidopsis thaliana. Plant J. 14, 121-128

81. Katsuhara, M., Sasano, S., Horie, T., Matsumoto, T., Rhee, J., Shibasaka, M. (2014) Functional and molecular characteristics of rice and barley NIP aquaporins transporting water, hydrogen peroxide and arsenite. Plant Biotech 31, 213-219

82. Khan, K., Agarwal, P., Shanware, A., Sane, V.A. (2015) Heterologous Expression of Two Jatropha Aquaporins Imparts Drought and Salt Tolerance and Improves Seed Viability in Transgenic *Arabidopsis thaliana*. PLoS One. 10, e0128866

83. Kruse, E., Uehlein, N., Kaldenhoff, R. (2006) The aquaporins. Genome Biol. 7, 206

84. Kukulski, W., Schenk, A.D., Johanson, U., Braun, T., de Groot, B.L., Fotiadis, D., Kjellbom, P., Engel, A. (2005) The 5A structure of heterologously expressed plant aquaporin SoPIP2;1. J. Mol. Biol. 350, 611-616

85. Lagrée, V., Froger, A., Deschamps, S., Hubert, J.F., Delamarche, C., Bonnec, G., Thomas, D., Gouranton, J., Pellerin, I. (1999) Switch from an aquaporin to a glycerol channel by two amino acids substitution. J Biol Chem. 274, 6817-6819

86. Lee, H.K., Cho, S.K., Son, O., Xu, Z., Hwang, I., Kim, W.T. (2009) Drought stress-induced Rma1H1, a RING membrane-anchor E3 ubiquitin ligase homolog, regulates aquaporin levels via

ubiquitination in transgenic Arabidopsis plants. Plant Cell. 21, 622-641

87. Leitao, L., Prista, C., Moura, T.F., Loureiro-Dias, M.C., Soveral, G. (2012) Grapevine aquaporins: gating of a tonoplast intrinsic protein (TIP2;1) by cytosolic pH. PLoS One. 7, e33219

88. Leonhardt, N., Kwak, J.M., Robert, N., Waner, D., Leonhardt, G., Schroeder, J.I. (2004) Microarray expression analyses of *Arabidopsis* guard cells and isolation of a recessive abscisic acid hypersensitive protein phosphatase 2C mutant. Plant Cell. 16, 596-615

89. Li, G., Boudsocq, M., Hem, S., Vialaret, J., Rossignol, M., Maurel, C., Santoni, V. (2015) The calcium-dependent protein kinase CPK7 acts on root hydraulic conductivity. Plant Cell Environ. 38, 1312-1320

90. Li, R., Wang, J., Li, S., Zhang, L., Qi, C., Weeda, S., Zhao, B., Ren, S., Guo, Y.D. (2016) Plasma Membrane Intrinsic Proteins SlPIP2;1, SlPIP2;7 and SlPIP2;5 Conferring Enhanced Drought Stress Tolerance in Tomato. Sci Rep. 6, 31814

91. Li, X., Wang, X., Yang, Y., Li, R., He, Q., Fang, X., Luu, D.T., Maurel, C., Lin, J. (2011a) Single-molecule analysis of PIP2;1 dynamics and partitioning reveals multiple modes of Arabidopsis plasma membrane aquaporin regulation. Plant Cell. 23, 3780-3797

92. Li, T., Choi, W.G., Wallace, I.S., Baudry, J., Roberts, D.M. (2011b) *Arabidopsis thaliana* NIP7;1: an anther-specific boric acid transporter of the aquaporin superfamily regulated by an unusual tyrosine in helix 2 of the transport pore. Biochemistry. 50, 6633-6641

93. Lian, H.-L., Yu, X., Ye, Q., Ding, X.-S., Kitagawa, Y., Kwak, S.-S., Su, W.-A., Tang, Z.-C. (2004) The role of aquaporin RWC3 in drought avoidance in rice. Plant Cell Physiol. 45, 481-489

94. Lin, W., Peng, Y., Li, G., Arora, R., Tang, Z., Su, W. Cai, W. (2007) Isolation and functional characterization of PgTIP1, a hormone-autotrophic cells-specific tonoplast aquaporin in ginseng. J. Exp. Bot. 58, 947-956

95. Lopez, D., Bronner, G., Brunel, N., Auguin, D., Bourgerie, S., Brignolas, F., Carpin, S., Tournaire-Roux, C., Maurel, C., Fumanal, B., Martin, F., Sakr, S., Label, P., Julien, J.L., Gousset-Dupont,

A., Venisse, J.S. (2012) Insights into Populus XIP aquaporins: evolutionary expansion, protein functionality, and environmental regulation. J Exp Bot. 63, 2217-2230

96. Loque, D., Ludewig, U., Yuan, L., von Wiren, N. (2005) Tonoplast Intrinsic Proteins AtTIP2;1 and AtTIP2;3 facilitate $NH_3$ transport into the vacuole. Plant Physiol 137, 671-680

97. Ludevid, D., Höfte, H., Himelblau, E., Chrispeels, M.J. (1992) The expression pattern of the tonoplast intrinsic protein γ-TIP in *Arabidopsis thaliana* is correlated with cell enlargement. Plant Physiol. 100, 1633-1639

98. Luu, D.T., Martinière, A., Sorieul, M., Runions, J., Maurel, C. (2012) Fluorescence recovery after photobleaching reveals high cycling dynamics of plasma membrane aquaporins in Arabidopsis roots under salt stress. Plant J. 69, 894-905

99. Ma, J.F., Tamai, K., Yamaji, N., Mitani, N., Konishi, S., Katsuhara, M., Ishiguro, M., Murata, Y., Yano, M. (2006) A silicon transporter in rice. Nature. 440, 688-691

100. Maeshima, M. (2001) Tonoplast transporters: organization and function. Annu Rev Plant Physiol Plant Mol Biol. 52, 469-497

101. Maeshima, M., Ishikawa, F. (2008) ER membrane aquaporin in plants. Eur J Physiol. 456, 709-716

102. Maggio, A., Joly, R.J. (1995) Effects of mercuric chloride on the hydraulic conductivity of tomato root systems. Evidence for a channel-mediated water pathway. Plant Physiol 109, 331-335

103. Martinière, A. Li, X., Runions, J. Lin, J., Maurel, C., Luu, D.-T. (2012) Salt stress triggers enhanced cycling of Arabidopsis root plasma-membrane aquaporins, Plant Signal. Behav. 7, 529-532

104. Martre, P., Morillon, R., Barrieu, F., North, G.B., Nobel, P.S., Chrispeels, M.J. (2002).Plasma membrane aquaporins play a significant role during recovery from water deficit. Plant Physiol. 130, 2101-2110

105. Maurel, C. (1997) Aquaporins and water permeability of plant membranes. Annu Rev Plant Physiol Plant Mol Biol 48, 399-429

106. Maurel, C., Boursiac, Y., Luu, D.T., Santoni, V., Shahzad, Z.,

Verdoucq, L. (2015) Aquaporins in Plants. Physiol Rev. 95, 1321-1358

107. Maurel, C., Kado, R.T., Guern, J., Chrispeels, M.J. (1995) Phosphorylation regulates the water channel activity of the seed-specific aquaporin α-TIP. EMBO J. 14, 3028-3035

108. Maurel, C., Reizer, J., Schroeder, J.I., Chrispeels, M.J. (1993) The vacuolar membrane protein γ-TIP creates water specific channels in *Xenopus* oocytes. EMBO J. 12, 2241-2247

109. Maurel, C. Verdoucq, L., Luu, D.-T., Santoni, V. (2008) Plant aquaporins: membrane channels with multiple integrated functions, Annu. Rev. Plant Biol. 59, 595-624

110. McLean, E.H., Ludwig, M., Grierson, P.F. (2011) Root hydraulic conductance and aquaporin abundance respond rapidly to partial root-zone drying events in a riparian *Melaleuca species*. New Phytol 192, 664-675

111. Miao, G., Hong, Z., Verma, D. (2004) Topology and phosphorylation of soybean nodulin-26, an intrinsic protein of the peribacteroid membrane. J. Cell Biol. 118, 481-490

112. Mikosch, M., Hurst, A.C., Hertel, B., Homann, U. (2006) Diacidic motif is required for efficient transport of the $K^+$ channel KAT1 to the plasma membrane. Plant Physiol.142, 923-930

113. Mitani, N., Yamaji, N., Ma, J.F. (2008) Characterization of substrate specificity of a rice silicon transporter, Lsi1. Pflügers Arch. 456, 679-686

114. Mizutani, M., Watanabe, S., Nakagawa, T., Maeshima, M. (2006) Aquaporin NIP2;1 is mainly localized to the ER membrane and shows root-specific accumulation in *Arabidopsis thaliana.* Plant Cell Physiol. 47, 1420-1426

115. Mori, I.C., Rhee, J., Shibasaka, M., Sasano, S., Kaneko, T., Horie, T., Katsuhara, M. (2014) $CO_2$ transport by PIP2 aquaporins of barley. Plant Cell Physiol. 55, 251-257

116. Moshelion, M., Halperin, O., Wallach, R., Oren, R., Way, D.A. (2015) Role of aquaporins in determining transpiration and photosynthesis in water-stressed plants: crop water-use efficiency, growth and yield. Plant Cell Environ. 38, 1785-1793

117. Moshelion, M., Moran, N., Chaumont, F. (2004) Dynamic changes in the osmotic water permeability of protoplast plasma membrane. Plant Physiol. 135, 2301-2317

118. Nardini, A., Salleo, S. (2005) Water stress-induced modifications of leaf hydraulic architecture in sunflower: co-ordination with gas exchange. J. Exp. Bot. 56, 3093-3101

119. Niemietz, C.M., Tyerman, S.D. (2002) New potent inhibitors of aquaporins: silver and gold compounds inhibit aquaporins of plant and human origin. FEBS Lett. 531, 443-447

120. Niittylä, T., Fuglsang, A.T., Palmgren, M.G., Frommer, W.B., Schulze, W.X. (2007) Temporal analysis of sucrose-induced phosphorylation changes in plasma membrane proteins of Arabidopsis. Mol Cell Proteomics. 6, 1711-1726

121. Nühse, T.S., Stensballe, A., Jensen, O.N., Peck, S.C. (2004) Phosphoproteomics of the Arabidopsis plasma membrane and a new phosphorylation site database. Plant Cell. 16, 2394-2405

122. Nyblom, M., Frick, A., Wang, Y., Ekvall, M., Hallgren, K., Hedfalk, K., Neutze, R., Tajkhorshid, E., Törnroth-Horsefield, S. (2009) Structural and functional analysis of SoPIP2;1 mutants add insight into plant aquaporin gating. J Mol Biol. 387, 653-668

123. Okubo-Kurihara, E., Sano, T., Higaki, T., Kutsuna, N., Hasezawa, S. (2009) Acceleration of vacuolar regeneration and cell growth by overexpression of an aquaporin *Nt*TIP1;1 in tobacco BY-2 cells. Plant and Cell Phys. 50, 151-160

124. Oliviusson, P., Salaj, J., Hakman, I. (2001) Expression pattern of transcripts encoding water channel-like proteins in Norway spruce (*Picea abies*). Plant Mol Biol. 46, 289-299

125. Otto, B., Uehlein, N., Sdorra, S., Fischer, M., Ayaz, M., Belastegui-Macadam, X., Heckwolf, M., Lachnit, M., Pede, N., Priem, N., Reinhard, A., Siegfart, S., Urban, M., Kaldenhoff, R. (2010) Aquaporin tetramer composition modifies the function of tobacco aquaporins. J Biol Chem. 285, 31253-31260

126. Parent, B., Hachez, C., Redondo, E., Simonneau, T., Chaumont, F., Tardieu, F. (2009) Drought and abscisic acid effects on aquaporin content translate into changes in hydraulic conductivity and leaf growth rate: a trans-scale approach. Plant Physiol. 149, 2000-2012

127. Phillips, A.L., Huttly, A.K. (1994) Cloning of two gibberellin-regulated cDNAs from Arabidopsis thaliana by subtractive hybridization: expression of the tonoplast water channel, γ-TIP, is increased by GA3. Plant Mol. Biol. 24, 603-15

128. Di Pietro, M., Vialaret, J., Li, G., Hem, S., Rossignol, M., Maurel, C., Santoni, V. (2013) Coordinated post-translational responses of aquaporins to abiotic and nutritional stimuli in Arabidopsis roots. Mol Cell Proteomics 12, 3886–3897

129. Postaire, O., Tournaire-Roux, C., Grondin, A., Boursiac, Y., Morillon, R., Schäffner, A.R., Maurel, C. (2010) A PIP1 aquaporin contributes to hydrostatic pressure-induced water transport in both the root and rosette of Arabidopsis. Plant Physiol. 152, 1418-1430

130. Prado, K., Boursiac, Y., Tournaire-Roux, C., Monneuse, J.-M., Postaire, O. Da Ines, O., Schäffner, A.R., Hem, S., Santoni, V., Maurel, C. (2013) Regulation of Arabidopsis leaf hydraulics involves light-dependent phosphorylation of aquaporins in veins, Plant Cell. 25, 1029-1039

131. Prak, S. Hem, S., Boudet, J., Viennois, G., Sommerer, N., Rossignol, M., Maurel, C., Santoni, V. (2008) Multiple phosphorylations in the C-terminal tail of plant plasma membrane aquaporins: role in subcellular trafficking of AtPIP2;1 in response to salt stress. Mol. Cell. Proteomics. 7, 1019-1030

132. Prasch, C.M., Ott, K.V., Bauer, H., Ache, P., Hedrich, R., Sonnewald, U. (2015) β-amylase1 mutant Arabidopsis plants show improved drought tolerance due to reduced starch breakdown in guard cells. J Exp Bot. 66, 6059-6067.

133. Preston, G.M. Carroll, T.P. Guggino, W.B., Agre, P. (1992) Appearance of water channels in *Xenopus* oocytes expressing red cell CHIP28 protein. Science. 256, 385-387

134. Qing, D., Yang, Z., Li, M., Wong, W.S., Guo, G., Liu, S., Guo, H., Li, N. (2015) Quantitative and Functional Phosphoproteomic Analysis Reveals that Ethylene Regulates Water Transport via the C-Terminal Phosphorylation of Aquaporin PIP2;1 in Arabidopsis. Mol Plant. 9, 158-174

135. Quemeneur, F., Sigurdsson, J.K., Renner, M., Atzberger, P.J., Bassereau, P., Lacoste, D. (2014) Shape matters in protein mobility

within membranes. Proc Natl Acad Sci U S A. 111, 5083-5087

136. Quigley, F., Rosenberg, J.M., Shachar-Hill, Y., Bohnert, H.J. (2001) From genome to function: the Arabidopsis aquaporins. Genome Biol. 3, 1-17
137. Reisen, D., Leborgne-Castel, N., Özalp, C., Chaumont, F., Marty, F. (2003) Expression of a cauliflower tonoplast aquaporin tagged with GFP in tobacco suspension cells correlates with an increase in cell size. Plant Mol Biol. 52, 387-400
138. Robinson, S., Burian, A.,Couturier, E., Landrein, B., Louveaux, M., Neumann, E.D., Peaucelle, A., Weber, A., Nakayama, N. (2013) Mechanical control of morphogenesis at the shoot apex. J Exp Bot. 64, 4729-4744
139. Rougé, P., Barre, A. (2008) A molecular modeling approach defines a new group of Nodulin 26-like aquaporins in plants. Biochem Biophys Res Commun. 367, 60-66
140. Sade, N., Shatil-Cohen, A., Attia, Z., Maurel, C., Boursiac, Y., Kelly, G., Granot, D., Yaaran, A., Lerner, S., Moshelion, M. (2014) The role of plasma membrane aquaporins in regulating the bundle sheath-mesophyll continuum and leaf hydraulics. Plant Physiol. 166, 1609-1620
141. Sade, N., Shatil-Cohen, A., Moshelion, M. (2015) Bundle-sheath aquaporins play a role in controlling Arabidopsis leaf hydraulic conductivity. Plant Signal Behav. 10, e1017177
142. Sade, N., Vinocur, B.J., Diber, A., Shatil, A., Ronen, G., Nissan, H., Wallach, R., Karchi, H., Moshelion, M. (2009) Improving plant stress tolerance and yield production: is the tonoplast aquaporin SlTIP2;2 a key to isohydric to anisohydric conversion? New Phytologist. 181, 651–661
143. Saito, C., Ueda, T., Abe, H.,Wada, Y., Kuroiwa, T., Hisada, A., Furuya, M., Nakano, A. (2002) A complex and mobile structure forms a distinct subregion within the continuous vacuolar membrane in young cotyledons of Arabidopsis. Plant J. 29, 245-255
144. Sakurai, J., Ishikawa, F., Yamaguchi, T., Uemura, M., Maeshima, M. (2005) Identification of 33 rice aquaporin genes and analyses of their expression and function. Plant and Cell Physiology. 45, 1568–1577

145. Santoni, V., Verdoucq, L., Sommerer, N., Vinh, J., Pflieger, D., Maurel, C. (2006) Methylation of aquaporins in plant plasma membrane. Biochem J. 400, 189-197

146. Santoni, V., Vinh, J., Pflieger, D., Sommerer, N., Maurel, C. (2003) A proteomic study reveals novel insights into the diversity of aquaporin forms expressed in the plasma membrane of plant roots. Biochem. J. 372, 289-296

147. Sarda, X., Tousch, D., Ferrare, K., Legrand, E., Dupuis, J.M., Casse-Delbart, F., Lamaze, T. (1997) Two TIP-like genes encoding aquaporins are expressed in sunflower guard cells. Plant J. 12, 1103-1111

148. Savage, D.F., O'Connell, J.D. 3rd, Miercke, L.J., Finer-Moore, J., Stroud, R.M. (2010) Structural context shapes the aquaporin selectivity filter. Proc Natl Acad Sci USA. 107, 17164-17169

149. Schüssler, M.D., Alexandersson, E., Bienert, G.P., Kichey, T., Laursen, K.H., Johanson, U., Kjellbom, P., Schjoerring, J.K., Jahn, T.P. (2008) The effects of the loss of TIP1;1 and TIP1;2 aquaporins in *Arabidopsis thalina*. Plant J. 56, 756-767

150. Shope, J.C., Mott, K.A. (2006) Membrane trafficking and osmotically induced volume changes in guard cells. J. Exp. Bot. 57, 4123-4131

151. Siefritz, F., Otto, B., Bienert, G.P., Van Der Krol, A., Kaldenhoff, R. (2004) The plasma membrane aquaporin NtAQP1 is a key component of the leaf unfolding mechanism in tobacco. Plant J. 37, 147-155

152. Siefritz, F., Tyree, M.T., Lovisolo, C., Schubert, A., Kaldenhoff, R. (2002) PIP1 plasma membrane aquaporins in tobacco: from cellular effects to function in plants. Plant Cell. 14, 869-876

153. Sjovall-Larsen, S., Alexandersson, E., Johansson, I., Karlsson, M., Johanson, U., Kjellbom, P. (2006) Purification and characterization of two protein kinases acting on the aquaporin SoPIP2;1, Biochim. Biophys. Acta Biomembr. 1758, 1157–1164

154. Smart, L.B., Moskal, W.A., Cameron, K.D., Bennett, A.B. (2001) MIP genes are down-regulated under drought stress in *Nicotiana glauca*. Plant Cell Physiol. 42, 686-693.

155. Sorieul, M., Santoni, V., Maurel, C., Luu, D.T. (2011) Mechanisms

and effects of retention of over-expressed aquaporin AtPIP2;1 in the endoplasmic reticulum. Traffic. 12, 473-482

156. Soto, G., Fox, R., Ayub, N., Alleva, K., Guaimas, F., Erijman, E.J., Mazzella, A., Amodeo, G., Muschietti, J. (2010) TIP5;1 is an aquaporin specifically targeted to pollen mitochondria and is probably involved in nitrogen remobilization in Arabidopsis thaliana. Plant J. 64, 1038-1047

157. Steudle, E. (2001) The cohesion-tension mechanism and the acquisition of water by plant roots. Annu Rev of Plant Phys and Plant Mol Biol. 52, 847-875

158. Steudle, E., Peterson, C. (1998) How does water get through the roots? J Exp Bot. 49, 775-788

159. Suga, S., Imagawa, S., Maeshima, M. (2001) Specificity of the accumulation of mRNAs and proteins of the plasma membrane and tonoplast aquaporins in radish organs. Planta. 212, 294-304

160. Suga, S., Komatsu, S., Maeshima, M. (2002) Aquaporin isoforms responsive to salt and water stresses and phytohormones in radish seedlings. Plant Cell Physiol. 43, 1229-1237

161. Suga, S., Maeshima, M. (2004) Water channel activity of radish plasma membrane aquaporins heterologously expressed in yeast and their modification by site-directed mutagenesis. Plant Cell Physiol. 45, 823-830

162. Sun, M.-H., Xu, W., Zhu, Y.-F., Su, W.-A., Tang, Z.-C. (2001) A simple method for in situ hybridization to RNA in guard cells of *Vicia faba L.*: The expression of aquaporins in guard cells. Plant Mol. Biol. Rep. 19, 129-135

163. Sutka, M., Li, G., Boudet, J., Boursiac, Y., Doumas, P., Maurel, C. (2011) Natural variation of root hydraulics in Arabidopsis grown in normal and salt stress conditions. Plant Physiol 155, 1264-1276

164. Sutter, J.U., Campanoni, P., Tyrrell, M., Blatt, M.R. (2006) Selective mobility and sensitivity to SNAREs is exhibited by the Arabidopsis KAT1 $K^+$ channel at the plasma membrane. Plant Cell. 18, 935-954

165. Takano, J., Wada, M., Ludewig, U., Schaaf, G., von Wiren, N., Fujiwara, T. (2006) The Arabidopsis major intrinsic protein NIP5;1 is essential for efficient boron uptake and plant development under boron limitation. Plant Cell. 18, 1498-1509

166. Tao, P., Zhong, X., Li, B., Wang, W., Yue, Z., Lei, J., Guo, W., Huang, X. (2014) Genome-wide identification and characterization of aquaporin genes (AQPs) in Chinese cabbage (*Brassica rapa ssp. pekinensis*). Mol. Genet. Genom. 289, 1131-1145

167. Tardieu, F., Simonneau, T. (1998) Variability among species of stomatal control under fluctuating soil water status and evaporative demand: modelling isohydric and anisohydric behaviours. J Exp Bot. 49, 419-432

168. Tester, M., Leigh, R.A. (2001) Partitioning of nutrient transport processes in roots. J Exp Bot. 52, 445-457

169. Tong, J., Briggs, M.M., McIntosh, T.J. (2012) Water permeability of aquaporin-4 channel depends on bilayer composition, thickness, and elasticity. Biophys J. 103, 1899-1908

170. Tong, J., Canty, J.T., Briggs, M.M., McIntosh, T.J. (2013) The water permeability of lens aquaporin-0 depends on its lipid bilayer environment. Exp Eye Res. 113, 32-40

171. Törnroth-Horsefield, S., Wang, Y., Hedfalk, K., Johanson, U., Karlsson, M., Tajkhorshid, E., Neutze, R., Kjellbom, P. (2006) Structural mechanism of plant aquaporin gating. Nature. 439, 688-694

172. Tournaire-Roux, C., Sutka, M., Javot, H., Gout, E., Gerbeau, P., Luu, D.F.T., Bligny, R., Maurel, C. (2003) Cytosolic pH regulates root water transport during anoxic stress through gating of aquaporins. Nature. 425, 393-397

173. Ueda, M., Tsutsumi, N., Fujimoto, M. (2016) Salt stress induces internalization of plasma membrane aquaporin into the vacuole in *Arabidopsis thaliana*. Biochem Biophys Res Commun. 474, 742-746

174. Vandeleur, R.K., Mayo, G., Shelden, M.C., Gilliham, M., Kaiser, B.N., Tyerman, S.D. (2009) The role of plasma membrane intrinsic protein aquaporins in water transport through roots: diurnal and drought stress responses reveal different strategies between isohydric and anisohydric cultivars of grapevine. Plant Physiol. 149, 445-460

175. Vander-Willigen, C., Postaire, O., Tournaire-Roux, C., Boursiac, Y., Maurel, C. (2006) Expression and inhibition of aquaporins in germinating Arabidopsis seeds. Plant Cell Physiol. 47, 1241-1250

176. Vera-Estrella, R., Barkla, B.J., Bohnert, H.J., Pantoja, O. (2004) Novel regulation of aquaporins during osmotic stress. Plant Physiol. 135, 2318–2329

177. Verdoucq, L., Grondin, A., Maurel, C. (2008) Structure-function analysis of plant aquaporin AtPIP2;1 gating by divalent cations and protons. Biochem J. 415, 409-416

178. Wallace, I.S., Choi, W.G., Roberts, D.M. (2006) The structure, function and regulation of the nodulin 26-like intrinsic protein family of plant aquaglyceroporins. Biochim. Biophys. Acta. 1758, 1165-1175

179. Wallace, I.S., Roberts, D.M. (2005) Distinct transport selectivity of two structural subclasses of the nodulin-like intrinsic protein family of plant aquaglyceroporin channels. Biochemistry. 44, 16826-16834

180. Wan, X., Steudle, E., Hartung, W. (2004) Gating of water channels (aquaporins) in cortical cells of young corn roots by mechanical stimuli (pressure pulses): effects of ABA and of $HgCl_2$. J Exp Bot. 55, 411-422

181. Wang, X., Cai, H., Li, Y., Zhu, Y., Ji, W., Bai, X., Zhu, D., Sun, X. (2014) Ectopic overexpression of a novel Glycine soja stress-induced plasma membrane intrinsic protein increases sensitivity to salt and dehydration in transgenic *Arabidopsis thaliana* plants. J Plant Res. 128, 103-113

182. Wang, L.L., Chen, A.P., Zhong, N.Q., Liu, N., Wu, X.M., Wang, F., Yang, C.L., Romero, M.F., Xia, G.X. (2013) The *Thellungiella salsuginea* tonoplast aquaporin TsTIP1;2 functions in protection against multiple abiotic stresses. Plant Cell Physiol. 55, 148-161

183. Van Wilder, V., Miecielica, U., Degand, H., Derua, R., Waelkens, E., Chaumont, F. (2008) Maize plasma membrane aquaporins belonging to the PIP1 and PIP2 subgroups are *in vivo* phosphorylated. Plant Cell Physiol. 49, 1364-1377

184. Wu, F. Q., Sheng, P. K., Tan, J. J., Chen, X. L., Lu, G. W., Ma, W. W., Heng, Y. Q., Lin, Q. B., Zhu, S. S., Wang, J. L., Wang, J., Guo, X. P., Zhang, X., Lei, C. L., Wan, J. M. (2015) Plasma membrane receptor-like kinase leaf panicle 2 acts downstream of the drought and salt tolerance transcription factor to regulate drought sensitivity in rice. J. Exp. Bot. 66, 271-281

185. Wu, X.N., Sánchez-Rodríguez, C., Pertl-Obermeyer, H., Obermeyer, G., Schulze, W.X. (2013) Sucrose-induced receptor kinase SIRK1 regulates a plasma membrane aquaporin in Arabidopsis. Mol Cell Proteomics. 12, 2856-2873

186. Wudick, M.M., Li, X., Valentini, V., Geldner, N., Chory, J., Lin, J., Maurel, C., Luu, D.T. (2015) Subcellular Redistribution of Root Aquaporins Induced by Hydrogen Peroxide. Mol Plant. 8, 1103-1114

187. Xu, Y., Hu, W., Liu, J., Zhang, J., Jia, C., Miao, H., Xu, B., Jin, Z. (2014) A banana aquaporin gene, MaPIP1;1, is involved in tolerance to drought and salt stresses. BMC Plant Biol. 14, 59

188. Yaneff, A., Sigaut, L., Marquez, M., Alleva, K., Pietrasanta, L.I., Amodeo, G. (2014) Heteromerization of PIP aquaporins affects their intrinsic permeability. Proc Natl Acad Sci USA. 111, 231-236

189. Yang, H.M., Zhang, X.Y., Tang, Q.L., Wang, G.X. (2006) Extracellular calcium is involved in stomatal movement through the regulation of water channels in broad bean. Plant Growth Regul. 50, 79-83

190. Ye, Q., Wiera, B., Steudle, E. (2004) A cohesion/tension mechanism explains the gating of water channels (aquaporins) in Chara internodes by high concentration. J Exp Bot. 55, 449-461

191. Zardoya, R. (2005) Phylogeny and evolution of the major intrinsic protein family. Biol Cell. 97, 397-414

192. Zardoya, R., Ding, X., Kitagawa, Y., Chrispeels, M.J. (2002) Origin of plant glycerol transporters by horizontal gene transfer and functional recruitment. Proc. Natl. Acad. Sci. USA. 99, 14893-14896

193. Zelazny, E., Borst, J.W., Muylaert, M., Batoko, H., Hemminga, M.A., Chaumont, F. (2007) FRET imaging in living maize cells reveals that plasma membrane aquaporins interact to regulate their subcellular localization. Proc Natl Acad Sci U S A. 104, 12359-12364

194. Zelazny, E., Miecielica, U., Borst, J.W., Hemminga, M.A., Chaumont, F. (2009) An N-terminal diacidic motif is required for the trafficking of maize aquaporins ZmPIP2;4 and ZmPIP2;5 to the plasma membrane. Plant J. 57, 346-355

195. Zhang, W.H., Tyerman, S.D. (1999) Inhibition of water channels by $HgCl_2$ in intact wheat root cells. Plant Physiol. 120, 849-858

196. Zhang, J., Yu, H., Zhang, Y., Wang, Y., Li, M., Zhang, J., Duan, L., Zhang, M., Li, Z. (2016) Increased abscisic acid levels in transgenic maize overexpressing *AtLOS5* mediated root ion fluxes and leaf water status under salt stress. J Exp Bot. 67, 1339-1355

197. Zhao, X.Q., Mitani, N., Yamaji, N., Shen, R.F., Ma, J.F. (2010) Involvement of silicon influx transporter *OsNIP2;1* in selenite uptake in rice. Plant Physiol 153, 1871-1877

198. Zhu, D., Wu, Z., Cao, G., Li, J., Wei, J., Tsuge, T., Gu, H., Aoyama, T., Qu, L.J. (2014) TRANSLUCENT GREEN, an ERF family transcription factor, controls water balance in Arabidopsis by activating the expression of aquaporin genes. Mol Plant. 7, 601-615

199. Zimmermann, H.M., Steudle, E. (1998) Apoplastic transport across young maize roots: effect of the exodermis. Planta. 206, 7-19

CHAPTER

# 2

# HOW TO PUT PLANT ROOT UPTAKE INTO A SOIL WATER FLOW MODEL

Xuejun Dong[1,2]

[1] North Dakota State University, Central Grassland Research Extension Center, Streeter, USA

[2] Texas A&M AgriLife Research, Uvalde, USA

## ABSTRACT

The need for improved crop water use efficiency calls for flexible modeling platforms to implement new ideas in plant root uptake and its regulation mechanisms. This paper documents the details of modifying a soil infiltration and redistribution model to include (a) dynamic root growth, (b) non-uniform root distribution and water uptake, (c) the effect of water stress on plant water uptake, and (d) soil evaporation. The paper also demonstrates strategies of using the modified model to simulate soil water dynamics and plant transpiration considering different sensitivity of plants to soil dryness and different mechanisms of root water uptake. In particular, the flexibility of simulating various degrees of compensated uptake (whereby plants tend to maintain potential transpiration under mild water stress) is emphasized. The paper also describes how to estimate unknown root distribution and rooting depth parameters by the use of a simulation-based searching method. The full documentation of the computer code will allow further applications and new development.

## INTRODUCTION

Increased biomass and yield of agricultural crops hinges on improved efficiency of root water uptake[1,2]. Biophysical mechanisms responsible for the improved uptake efficiency in turn rely on linkages between plant physiological processes and the physics of soil water movement; the development of the latter has been well illustrated in 3,4 and 5. Research literature in this and closely related areas has been profound in past decades, though roughly two categories of approach can be identified: one that is mechanistic and one that is phenomenological in nature. The mechanistic approach has improved our understanding of the soil-root interactions by identifying major resistances and driving forces of water flow from soil to roots[3,4], as well as the scale-dependent interactions and geometry among soil organisms responsible for water and nutrient uptake[6,7]. Yet effective mechanistic modeling needs a large amount of empirical data support, which cannot always be met due to various factors constraining research activities. On the other hand, an intuitive, phenomenological approach can be implemented with less empirical data support. Examples of this latter approach may include constraining root water uptake capacity by soil water potential[8], by root length distribution[9–11], and through additional modifications of the root uptake sensitivity to local soil water potential and root density[12–16], which allows compensated uptake to maintain potential transpiration[17,18].

One of the major goals of agronomic management for increased water use efficiency is to channel water loss through transpiration[19]. This to some extent can be aided by the compensated root uptake, which has been implemented in software such as HYDRUS through the use of a threshold root uptake adaptation factor, whereby reduced water uptake in water-stressed parts of the root zone is fully compensated for by increased uptake in other soil regions where water is available[20].

However, root water uptake compensation has been implemented in different studies using different models, which naturally calls for the flexibility of implementation of various forms of the compensation mechanism in root uptake studies. The first objective of this paper is to show that this flexibility can be obtained by using a water flow model extended from an original model as described in 5. This extended model was used in generating numerical solutions of water flow problems in 16.

The second objective of this paper is to provide a tool for quantifying plant water uptake under conditions where information of root depth and distribution pattern is not known. This is relevant to agricultural applications as fine root information remains difficult to obtain despite the progress made in advanced underground sensing[21]. We take advantage of inverse modeling, similar to 22 and 10, to find optimal solutions of unknown root parameters based on good measurement of soil water content, which can be obtained through the use of various soil moisture sensors. Despite excellent documentation of the methods for solving the soil water flow equations relevant to agricultural water management[3,5], the author has not seen a paper documenting the details of how to put the root uptake component into a soil water flow model, as well as providing the details of a search-based optimization procedure for root parameter estimation.

# METHODS

## Implementation

The one dimensional Richards equation of soil water flow in the vertical direction can be written as

$$c\frac{\partial h}{\partial t}=\frac{\partial}{\partial z}\left[K\frac{\partial h}{\partial z}\right]-\frac{\partial K}{\partial z}, \qquad (1)$$

where $h$ is water pressure head (m), $t$ is time (s), $z$ space (m, positive downward and soil surface as zero), $K$ hydraulic conductivity (m/s) and $c=\partial\theta/\partial h$. The reason why there is a negative sign in Equation 1 is that, when we take the space derivative for the gravity potential head $z$, namely, in $\partial z/\partial z$, the values in the numerator change in the opposite direction as that in the denominator: gravity head always decreases downwards but our chosen space $z$ increases downwards.

The model described by Equation 1 does not consider a sink of plant root water uptake. It can be solved using the implicit finite difference method as described by 5. Here is a summary of key points with a few intuitive comments. Let's for now assume surface fluxes as precipitation, irrigation, or surface evaporation. As Equation 1 considers soil water dynamics both in space (vertical axis) and time, we need to properly discretize the equation considering both space and time. Here we let all $h$'s that take space derivatives be approximated using values at the old ($t = n$) and the new ($t = t_{n+1}$) times, but let $K$'s be approximated in old time ($t = n$). All other variables not taking partial derivatives are to be approximated in the old time. Assume the vertical axis is equally divided by $\Delta z$, to approximate Equation 1 at $z = z_i$ and $t_{n+1} = t_n + \Delta t$, we have the following formula to use:

$$\frac{c_{i,n}(h_i^* - h_{i,n})}{\Delta t} = \frac{K_{i+0.5,n}(h_{i+1}^* - h_i^*)}{(\Delta z)^2} - \frac{K_{i-0.5,n}(h_i^* - h_{i-1}^*)}{(\Delta z)^2} - \frac{(K_{i+0.5,n} - K_{i-0.5,n})}{\Delta z}. \tag{2}$$

Another interesting aspect of Equation 2 is that the space index for $K$ is shifted to the upper edge of each of the blocks, while the index of $h$ or $\theta$ is located at the center of the blocks. This is the characteristic of the *block-centered* finite difference scheme and it will enable the calculation of fluxes between two adjacent soil layers, or blocks, as seen in the right sub-figure of Figure 1, which is based on Figure 5–13 of Warrick's book (on page 195). In Figure 1, we reinforce the idea that, similar to the situation of conductivity, $K$, the index for flux density also is located at the edge of the blocks. Intuitively, this makes sense, because in the head-based equation of soil water flow, $K$ and $J$ have the same unit (both are in m/s). As a result, we can compare the magnitude of $K$ and $J$ directly. For example, if the water conductivity of the surface soil is similar to the rate of precipitation, we can roughly say that all the rain water would have time to fully infiltrate into the soil before ponding on the soil surface.

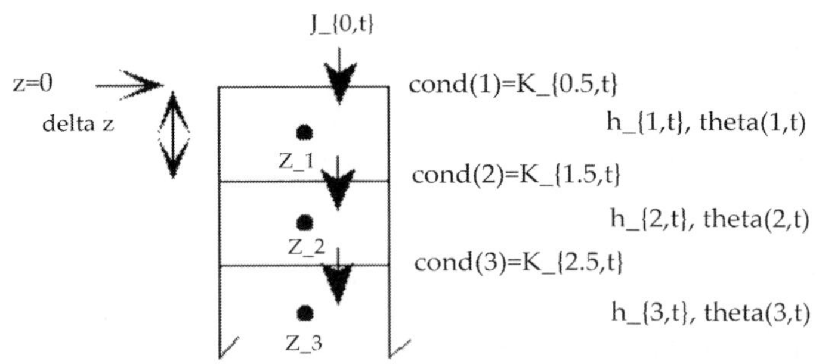

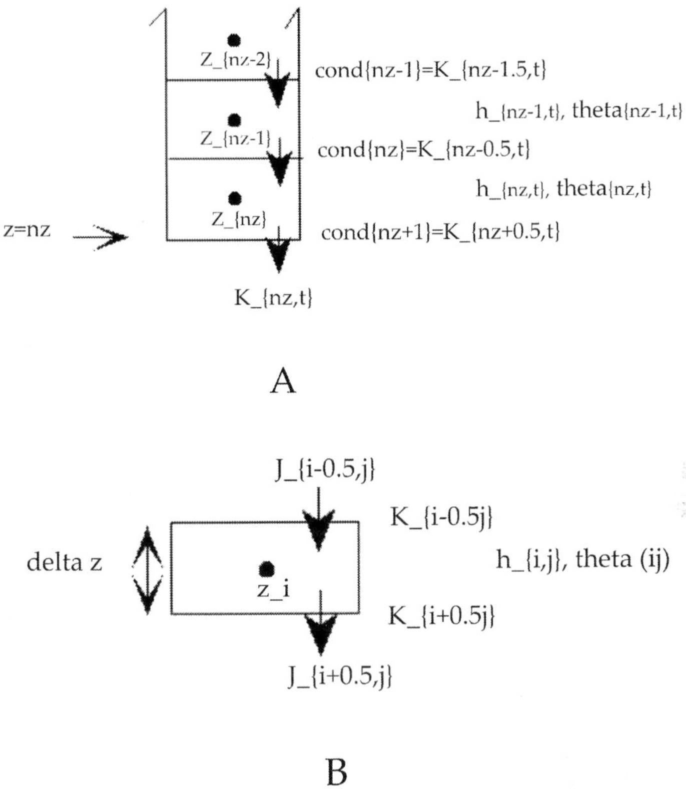


**Figure 1.** Block-centered finite difference scheme for numerical solution of 1-D Richards equation. A. Pressure heads and water conductivities are indexed differently. B. Highlight of one internal block. Modified based on Figure 5–13 of 5.

The terms of the implicit scheme can be collected to form a *tridiagonal* system of linear equations, with the water potentials for the new time (starred) arranged at the left side and those for the old time on the right side of the equation. Then, terms for different space segments can be collected to form the following equation:

$$A_i h_{i-1}^* + B_i h_i^* + C_i h_{i+1}^* = D_i, \quad i = 1, \ldots, nz, \tag{3}$$

which can be written in matrix form as

$$\begin{bmatrix} B_1 & C_1 & 0 & \cdots & 0 & 0 & 0 \\ A_2 & B_2 & C_2 & \cdots & 0 & 0 & 0 \\ \vdots & \vdots & \vdots & \ddots & \vdots & \vdots & \vdots \\ 0 & 0 & 0 & \cdots & A_{nz-1} & B_{nz-1} & C_{nz-1} \\ 0 & 0 & 0 & \cdots & 0 & A_{nz} & B_{nz} \end{bmatrix} \times \begin{bmatrix} h_1^* \\ h_2^* \\ \vdots \\ h_{nz-1}^* \\ h_{nz}^* \end{bmatrix} = \begin{bmatrix} D_1 \\ D_2 \\ \vdots \\ D_{nz-1} \\ D_{nz}. \end{bmatrix} \quad (4)$$

Equation 4 is called tridiagonal because, for the coefficient of the *nz* by *nz* square matrix, the elements are all zero except for those located along the three diagonals. From Equation 4, we can see that $A_1 = C_{nz} = 0$. Actually, values of the coefficients on the left side of Equation 4 should be determined considering the boundary conditions at the soil surface and at the bottom of the root zone.

Detailed solution procedures of the above model "Austere-Layered" (hereafter referred to as AL.f) are provided in 5. Here in order to let this soil water flow model simulate root uptake, the following new components have been added: (a) dynamic root growth, (b) non-uniform root distribution and water uptake, (c) effect of water stress on plant water uptake, and (d) soil evaporation. Following Warrick's original model, the extended model (ALS.f) was also programmed using Fortran 77. To facilitate further development, variables (array variables and non-array variables) used in the original program AL.f. The subroutine SINK() in ALS.f encapsulates various options for root uptake mechanisms. Similar to some other external subroutines, such as THOMAS(), variables inside the subroutine are local. In the following we document major changes made to the program ALS.f:

## Modeling Water Flow in Variably Saturated Soil

The model AL.f was modified to include the flux update scheme proposed by 23 and was applied in program a2&3.for of 5. The main idea is summarized in the following (variables referred to are included in ALS.f):

- Assume we use the flux upper boundary condition (B_TYPE= 2). At each new time step, the old hydraulic conductivities and soil water capacities are used to find the needed coefficients for solving the tridiagonal system of equations. However, the surface flux is temporarily not considered (C2=0). Then, these coefficients are provided to the Thomas subroutine to find the updated pressure head (HSTAR). The trial soil water content at the new time (THETA) is found using the newly computed conductivities based on values of HSTAR as well as the surface flux.

- The trial value of THETA is accepted only if the soil water content at a depth interval, or that for intervals immediately adjacent to this current depth interval, is unsaturated. In such a case, the soil water retention curve is used to update the new pressure head H based on the trial value of THETA.
- Otherwise, the trial value of THETA is rejected, the value of HSTAR is accepted as the updated value for H, and the updated THETA is computed from the soil water retention curve based on HSTAR.

## Testing Different Ending Soil Water Potentials for Optimal Root Uptake

Detailed information regarding the root uptake sink term is included. In particular, Equation illustrates the importance of four water pressure heads for root uptake: $h_1$ marks the anaerobiosis point; $h_4$ the pressure head at permanent wilting point, and $h_2$ and $h_3$ marks the range of pressure head for optimal root uptake. Here we refer to $h_3$ as the ending water potential for optimal uptake, because root uptake starts to reduce when the soil water potential becomes lower than this point. As different plants might have different sensitivity to a mild water stress, it is useful to test the model's sensitivity to $h_3$.

Following 14 and 24, $h_3$ may be assumed as $h^1_3=-11$ m for a slow transpiration of $T^1_p=1$ mm day$^{-1}$ and $h^2_3=-5$ m for a fast transpiration of $T_{2p}=5$ mm day$^{-1}$. We tested two more possibilities of the $h_3$ values, namely, $(h^1_3,h^2_3)=(-14,-8)$ m, and $(h^1_3,h^2_3)=(-16,-10)$, for $(T^1_p,T^2_p)=(1,5)$ mm day$^{-1}$. We also assumed that $h_3$changed in a linear fashion for intermediate transpiration rates between $T^1_p=1$ and $T^2_p=5$ mm day$^{-1}$), using different $h^1_3$ and $h^2_3$ values:

$$h_3 = aT_p + h_3^1 - aT_p^1, \tag{5}$$

where $a=(h^1_3-h^2_3)/(T^1_p-T^2_p)$. This change is reflected in the new subroutine HTHREE, as shown in ALS.f.

## Testing Different Functions for Root Water Uptake

The extended program ALS.f also allows testing different root uptake functions with or without compensation mechanisms, such as those demonstrated in 14,15,25. In particular, in subroutine SINK(), the model by 12 is used when ICPS= 1; a function by 9 is used when ICPS= 3; and the compensation function of 14 (in combination with Wu's function) is used when ICPS= 2. The normalized root density function ($L_{nrd}$) of 9 has four parameters:

The extended program ALS.f also allows testing different root uptake functions with or without compensation mechanisms, such as those demonstrated in 14,15,25. In particular, in subroutine SINK(), the model by 12 is used when ICPS= 1; a function by 9 is used when ICPS= 3; and the compensation function of 14 (in combination with Wu's function) is used when ICPS= 2. The normalized root density function ($L_{nrd}$) of 9 has four parameters:

$$L_{nrd}(z_r) = R_1 + R_2 z_r + R_3 z_r^2 + R_4 z_r^3, \tag{6}$$

where $z_r = z/Z_r(t)$ is the normalized root density at time $t$ with $Z_r(t)$ the maximum rooting depth (m) at time $t$. The coefficients $R_1$ through $R_4$ represent experimentally fitted values for wheat, with $R_1 = 2.21$, $R_2 = -3.72$, $R_3 = 3.46$, and $R_4 = -1.87$ according to 9. The compensation function proposed by 14 at time $t$ has the following form:

$$S_i = \frac{\alpha_i^2 L_{nrd,i}^{\lambda} T_p}{\Delta z_i \sum_{i=1}^{n} \alpha_i L_{nrd,i}^{\lambda}}, \tag{7}$$

where $S_i$ is the actual water uptake from $i$th soil depth increment, $\alpha_i$ is the reduction function for this $i$th depth increment, $L_{nrd,i}$ the normalized root density function for the $i$th depth increment, $\Delta z_i$ the thickness (m) of the $i$th depth increment, and $\lambda$ is a fitting parameter varying from 0.01 to 2.0 according to 14.

## Details of Simulation-Based Searching Program

In order to cope with situations in which the information of rooting depth and root distribution pattern is unknown, the program ALS.f was converted into a subroutine and is callable by a searching main program, ALSS.f. This conversion was quite straight forward in Fortran 77 in that we only did three things: (a) we deleted parameters BETA and ZRMAX from the parameter listing; (b) we converted the program ALS.f from a main program to a subroutine using 2 input dummy variables BETA and ZRMAX and 7 output dummy variables; and (c) we let this subroutine be called multiple times using different values of true BETA and true ZRMAX from arrays BETAS() and ZRMAX() within the main program unit and made sure that only the values of selected 7 variables were returned to the main program after each run but all the other information from within the subroutine ALS() is lost.

## Operation

The provided two Fortran programs, ALS.f and ALSS.f, can be run from one of the freely available compilers, such as g95.

To run ALS.f and ALSS.f, the following input information is needed:

- Soil physical properties and in particular soil water retention curve and saturated hydraulic conductivity. A method for predicting soil water retention curves based on easily measured soil texture and bulk density data is illustrated in 26.
- Initial soil water content profile. This must be provided and put to exact format as shown in INITF.DAT, following the template provided by 5. The measured soil water content must be interpolated so that every depth increment of soil has a water content value in it. One method of data interpolation is outlined in 16 using routines provided in 27 and 28.
- Measured terminating soil water content. This can be the measured soil water content on last day of simulation and must be entered in the format as shown in TERMF.DAT. No interpolation is needed for this file.
- Daily weather data. Daily evapo-transpiration rate, daily average air temperature, daily rainfall, relatively humidity, as

well as leaf area index, must be entered in data file WEA2.DAT in the order as shown in the data file. Daily leaf area index can be interpolated based on field measured result, as shown in 16.

There are four output files from ALS.f:

- OUTS.DAT – Mirrors output from AL.f of 5.
- OUTS2.DAT – Daily soil water potential and water contents for interested soil depths.
- INFS.DAT – Mirrors output from AL.f of 5.
- WUE.DAT – Summary of daily cumulative water loss.

There are seven output files from ALSS.f. They provide the following information based on particular realization of combinations of root distribution shape factor BETAS and the maximum rooting depth ZRMAXS:

- ARD.DAT – Average relative discrepancy between measured and simulated soil water content throughout the simulation period.
- C_AEVA.DAT – Cumulative soil evaporation throughout the simulation period.
- C_ATRA.DAT – Cumulative transpiration throughout the simulation period.
- WBELOW.DAT – Amount of water to deep drainage out of the specified root zone. This was inherited from AL.f
- WADD.DAT – Amount of water added to the system. This was also inherited from AL.f
- BIGI.DAT – Cumulative infiltration. This was also inherited from AL.f
- PCTDIF.DAT – Percent difference in water balance, defined in 5 as (BIGI-WADD)/WADD*100.

The main difference between the output files of ALS.f and those of ALSS.f is that the outputs from the former are a summary of water balance from the simulation of soil water balance based on one set of root parameters (rooting depth and shape factors), while those from the latter (ALSS.f) contain the simulation outcomes using many sets of root parameters (see next section for further details).

## USE CASES

Programs ALS.f and ALSS.f in their current forms were customized to simulate the soil water dynamics for 111 days from May 14 to September 1, 2009. This is reflected both in the programs and the input data files. To use the model in other locations and other durations, appropriate changes must be made both in a few parameters of the programs and input data files. The two potential uses of the programs are outlined in the following:

- If root distribution parameter BETA and maximum rooting depth ZRMAX are known, then program ALS.f may be used to simulate soil water dynamics with root uptake. Make sure to provide updated values for the major parameter values, such as JMATUR (Julian day number of growth maturity), JTHAW (Julian day number of growth initiation), BETA, and ZRMAX.
- If root distribution parameter BETA and maximum rooting depth ZRMAX are unknown, then it is recommended to first use ALSS.f to find the optimal values for BETA and ZRMAX before using ALS.f to simulate soil water dynamics. Before running ALSS.f, three things must be done:
  - Check to see if the current value of $h_3$ is correct by changing, if necessary, the value of parameter IHTH (current default value is set to 1, meaning ($h^1_3$, $h^2_3$)=(−11,−5)). See the preamble of subroutine SINK() for two other options.
  - Check to see if the parameter of ICPSN is correctly specified. currently, the default value is 1, meaning to use Ojha's compensation function. Setting the value of ICPSN to 2 will switch to Li's compensation function, and a value of 3 will invoke Wu's uptake function (without compensation).
  - Check the nested DO LOOPs structure to adjust the admissible values for index variable II (currently it is set to have 16 different values for BETA) and JJ (currently it is set to have 22 different values for ZRMAX). At the same time, the following lines in the main program of ALSS.f must be changed to be compatible with the range of index values in the nested DO LOOPS.

```
BETAS(1)=0.5
DO I=1,15
  BETAS(I+1)=BETAS(I)+0.25
END DO

ZRMAXS(1)=0.75
DO I=1,21
  ZRMAXS(I+1)=ZRMAXS(I)+0.05
END DO
```

- Finally, it is recommended to put both the program files and the required input data files in the same folder before running the programs.

## SUMMARY

While plant root water uptake has been implemented in commercial software packages, very few researchers have found the time to document important details of how root uptake can be included in a soil water flow model. This paper tries to fill this knowledge gap by using a simple computer program aided with sample input data sets. Specifically, we modified and extended a soil water flow model of 5 to include plant root water uptake capability. We showed that the extended program has the flexibility of considering different sensitivity of root uptake, as well as different water uptake mechanisms. We also modified the model to search for optimal root parameters based on measured soil water content data. This is relevant for new application in that field data of root development usually is expensive and sometimes difficult to obtain, and the utilization of a searching-based modeling approach will help to identify root parameters for improved water balance characterization, which in turn may help to identify ways to improve crop water use efficiency under field conditions. No menu-driven user interface is provided but the author believes that, under some circumstances of knowledge discovery, it is more efficient to work directly with a well-structured computer program, such as the one described in this paper, than relying on a packaged one.

## ACKNOWLEDGMENTS

The author thanks Bob Patton, Paul and Anne Nyren and Lyle Prunty at North Dakota State University for fruitful collaboration leading to this paper.

## REFERENCES

1. Campbell GS, Diaz R: Simplified soil-water balance models to predict crop transpiration. In F. R. Bidinger and C. Johansen, editors, *Drought research priorities for the dryland tropics.* International Crops Institute for the Semi-Arid Tropics, Parancheru, A. P. 502 324, India, 1988; 15–26.
2. Passioura JB, Angus JF: Improving productivity of crops in water-limited environments. In D. L. Sparks, editor, *Advances in Agronomy.* Elsevier Science, USA, 2010; **106**: 37–75.
3. Campbell GS: Soil Physics with Basic: Transport Models for Soil–Plant Systems. Developments in Soil Science 14. Elsevier, New York, 1985.
4. Hopmans JW, Bristow KL: Current capabilities and future needs of root water and nutrient uptake modeling. In D. L. Sparks, editor, *Advances in Agronomy.* Elsevier Science, USA, 2002; **77**: 103–183.
5. Warrick AW: Soil Water Dynamics. Oxford University Press, 2003.
6. Roose T, Schnepf A: Mathematical models of plant-soil interaction. *Philos Trans A Math Phys Eng Sci.* 2008; **366**(1885): 4597–4611.
7. Leitner D, Schnepf A, Klepsch S, *et al.*: Comparison of nutrient uptake between three-dimensional simulation and an averaged root system model. *Plant Biosystem.* 2010; **144**(2): 443–447.
8. Feddes RA, Kowalik PJ, Zaradny H: Simulation of field water use and crop yield. Center for Agricultural Publishing and Documentation, Wageningen, The Netherlands, 1978.
9. Wu J, Zhang R, Gui S: Modeling soil water movement with water uptake by roots. *Plant Soil.* 1999; **215**(1): 7–17.
10. Zuo Q, Zhang R: Estimating root-water-uptake using an inverse method. *Soil Sci.* 2002; **167**(9): 561–571.
11. Zuo Q, Jie F, Zhang R, *et al.*: A generalized function of wheat's root length density distributions. *Vadose Zone J.* 2004; **3**: 271–277.
12. Ojha CS, Rai AK: Nonlinear root water uptake model. *J Irrig Drain Eng.* 1996; **122**(4): 198–201.
13. Adiku SG, Rose CW, Braddock RD, *et al.*: On the simulation of root water extraction: examination of a minimum energy hypothesis. *Soil Sci.* 2000; **165**(3): 226–236.

14. Li KY, De Jong R, Boisvert JB: An exponential root water uptake model with water stress compensation. *J Hydrol.* 2001; **252**(1–4): 189–204.
15. Yadav BK, Mathur S, Siebel MA: Soil moisture dynamics modeling considering the root compensation mechanism for water uptake by plants. *J Hydrol Eng.* 2009; **14**(9): 913–922.
16. Dong X, Patton BD, Nyren AC, *et al.*: Quantifying root water extraction by rangeland plants through soil water modeling. *Plant Soil.* 2010; **335**(1): 181–198.
17. Heinen M: Compensation in root water uptake models combined with three-dimensional root length density distribution. *Vadose Zone J.* 2014; **13**(2).
18. Verma P, Loheide SP, Eamus D, *et al.*: Root water compensation sustains transpiration rates in an Australian woodland. *Adv Water Resour.* 2014; **74**: 91–101.
19. Bodner G, Nakhforoosh A, Kaul HP: Management of crop water under drought: A review. *Agron Sustain Dev.*2015; **35**(2): 401–442.
20. Simunek J, Hopmans JW: Modeling compensated root water and nutrient uptake. *Ecol Model.* 2009; **220**(4): 505–521.
21. Jayawickreme DH, Jobbágy EG, Jackson RB: Geophysical subsurface imaging for ecological applications.*New Phytol.* 2014; **201**(4): 1170–1175.
22. Musters PA, Bouten W: A method for identifying optimum strategies of measuring soil water contents for calibrating a root water uptake model. *J hydrol.* 2000; **227**(1–4): 273–286.
23. Kirkland MR, Hills RG, Wierenga PJ: Algorithms for solving Richards' equation for variably saturated soils.*Water Resour Res.* 1992; **28**(8): 2049–2058.
24. Yang D, Zhang T, Zhang K, *et al.*: An easily implemented agro-hydrological procedure with dynamic root simulation for water transfer in the crop-soil system: validation and application. *J Hydrol.* 2009; **370**(1–4): 177–190.
25. Lai CT, Katul G: The dynamic role of root-water uptake in coupling potential to actual transpiration. *Adv Water Resour.* 2000; **23**(4): 427–439.

26. Dong X, Patton BD: Predicting soil water retention curves based on particle-size distribution using a Minitab macro. *Afr J Soil Sci.* 2015; **3**(1): 079–085.
27. Doughty S: Solve It: Computer Aided Mathematics for Science and Engineering. Gulf Publishing Company, 1995.
28. Hanna OT, Sandall OC: Computational Methods in Chemical Engineering. Prentice Hall PTR, 1995.
29. Dong X: A plant root water uptake model. *Zenodo.* 2015.
30. Dong X: A program for searching optimal root parameters in plant water uptake modeling. *Zenodo.* 2015.

CHAPTER

# 3

# PLANT ADAPTATION TO DROUGHT STRESS

Supratim Basu, Venkategowda Ramegowda, Anuj Kumar, Andy Pereira

Crop, Soil, and Environmental Sciences, University of Arkansas, Fayetteville, Arkansas, 72701, USA

## ABSTRACT

Plants in their natural habitats adapt to drought stress in the environment through a variety of mechanisms, ranging from transient responses to low soil moisture to major survival mechanisms of escape by early flowering in absence of seasonal rainfall. However, crop plants selected by humans to yield products such as grain, vegetable, or fruit in favorable environments with high inputs of water and fertilizer are expected to yield an economic product in response to inputs. Crop plants selected for their economic yield need to survive drought stress through mechanisms that maintain crop yield. Studies on model plants for their survival under stress do not, therefore, always translate to yield of crop plants under stress, and different aspects of drought stress response need to be emphasized. The crop plant model rice (*Oryza sativa*) is used here as an example to highlight mechanisms and genes for adaptation of crop plants to drought stress.

## INTRODUCTION

Drought stress is the most prevalent environmental factor limiting crop productivity[1], and global climate change is increasing the frequency of severe drought conditions[2]. The sheer diversity of plant species grown across climatic regions that include extreme dry conditions suggests that, in nature, plants have evolved to endure drought stress with an array of morphological, physiological, and biochemical adaptations[3]. 'Drought resistance' (DR) is a broader term applied to plant species with adaptive features that enable them to escape, avoid, or tolerate drought stress[4]. 'Drought escape' is the ability of a plant species to complete its life cycle before the onset of drought. Thereby, plants do not experience drought stress, as they are able to modulate their vegetative and reproductive growth according to water availability, essentially through two different mechanisms: rapid phenological development and developmental plasticity[5]. Rapid phenological development involves rapid plant growth, producing a minimal number of seeds before the soil water depletes, and these plants are considered not to have any special morphological, physiological, or biochemical adaptations. Plants with mechanisms of developmental plasticity show little growth during the dry season, with very few flowers and seeds, but in wet seasons they grow indeterminately, producing a large amount of seed. 'Drought avoidance' is the ability of plants to maintain (relatively) higher tissue water content despite reduced water content in the soil[4]. This is achieved through a variety of adaptive traits involving the minimization of water loss (water savers) and optimization of water uptake (water spenders). Water spenders achieve higher tissue water status by maintaining the water uptake through increased rooting, hydraulic conductance, etc. under drought stress. In contrast, water savers use water effectively through reduced loss of water by reducing transpiration, transpiration area, radiation absorption, etc. under drought stress. 'Drought tolerance' (DT) is the ability of plants to endure low tissue water content through adaptive traits. These adaptive traits involve maintenance of cell turgor through osmotic adjustment and cellular elasticity, and increasing protoplasmic resistance[6].

Improvement of yield and maintaining yield stability of crops, under normal as well as drought stress conditions, is essential for the food security of the growing global population. It is difficult to resolve the role of different components of DR in the stability of the crop yield as the major objective. However, there exist a variety of different mechanisms for drought escape, avoidance, or tolerance in natural populations that can improve DR and maintain grain yield in crop plants. In nature, extreme DR is found in resurrection plants[7,8] which possess strong drought escape mechanisms. Resurrection plants can be exposed to severe drought for months, extending up to years, forcing them to optimize their growth for survival, but not for seed production, in the long term[9]. Therefore, the DR mechanisms that enable plants to merely survive longer lead to subsistence yield, which is much lower than that which is observed under normal conditions. Crop plants, on the other hand, are grown by humans in environments under conditions for high agricultural production and will be exposed to only a random short-term drought stress of days to weeks, from which they must quickly respond to limit the damage caused by short-term drought stress while they continue to grow and yield in the stressful environments. Therefore, bringing in the drought adaptive mechanisms from plants adapted to grow in extreme dry conditions may not be a feasible option, as it may result in growth and/or yield penalty in crop plants under drought as well as normal conditions.

Although plant survival is very critical in the early growth stages, the mechanisms have little relevance to increasing grain yield directly. The emphasis to improve DR of crop plants should therefore be based on stability of yield components and not on plant survival alone. So far, most of the efforts to improve grain yield under drought stress were focused on secondary traits such as root architecture, leaf water potential, osmotic adjustment, and relative water content at the vegetative stage, which are often not highly correlated with grain yield[10,11]. Looking forward in crops, the effective drought improvement approach should be selection for yield and its component traits under reproductive-stage drought stress[12]. Additionally, little importance has also been given to simultaneous improvement of grain yield under normal and drought conditions. Selection for DT has been suggested to have a yield drag under normal conditions. It has been proposed that the yield potential of crop plants should be simultaneously selected for under favorable and environmental stress conditions, as there is a positive correlation between yield potential under normal and drought stress conditions[13]. Combining high yield potential under normal conditions with good yield under drought stress is the ideal trait. Identification of mechanisms, traits, and genes regulating yield under drought stress that are free from yield drag under normal conditions should be the focus. For example, regulation of yield under normal as well as drought stress conditions has been shown for three NAC family transcription factors (TFs). Transgenic plants expressing *OsNAC5*, *OsNAC9*, and *OsNAC10* TFs showed an increase in grain yield of 5-26% under normal conditions[14–16]. Nevertheless, in these studies, yield under normal conditions has been overlooked with more emphasis given to yield under drought stress. Two of our recent studies show the potential of simultaneously improving and stabilizing grain yield, both under normal as well as drought stress conditions, using two regulatory genes, namely *GUDK* and *HYR* in rice[17,18]. These studies indicate that it might be advantageous to identify mechanisms and genes for increasing grain yield that are also stable or maintained under drought stress conditions.

Despite the complexity of DR, tremendous progress has been made in understanding the drought-adaptive mechanisms of plants[1,19,20]. Adaptation through DR mainly involves morpho-physiological alterations. These alterations in adaptive processes are controlled by molecular mechanisms that regulate the expression of genes[21]. There exists a large diversity in drought adaptation within a crop species, as some genotypes are able to cope with drought better than others. Genotypes that differ in drought adaptive mechanisms serve as an important resource to study the variation in drought adaption in crop plants. This natural variation needs to be exploited to simultaneously improve DR and yields of cultivated varieties through better understanding of the underlying mechanisms and to aid in selection for these traits[22]. In the following sections, we describe the widely known morpho-physiological processes and recent molecular advances in regulating these drought-adaptive processes leading to increased yield in crop plants.

## PHOTOSYNTHESIS

Drought stress is known to reduce photosynthesis by decreasing both leaf area and photosynthetic rate per unit leaf area. Reduced photosynthetic rate is mainly through stomatal closure or metabolic impairment[23]. Continued photosynthetic light reactions during drought stress under limited intercellular $CO_2$ concentration results in the accumulation of reduced photosynthetic electron transport components, which can potentially reduce molecular oxygen, resulting in the production of reactive oxygen species (ROS). ROS can cause severe damage to the photosynthetic apparatus[24]. The adaptive responses that plants have developed to reduce drought-induced damage to photosynthesis include thermal dissipation of light energy, the xanthophyll cycle, the water-water cycle, and dissociation of the light-harvesting complexes from photosynthetic reaction centers[25–27]. The metabolic impairment during drought stress is mainly caused by changes in photosynthetic carbon metabolism[24]. The biochemical efficiency of photosynthesis under drought stress mainly depends on ribulose-1,5-bisphosphate (RuBP) regeneration and the activity of ribulose-1,5-bisphosphate carboxylase/oxygenase (RuBisCO)[28,29]. Considerable progress has been made in improving the stomatal components for $CO_2$ diffusion, photosynthetic light reaction, and metabolic changes, including the expression of photosynthesis-related genes to regulate photosynthesis under drought towards the improvement of grain yield[30].

The C4 pathway of carbon assimilation has been suggested to be the major adaptation of the C3 pathway to limit water loss, reduce photorespiration, and improve photosynthetic efficiency under drought stress[31]. However, many important crops—including rice, wheat, soybean, and potato—use the C3 pathway of photosynthesis. Although the transfer of the C4 pathway into C3 crops is underway, so far its contribution to increased grain yield is very limited[32]. Photosynthetic adaptation of plants to drought stress involves a complex interaction of hormones, ROS, sugars, and other metabolic events[33]. Combinations of computational models, which integrate the physiological and metabolic processes with gene expression data, along with modern breeding and transgenic technologies hold promise in improving photosynthesis and hence crop yield under normal as well as drought stress conditions.

In recent studies, we used a rice gene regulatory network to identify a TF termed HYR (HIGHER YIELD RICE), which was highly associated with primary carbon metabolism[17], and on overexpression in rice enhanced photosynthesis under normal conditions as well as under drought and high temperature stress. HYR regulates several morpho-physiological processes leading to higher yield under normal and environmental stress conditions. Our study showed that HYR is a master regulator of photosynthesis, directly activating photosynthesis genes, cascades of

TFs, and other downstream genes involved in photosynthetic carbon metabolism, resulting in improved yield.

## HORMONAL REGULATION

Major phytohormones, such as abscisic acid (ABA), cytokinin (CK), gibberellic acid (GA), auxin, and ethylene, regulate diverse processes which enable plant adaptation to drought stress[34]. Upon exposure of plants to drought stress, ABA is the major hormone synthesized in roots and translocated to leaves to initiate adaptation of plants to drought stress through stomatal closure and reduced plant growth[35]. However, modulating the ABA-induced drought adaptation of plants for better yield remains a greater challenge because of the potential inadvertent reduction in carbon gain upon stomatal closure and ABA-induced senescence, especially if the drought occurs at the reproductive stage[36]. There are ABA signaling genes, such as *OsNAP*, *OsNAC5*, and *DSM2*, which promote improved yield under reproductive drought[37–40]. These ABA-induced non-stomatal adaptations of plants under drought stress can be exploited to improve grain yield under reproductive drought (Figure 1).

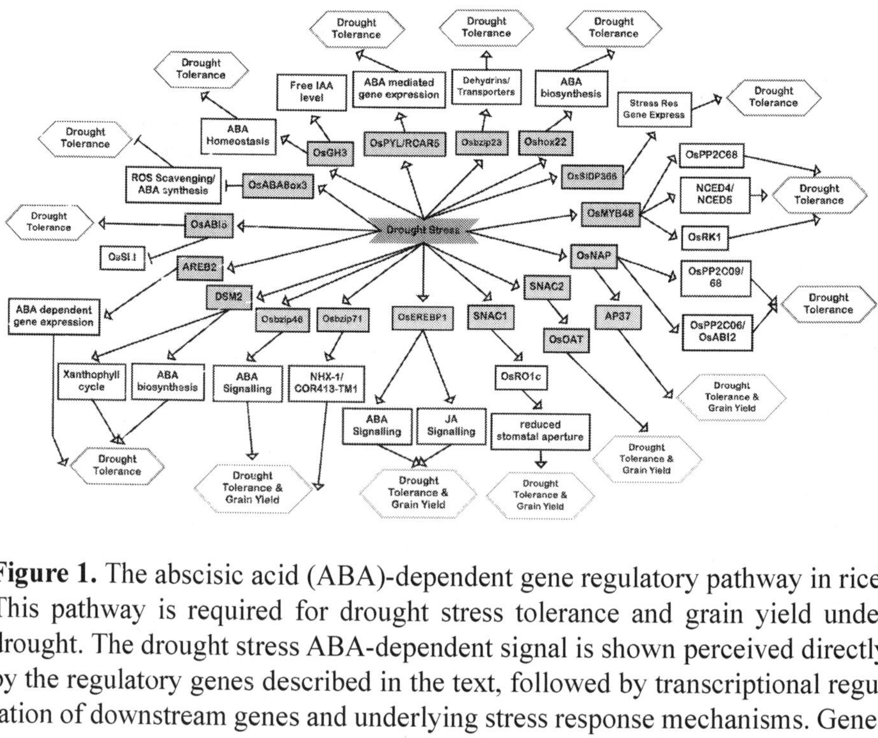


**Figure 1.** The abscisic acid (ABA)-dependent gene regulatory pathway in rice. This pathway is required for drought stress tolerance and grain yield under drought. The drought stress ABA-dependent signal is shown perceived directly by the regulatory genes described in the text, followed by transcriptional regulation of downstream genes and underlying stress response mechanisms. Genes

regulating drought tolerance (DT) at the vegetative stage are shaded green, and genes regulating DT and grain yield under drought are shaded orange. The resulting phenotypes are represented for DT at the vegetative level only (green diamonds), or DT and grain yield (orange diamonds). The genes described here are *OsGH3*[92], *OsNAP*, *OsABI2*[39], *AP37*[93], *OsPP2C09*, *OsPP2C06*[94], *OsPYL/RCAR5*, *OsSIDP366*[95], *OsMYB48*[96], *OsRK1*[97], *Oshox22*[98], *SNAC2*[99,100], *OsOAT*[101], *OsbZIP23*[102], SNAC1[99], OsEREBP1[103], *OsbZIP71*[104], *OsbZIP46*[105], *OsABI5*[106], *DSM2*[40], *AREB2*[107], *OsSRO1c*[108], and *OsABA8OX3*[109].

Under drought stress, CKs are known to delay premature leaf senescence and death, adaptive traits very useful for increasing grain yield. An increase in the endogenous levels of CK through expression of *isopentenyltransferase* (*IPT*), a CK biosynthetic pathway gene, leads to stress adaptation by delaying drought-induced senescence and an increase in yield[41,42]. Generally, auxin has been shown to negatively regulate drought adaptation in plants. Decrease in indole-3-acetic acid (IAA) content was shown to be associated with up-regulation of genes encoding late embryogenesis abundant (LEA) proteins, leading to drought adaptation in plants[43,44]. Recently, the *DEEPER ROOTING 1* (*DRO1*) gene determining a quantitative trait locus (QTL) controlling root growth angle was shown to be negatively regulated by auxin. Higher expression of *DRO1* in a shallow-rooting rice cultivar resulted in drought avoidance and high yield under drought[45]. GA is suggested to positively regulate plant adaptation to drought stress. A rapid decline in levels of endogenous GA was observed in plants subjected to drought stress, resulting in growth inhibition[46]. The role of GA in regulating grain yield of crop plants is thus an important area that can be further explored. Ethylene is a negative regulator of drought stress response by promoting leaf senescence and inhibiting root growth and development, shoot/leaf expansion, and photosynthesis[47–51]. Ethylene can also directly affect yield by increasing embryo and grain abortion and reducing the grain-filling rate[52]. In addition to the major hormones, other hormones such as brassinosteroids, jasmonic acid (JA), salicylic acid (SA), and strigolactone also have an equally important role in plant growth and development. However, their function under drought stress is relatively less characterized. Tillering in rice has been suggested to be the outcome of an interaction among three hormones, CK, auxin, and strigolactone, with CK promoting branching and the other two inhibiting it[53,54], suggesting that all hormones do not act in isolation but instead interact and modulate each other's biosynthesis and responses. Therefore, the net outcome of the drought stress response is regulated by a balance between hormones that promote and those that inhibit the trait, rather than individual hormones.

## TRANSPIRATION AND STOMATAL CONDUCTANCE

The immediate response of plants on being exposed to drought stress is stomatal closure. However, stomatal closure not only diminishes water loss through transpiration but also reduces $CO_2$ and nutrient uptake, and hence alters metabolic pathways such as photosynthesis[55]. Plants growing in dry areas have developed xeromorphic traits to reduce transpiration under drought stress. Reduction in transpiration under drought stress conditions can also be achieved through leaf shedding (i.e. deciduous species in drought) as well as decrease in leaf number, leaf size, and branching. Another adaptation to counter drought stress is sclerophylly, where plants form hard leaves that will not suffer permanent damage due to wilting and can be restored to full functionality when normal conditions resume[56]. Recent research has shown that decreased stomatal conductance in response to drought stress is related not only to reduced expression of aquaporin genes but also to anatomical traits leading to reduction of chloroplast surface area exposed to intercellular space per unit leaf area[57,58]. Several other factors, including leaf developmental stage and light availability, are also known to interact with drought in modulating mesophyll and chloroplast differentiation, ultimately affecting conductance and photosynthetic capacity[58]. Reduction in stomatal size and number on exposure to drought is another adaptation for survival under drought conditions. Previous studies have shown that while there is an increase in stomatal density under mild drought stress, there is a decrease during severe drought[59]. Thus, all these adaptations in plants reduce the negative impacts of drought stress on photosynthesis and thereby have a positive effect on water use efficiency (WUE), which in turn will result in high yield potential and high yield[60]. Such an adaptation was shown in rice by overexpression of the Arabidopsis AP2/ERF TF *HARDY* that improved WUE (the ratio of biomass produced to water used) by enhancing photosynthesis and reducing transpiration[61].

The above reported traits therefore exemplify adaptive mechanisms in plants to survive under drought stress without loss of productivity or yield.

## ROOT MORPHOLOGY

In many agriculturally important crops, drought stress is perceived first by the root system, which continues to grow underneath the soil even though shoot growth is inhibited under these conditions[62]. Although the growth of the primary root is not affected by drought stress, the growth of lateral roots is significantly reduced, mainly by suppression of the activation of the lateral root meristems[63]. The Arabidopsis R2R3-type MYB TF *MYB96* has been shown to regulate activation of lateral root meristem through an ABA signaling cascade, with an activation-tagged mutant showing enhanced DR with reduced lateral root formation[64]. The plant microRNA miR393 has also been shown to play a role in root-mediated adaptation to drought stress response through attenuation of auxin signaling[65]. In addition to the lateral roots, the presence of small roots is also considered as an adaptive strategy to increase water uptake by providing more absorptive surface. Presence of specialized tissues like rhizodermis, with a thickened outer cell wall or suberized exodermis, or reduction in the number of cortical layers are considered an adaptive advantage for drought stress survival. Hydrotropism is another adaptive measure taken by plants to counter stress, where studies have shown that degradation of amyloplasts in the columella cells of plant roots on exposure to drought stress increases hydrotropism[66,67]. Hormonal cross-talk mediated by auxin, CK, GA, and ABA has been implicated as a potential chemical signal in response to water stress to modulate root system architecture[68].

The expression of enzymes related to root morphology (e.g. xyloglucan endotransglucosylase) is induced upon mild drought stress, while other structural proteins are down-regulated, which is strongly correlated with root growth and hence an augmentation in the surface area for water uptake. The alterations in the expression of these proteins correlate positively with lateral development that in turn also affects photosynthesis[69]. More lateral root and root hair formation was found in lines possessing a QTL, qDTY12.1, only when under drought[70]. Such traits, which are expressed only under drought stress, have higher potential to increase grain yield under drought. Moreover, it has also been shown that drought stress triggers a wide variety of anatomical traits expressed to different levels and patterns in different species and even in different cultivars within species[71–73]. For example, suberization and compaction of sclerenchyma layer cells were shown to decrease in rice under drought, which increases retention of water under drought stress[71].

## OSMOTIC ADJUSTMENT

Osmotic adjustment (OA) is defined as a process of solute accumulation in dividing cells when the water potential is reduced, and thereby helps in maintaining the turgor[74]. Cell enlargement and growth in plants is highly dependent on water availability and helps in maintaining the turgor. Turgor measurement in growing regions of plants, especially the leaves and stems, shows little or no reduction, though cell enlargement is inhibited during drought stress and is believed to be due to OA[75,76]. Under conditions of drought stress, OA has been implicated in maintaining stomatal conductance, photosynthesis, leaf water volume, and growth[74,77]. At times of drought stress, in addition to the reduction in water content, there are also other associated changes, such as increases in salt concentration and mechanical impedance[78]. Inorganic cations, organic acids, carbohydrates, and free amino acids are the known predominant solutes that accumulate in response to water stress. Previous studies have shown that drought-resistant wheat varieties, with yield stability under drought stress, have a greater capacity for osmoregulation than less resistant varieties[76]. The accumulation of compatible solutes such as proline and glycine betaine help in protecting the plants from detrimental effects of drought stress not only by OA but also by detoxification of ROS, protection of membrane integrity, and stabilization of enzymes or proteins[79]. Enzymes such as betaine aldehyde dehydrogenase (BADH), pyrroline-5-carboxylate reductase (P5CR), and ornithine δ-aminotransferase (OAT) have been shown to play major roles in OA. Overexpression of Arabidopsis *EDT1/HDG11* was shown to increase DT of poplar and cotton through increased accumulation of solutes such as proline and soluble sugars and also increase the yield of cotton in the field[80]. However, there are some plants in which sugars are the main osmolytes that play a significant role in OA, including sucrose, trehalose, glucose, and fructose. Previous studies have shown that overexpression of the *sucrose:fructan-6-fructosyltransferase*

*(6-SFT)* gene from *Psathyrostachys huashanica* in tobacco and the trehalose-6-phosphate phosphatase gene *OsTPP1* in rice confers abiotic stress tolerance[81,82]. Researchers have also identified a QTL for OA on chromosome 8 in rice that is homeologous with a segment of wheat chromosome 7[83].

## SOURCE-SINK RELATIONSHIPS

Source-sink relationships largely determine the grain yield of cereal crops, with developing grains being primary sinks while the top two leaves, the flag leaf in particular, are the primary source[84,85]. Drought stress affects the source-sink relationship by reducing the source strength, leading to yield reduction. Sufficient sugar supply through photosynthesis, transport, and conversion of sugars is regarded as the most critical component in determining the viability of reproductive organs in rice[86–88]. Drought stress dramatically affects pollen viability due to abnormal starch accumulation[89]. Insufficient starch synthesis and arrested pollen development have been linked to reduced invertase activity under drought stress[88]. In addition to invertases, active grain filling involves other key enzymes, such as sucrose synthase, ADP glucose pyrophosphorylase (AGPase), and starch synthase as well as starch branching and debranching enzymes, which are also affected by drought stress[88]. Therefore, enhancing rice yield through source-sink relationships involves not only sugar metabolism but also the regulated mobilization of metabolic resources from source to sink tissue.

## FUTURE RESEARCH PERSPECTIVES

Improving DR in crop plants is a challenge for plant breeders and crop physiologists, as it is a complex genetic trait with multiple pathways involved. Effective development of drought-resistant crop plants thus requires the pyramiding and interaction of many mechanisms, traits, and genes that are appropriate to individual crops and their growing environments. Success in this direction not only extends the growing area of crop plants but also achieves stable yield in drought-prone areas. Identifying genetic variation for DR is the first step towards development of drought-resistant crop plants. Such variation is often present in wild species and adapted genotypes that have evolved under natural selection and these are the best source of DR traits. Evaluation of these resources through an integrated phenotyping and genotyping approach under field conditions alongside identification of traits that are directly associated with yield is key to improving DR[90]. Comparative omics analyses between the diverse germplasm could aid in bettering our understanding of the variety of crop adaptations to drought stress. In addition, analysis of specific genes focused on increasing DR while stabilizing the yield is crucial for understanding the broad basis of complex traits such as DR[91]. The development of high-yielding resilient crops[74] that maintain yield stability under drought and other environmental stresses due to climate change is also currently needed. Drought-resistant plants should combine a better root system, stomatal regulation, WUE, and hormonal balance while avoiding the negative effects on grain yield under both normal and drought stress conditions. Therefore, a holistic crop improvement strategy should involve the deployment of high crop yield potential and the utilization of a combination of morpho-physiological, biochemical, and anatomical adaptive responses to drought stress.

## AUTHOR CONTRIBUTIONS

All authors contributed to writing the manuscript. Supratim Basu and Andy Pereira prepared Figure 1.

## REFERENCES

1. Bray EA: Plant responses to water deficit. *Trends Plant Sci.* 1997; 2(2): 48–54.
2. Dai A: Increasing drought under global warming in observations and models. *Nat Clim Chang.* 2012; 3: 52–8.
3. Bohnert HJ, Nelson DE, Jensen RG: Adaptations to Environmental Stresses. *Plant Cell.* 1995; 7(7): 1099–111.
4. Levitt J: Responses of plants to environmental stresses. Volume II. Water, radiation, salt, and other stresses. (Academic Press), 1980. Reference Source
5. Jones MM, Turner NC, Osmond CB: Mechanisms of drought resistance. The physiology and biochemistry of drought resistance in plants. 1981; 1: 15–37.

6. Morgan JM: Osmoregulation and Water Stress in Higher Plants. *Annu Rev Plant Physiol.* 1984; 35: 299–319.
7. Gaff DF: Desiccation tolerant plants in South America. *Oecologia.* 1987; 74(1): 133–6.
8. Gaff DF: Desiccation-tolerant flowering plants in southern Africa. *Science.* 1971; 174(4013): 1033–4.
9. Porembski S: Granitic and gneissic outcrops (inselbergs) as centers of diversity for desiccation-tolerant vascular plants. *Plant Ecol.* 2000; 151(1): 19–28.
10. Jongdee B, Fukai S, Cooper M: Leaf water potential and osmotic adjustment as physiological traits to improve drought tolerance in rice. *Field Crops Res.* 2002; 76(2–3): 153–63.
11. Pantuwan G, Fukai S, Cooper M, *et al.*: Yield response of rice (*Oryza sativa* L.) genotypes to drought under rainfed lowland: 3. Plant factors contributing to drought resistance. *Field Crops Res.* 2002; 73(2–3): 181–200.
12. Venuprasad R, Lafitte HR, Atlin GN: Response to Direct Selection for Grain Yield under Drought Stress in Rice. *Crop Sci.* 2007; 47(1): 285–293.
13. Guan YS, Serraj R, Liu SH, *et al.*: Simultaneously improving yield under drought stress and non-stress conditions: a case study of rice (*Oryza sativa* L.). *J Exp Bot.* 2010; 61(15): 4145–56.
14. Jeong JS, Kim YS, Redillas MC, *et al.*: *OsNAC5* overexpression enlarges root diameter in rice plants leading to enhanced drought tolerance and increased grain yield in the field. *Plant Biotechnol J.* 2013; 11(1): 101–14.
15. Redillas MC, Jeong JS, Kim YS, *et al.*: The overexpression of *OsNAC9* alters the root architecture of rice plants enhancing drought resistance and grain yield under field conditions. *Plant Biotechnol J.* 2012; 10(7): 792–805.
16. Jeong JS, Kim YS, Baek KH, *et al.*: Root-specific expression of *OsNAC10* improves drought tolerance and grain yield in rice under field drought conditions. *Plant Physiol.* 2010; 153(1): 185–97.
17. Ambavaram MM, Basu S, Krishnan A, *et al.*: Coordinated

regulation of photosynthesis in rice increases yield and tolerance to environmental stress. *Nat Commun.* 2014; 5: 5302.

18. Ramegowda V, Basu S, Krishnan A, *et al.*: Rice *GROWTH UNDER DROUGHT KINASE* is required for drought tolerance and grain yield under normal and drought stress conditions. *Plant Physiol.* 2014; 166(3): 1634–45.
19. Yordanov I, Velikova V, Tsonev T: Plant responses to drought, acclimation, and stress tolerance.*Photosynthetica.* 2000; 38(2): 171–86.
20. Bray EA: Molecular Responses to Water Deficit. *Plant Physiol.* 1993; 103(4): 1035–40.
21. Osakabe Y, Osakabe K, Shinozaki K, *et al.*: Response of plants to water stress. *Front Plant Sci.* 2014; 5: 86.
22. Reynolds M, Tuberosa R: Translational research impacting on crop productivity in drought-prone environments. *Curr Opin Plant Biol.* 2008; 11(2): 171–9.
23. Tezara W, Mitchell VJ, Driscoll SD, *et al.*: Water stress inhibits plant photosynthesis by decreasing coupling factor and ATP. *Nature.* 1999; 401: 914–7.
24. Lawlor DW, Cornic G: Photosynthetic carbon assimilation and associated metabolism in relation to water deficits in higher plants. *Plant Cell Environ.* 2002; 25(2): 275–94.
25. Niyogi KK: PHOTOPROTECTION REVISITED: *Genetic and Molecular Approaches. Annu Rev Plant Physiol Plant Mol Biol.* 1999; 50: 333–59.
26. Demmig-Adams B, Adams WW III: The role of xanthophyll cycle carotenoids in the protection of photosynthesis. *Trends Plant Sci.* 1996; 1(1): 21–6.
27. Demmig-Adams B, Adams WW 3rd: Photoprotection in an ecological context: the remarkable complexity of thermal energy dissipation. *New Phytol.* 2006; 172(1): 11–21.
28. Lawlor DW: Limitation to photosynthesis in water-stressed leaves: stomata *vs*. metabolism and the role of ATP. *Ann Bot.* 2002; 89(Spec No): 871–85.
29. Medrano H, Parry MAJ, Socias X, *et al.*: Long term water stress

inactivates Rubisco in subterranean clover.*Ann Appl Biol.* 1997; 131(3): 491–501.

30. Chaves MM, Flexas J, Pinheiro C: Photosynthesis under drought and salt stress: regulation mechanisms from whole plant to cell. *Ann Bot.* 2009; 103(4): 551–60.
31. Edwards G, Walker D: C3, C4: mechanisms, and cellular and environmental regulation, of photosynthesis. Univ of California Press, 1983. Reference Source
32. Gowik U, Westhoff P: The path from $C_3$ to $C_4$ photosynthesis. *Plant Physiol.* 2011; 155(1): 56–63.
33. Pinheiro C, Chaves MM: Photosynthesis and drought: can we make metabolic connections from available data? *J Exp Bot.* 2011; 62(3): 869–82.
34. Wilkinson S, Kudoyarova GR, Veselov DS, *et al.*: Plant hormone interactions: innovative targets for crop breeding and management. *J Exp Bot.* 2012; 63(9): 3499–509.
35. Wilkinson S, Davies WJ: Drought, ozone, ABA and ethylene: new insights from cell to plant to community.*Plant Cell Environ.* 2010; 33(4): 510–25.
36. Ji X, Dong B, Shiran B, *et al.*: Control of abscisic acid catabolism and abscisic acid homeostasis is important for reproductive stage stress tolerance in cereals. *Plant Physiol.* 2011; 156(2): 647–62.
37. Sperotto RA, Ricachenevsky FK, Duarte GL, *et al.*: Identification of up-regulated genes in flag leaves during rice grain filling and characterization of *Os*NAC5, a new ABA-dependent transcription factor. *Planta.* 2009; 230(5): 985–1002.
38. Liang C, Wang Y, Zhu Y, *et al.*: OsNAP connects abscisic acid and leaf senescence by fine-tuning abscisic acid biosynthesis and directly targeting senescence-associated genes in rice. *Proc Natl Acad Sci U S A.* 2014; 111(27): 10013–8.
39. Chen X, Wang Y, Lv B, *et al.*: The NAC family transcription factor OsNAP confers abiotic stress response through the ABA pathway. *Plant Cell Physiol.* 2014; 55(3): 604–19.
40. Du H, Wang N, Cui F, *et al.*: Characterization of the beta-carotene hydroxylase gene *DSM2* conferring drought and oxidative stress

resistance by increasing xanthophylls and abscisic acid synthesis in rice. *Plant Physiol.*2010; 154(3): 1304–18.

41. Peleg Z, Reguera M, Tumimbang E, *et al.*: Cytokinin-mediated source/sink modifications improve drought tolerance and increase grain yield in rice under water-stress. *Plant Biotechnol J.* 2011; 9(7): 747–58.
42. Peleg Z, Blumwald E: Hormone balance and abiotic stress tolerance in crop plants. *Curr Opin Plant Biol.* 2011; 14(3): 290–5.
43. Xie Z, Jiang D, Cao W, *et al.*: Relationships of endogenous plant hormones to accumulation of grain protein and starch in winter wheat under different post-anthesis soil water statusses. *Plant Growth Regul.* 2003; 41(2): 117–27.
44. Zhang S, Li C, Cao J, *et al.*: Altered architecture and enhanced drought tolerance in rice via the down-regulation of indole-3-acetic acid by *TLD1/OsGH3.13* activation. *Plant Physiol.* 2009; 151(4): 1889–901.
45. Uga Y, Sugimoto K, Ogawa S, *et al.*: Control of root system architecture by *DEEPER ROOTING 1* increases rice yield under drought conditions. *Nat Genet.* 2013; 45(9): 1097–102.
46. Wang C, Yang A, Yin H, *et al.*: Influence of water stress on endogenous hormone contents and cell damage of maize seedlings. *J Integr Plant Biol.* 2008; 50(4): 427–34.
47. Munné-Bosch S, Alegre L: Die and let live:Leaf senescence contributes to plant survival under drought stress.*Funct Plant Biol.* 2004; 31(3): 203–216.
48. Fukao T, Xu K, Ronald PC, *et al.*: A variable cluster of ethylene response factor-like genes regulates metabolic and developmental acclimation responses to submergence in rice. *Plant Cell.* 2006; 18(8): 2021–34.
49. Perata P, Voesenek LA: Submergence tolerance in rice requires *Sub1A*, an ethylene-response-factor-like gene.*Trends Plant Sci.* 2007; 12(2): 43–6.
50. Rajala A, Peltonen-Sainio P: Plant Growth Regulator Effects on Spring Cereal Root and Shoot Growth. *Agron J.*2001; 93(4): 936–943.

51. Sharp RE: Interaction with ethylene: changing views on the role of abscisic acid in root and shoot growth responses to water stress. *Plant Cell Environ.* 2002; 25(2): 211–22.

52. Yang JC, Zhang JH, Ye YX, *et al.*: Involvement of abscisic acid and ethylene in the responses of rice grains to water stress during filling. *Plant Cell Environ.* 2004; 27(8): 1055–64.

53. Gomez-Roldan V, Fermas S, Brewer PB, *et al.*: Strigolactone inhibition of shoot branching. *Nature.* 2008; 455(7210): 189–94.

54. Xing Y, Zhang Q: Genetic and molecular bases of rice yield. *Annu Rev Plant Biol.* 2010; 61: 421–42.

55. Xiong L, Zhu J: Molecular and genetic aspects of plant responses to osmotic stress. *Plant Cell Environ.* 2002; 25(2): 131–9.

56. Micco VD, Aronne G: in Plant Responses to Drought Stress. (ed Aroca. Ricardo). Springer, 2002.

57. Miyazawa S, Yoshimura S, Shinzaki Y, *et al.*: Deactivation of aquaporins decreases internal conductance to $CO_2$ diffusion in tobacco leaves grown under long-term drought. *Funct Plant Biol.* 2008; 35(7): 553–564.

58. Tosens T, Niinemets U, Vislap V, *et al.*: Developmental changes in mesophyll diffusion conductance and photosynthetic capacity under different light and water availabilities in *Populus tremula*: how structure constrains function. *Plant Cell Environ.* 2012; 35(5): 839–56.

59. Xu Z, Zhou G: Responses of leaf stomatal density to water status and its relationship with photosynthesis in a grass. *J Exp Bot.* 2008; 59(12): 3317–25.

60. Blum A: Drought resistance, water-use efficiency, and yield potential—are they compatible, dissonant, or mutually exclusive? *Aust J Agric Res.* 2005; 56(11): 1159–1168.

61. Karaba A, Dixit S, Greco R, *et al.*: Improvement of water use efficiency in rice by expression of *HARDY*, an *Arabidopsis* drought and salt tolerance gene. *Proc Natl Acad Sci U S A.* 2007; 104(39): 15270–5.

62. Spollen WG, Sharp RE: Spatial distribution of turgor and root growth at low water potentials. *Plant Physiol.* 1991; 96(2): 438–43.

63. Deak KI, Malamy J: Osmotic regulation of root system architecture. *Plant J.* 2005; 43(1): 17–28.

64. Seo PJ, Park C: Auxin homeostasis during lateral root development under drought condition. *Plant Signal Behav.* 2009; 4(10): 1002–4.

65. Chen H, Li Z, Xiong L: A plant microRNA regulates the adaptation of roots to drought stress. *FEBS Lett.* 2012; 586(12): 1742–7.

66. Jaffe MJ, Takahashi H, Biro RL: A pea mutant for the study of hydrotropism in roots. *Science.* 1985; 230(4724): 445–7.

67. Takahashi N, Yamazaki Y, Kobayashi A, *et al.*: Hydrotropism interacts with gravitropism by degrading amyloplasts in seedling roots of Arabidopsis and radish. *Plant Physiol.* 2003; 132(2): 805–10.

68. Blilou I, Xu J, Wildwater M, *et al.*: The PIN auxin efflux facilitator network controls growth and patterning in *Arabidopsis* roots. *Nature.* 2005; 433(7021): 39–44.

69. Sengupta D, Reddy AR: Water deficit as a regulatory switch for legume root responses. *Plant Signal Behav.* 2011; 6(6): 914–7.

70. Dixit S, Kumar Biswal A, Min A, *et al.*: Action of multiple intra-QTL genes concerted around a co-localized transcription factor underpins a large effect QTL. *Sci Rep.* 2015; 5: 15183.

71. Henry A, Cal AJ, Batoto TC, *et al.*: Root attributes affecting water uptake of rice (*Oryza sativa*) under drought.*J Exp Bot.* 2012; 63(13): 4751--63.

72. Huang B, Fry JD: Root Anatomical, Physiological, and Morphological Responses to Drought Stress for Tall Fescue Cultivars. *Crop Sci.* 1998; 38(4): 1017–1022.

73. Steudle E: Water uptake by roots: effects of water deficit. *J Exp Bot.* 2000; 51(350): 1531–42.

74. Chaves MM, Oliveira MM: Mechanisms underlying plant resilience to water deficits: prospects for water-saving agriculture. *J Exp Bot.* 2004; 55(407): 2365–84.

75. Meyer RF, Boyer JS: Sensitivity of cell division and cell elongation to low water potentials in soybean hypocotyls. *Planta.* 1972; 108(1): 77–87.

76. Serraj R, Sinclair TR: Osmolyte accumulation: can it really help

increase crop yield under drought conditions?*Plant Cell Environ.* 2002; 25(2): 333–41.

77. Oosterhuis DM, Wullschleger SD: Osmotic Adjustment in Cotton (*Gossypium hirsutum* L.) Leaves and Roots in Response to Water Stress. *Plant Physiol.* 1987; 84(4): 1154–7.

78. Sauter A, Davies WJ, Hartung W: The long-distance abscisic acid signal in the droughted plant: the fate of the hormone on its way from root to shoot. *J Exp Bot.* 2001; 52(363): 1991–7.

79. Ashraf M, Foolad MR: Roles of glycine betaine and proline in improving plant abiotic stress resistance.*Environ Exp Bot.* 2007; 59(2): 206–16.

80. Yu LH, Wu SJ, Peng YS, *et al.*: *Arabidopsis EDT1/HDG11* improves drought and salt tolerance in cotton and poplar and increases cotton yield in the field. *Plant Biotechnol J.* 2016; 14(1): 72–84.

81. He X, Chen Z, Wang J, *et al.*: A sucrose: Fructan-6-fructosyltransferase (6-SFT) gene from *Psathyrostachys huashanica* confers abiotic stress tolerance in tobacco. *Gene.* 2015; 570(2): 239–47.

82. Ge L, Chao D, Shi M, *et al.*: Overexpression of the trehalose-6-phosphate phosphatase gene *OsTPP1* confers stress tolerance in rice and results in the activation of stress responsive genes. *Planta.* 2008; 228(1): 191–201.

83. Ahn S, Anderson JA, Sorrells ME, *et al.*: Homoeologous relationships of rice, wheat and maize chromosomes.*Mol Gen Genet.* 1993; 241(5–6): 483–90.

84. Hirota O, Oka M, Takeda T: Sink activity estimation by sink size and dry matter increase during the ripening stage of barley (*Hordeum vulgare*) and rice (*Oryza sativa*). *Ann Bot.* 1990; 65(4): 349–353. Reference Source

85. Gladun IV, Karpov EA: Production and partitioning of assimilates between the panicle and vegetative organs of rice after flowering. *Russ J Plant Physiol.* 1993; 40(5): 629-633. Reference Source

86. Li X, Lawas LMF, Malo R, *et al.*: Metabolic and transcriptomic signatures of rice floral organs reveal sugar starvation as a factor in reproductive failure under heat and drought stress. *Plant Cell Environ.* 2015; 38(10): 2171–92.

87. Nguyen GN, Hailstones DL, Wilkes M, *et al.*: DROUGHT STRESS: Role of Carbohydrate Metabolism in Drought-Induced Male Sterility in Rice Anthers. *J Agron Crop Sci.* 2010; 196(5): 346–57.

88. Sheoran IS, Saini HS: Drought-induced male sterility in rice: Changes in carbohydrate levels and enzyme activities associated with the inhibition of starch accumulation in pollen. *Sexual Plant Reprod.* 1996; 9(3): 161–9.

89. Jin Y, Yang H, Wei Z, *et al.*: Rice male development under drought stress: phenotypic changes and stage-dependent transcriptomic reprogramming. *Mol Plant.* 2013; 6(5): 1630–45.

90. Tuberosa R: Phenotyping for drought tolerance of crops in the genomics era. *Front Physiol.* 2012; 3: 347.

91. Nguyen TT, Klueva N, Chamareck V, *et al.*: Saturation mapping of QTL regions and identification of putative candidate genes for drought tolerance in rice. *Mol Genet Genomics.* 2004; 272(1): 35–46.

92. Du H, Wu N, Fu J, *et al.*: A GH3 family member, OsGH3-2, modulates auxin and abscisic acid levels and differentially affects drought and cold tolerance in rice. *J Exp Bot.* 2012; 63(18): 6467–80.

93. Oh SJ, Kim YS, Kwon CW, *et al.*: Overexpression of the Transcription Factor *AP37* in Rice Improves Grain Yield under Drought Conditions. *Plant Physiol.* 2009; 150(3): 1368–79.

94. He Y, Hao Q, Li W, *et al.*: Identification and Characterization of ABA Receptors in *Oryza sativa*. *PLoS One.*2014; 9(4): e95246.

95. Guo C, Luo C, Guo L, *et al.*: *OsSIDP366*, a DUF1644 gene, positively regulates responses to drought and salt stresses in rice. *J Integr Plant Biol.* 2016; 58(5): 492–502.

96. Xiong H, Li J, Liu P, *et al.*: Overexpression of *OsMYB48-1*, a Novel MYB-Related Transcription Factor, Enhances Drought and Salinity Tolerance in Rice. *PLoS One.* 2014; 9(3): e92913.

97. Chae M, Lee JS, Nam MH, *et al.*: A rice dehydration-inducible SNF1-related protein kinase 2 phosphorylates an abscisic acid responsive element-binding factor and associates with ABA signaling. *Plant Mol Biol.* 2007; 63(2): 151–69.

98. Zhang S, Haider I, Kohlen W, *et al.*: Function of the HD-Zip I gene *Oshox22* in ABA-mediated drought and salt tolerances in rice. *Plant Mol Biol.* 2012; 80(6): 571–85.

99. Hu H, Dai M, Yao J, *et al.*: Overexpressing a NAM, ATAF, and CUC (NAC) transcription factor enhances drought resistance and salt tolerance in rice. *Proc Natl Acad Sci U S A.* 2006; 103(35): 12987–92.

100. Hu H, You J, Fang Y, *et al.*: Characterization of transcription factor gene *SNAC2* conferring cold and salt tolerance in rice. *Plant Mol Biol.* 2008; 67(1–2): 169–81.

101. You J, Hu H, Xiong L: An ornithine $\delta$-aminotransferase gene *OsOAT* confers drought and oxidative stress tolerance in rice. *Plant Sci.* 2012; 197: 59–69.

102. Xiang Y, Tang N, Du H, *et al.*: Characterization of OsbZIP23 as a key player of the basic leucine zipper transcription factor family for conferring abscisic acid sensitivity and salinity and drought tolerance in rice.*Plant Physiol.* 2008; 148(4): 1938–52.

103. Jisha V, Dampanaboina L, Vadassery J, *et al.*: Overexpression of an AP2/ERF Type Transcription Factor *OsEREBP1* Confers Biotic and Abiotic Stress Tolerance in Rice. *PLoS One.* 2015; 10(6): e0127831.

104. Liu C, Mao B, Ou S, *et al.*: OsbZIP71, a bZIP transcription factor, confers salinity and drought tolerance in rice. *Plant Mol Biol.* 2014; 84(1–2): 19–36.

105. Tang N, Zhang H, Li X, *et al.*: Constitutive activation of transcription factor OsbZIP46 improves drought tolerance in rice. *Plant Physiol.* 2012; 158(4): 1755–68.

106. Zou M, Guan Y, Ren H, *et al.*: A bZIP transcription factor, OsABI5, is involved in rice fertility and stress tolerance. *Plant Mol Biol.* 2008; 66(6): 675–83.

107. Yoshida T, Fujita Y, Sayama H, *et al.*: AREB1, AREB2, and ABF3 are master transcription factors that cooperatively regulate ABRE-dependent ABA signaling involved in drought stress tolerance and require ABA for full activation. *Plant J.* 2010; 61(4): 672–85.

108. You J, Zong W, Li X, *et al.*: The SNAC1-targeted gene *OsSRO1c* modulates stomatal closure and oxidative stress tolerance by regulating hydrogen peroxide in rice. *J Exp Bot.* 2013; 64(2): 569–83.

109. Cai S, Jiang G, Ye N, *et al.*: A key ABA catabolic gene, *OsABA8ox3*, is involved in drought stress resistance in rice. *PLoS One.* 2015; 10(2): e0116646.

CHAPTER

# 4

# WATER USE PATTERNS OF FOUR TROPICAL BAMBOO SPECIES ASSESSED WITH SAP FLUX MEASUREMENTS

Tingting Mei[1], Dongming Fang[1], Alexander Röll[1], Furong Niu[1], Hendrayanto[2] and Dirk Hölscher[1]

[1]Tropical Silviculture and Forest Ecology, Georg-August-University of Göttingen, Göttingen, Germany

[2]Department of Forest Management, Institut Pertanian Bogor, Bogor, Indonesia

Bamboos are grasses (Poaceae) that are widespread in tropical and subtropical regions. We aimed at exploring water use patterns of four tropical bamboo species (*Bambusa vulgaris, Dendrocalamus asper*, *Gigantochloa atroviolacea*, and *G. apus*) with sap flux measurement techniques. Our approach included three experimental steps: (1) a pot experiment with a comparison of thermal dissipation probes (TDPs), the stem heat balance (SHB) method and gravimetric readings using potted *B. vulgaris* culms, (2) an *in situ* calibration of TDPs with the SHB method for the four bamboo species, and (3) field monitoring of sap flux of the four bamboo species along with three tropical tree species (*Gmelina arborea*, *Shorea leprosula*, and *Hevea brasiliensis*) during a dry and a wet period. In the pot experiment, it was confirmed that the SHB method is well suited for bamboos but that TDPs need to be calibrated. *In situ*, species-specific parameters for such calibration formulas were derived. During field monitoring we found that some bamboo species reached high maximum sap flux densities. Across bamboo species, maximal sap flux density increased with decreasing culm diameter. In the diurnal course, sap flux densities in bamboos peaked much earlier than radiation and vapor pressure deficit (VPD), and also much earlier than sap flux densities in trees. There was a pronounced hysteresis between sap flux density and VPD in bamboos, which was less pronounced in trees. Three of the four bamboo species showed reduced sap flux densities at high VPD values during the dry period, which was associated with a decrease in soil moisture content. Possible roles of internal water storage, root pressure and stomatal sensitivity are discussed.

**Keywords**: Calibration, environmental drivers, hysteresis, stem heat balance, thermal dissipation probes, trees, bamboos

## INTRODUCTION

Bamboos (Poaceae, Bambuseae) are abundant in the natural vegetation of tropical and subtropical regions. They have been used by people for millennia and are still used as food and construction materials. In addition, a large variety of bamboo usages have been developed in recent decades, for example for pulp, paper, or clothing production (International Network for Bamboo and Rattan [INBAR], 2014). The increasing economic exploitation of bamboos goes along with a considerable expansion of bamboo plantations in some regions (Chen et al., 2009; Food Agriculture Organization [FAO], 2010), which may lead to changes in ecological processes such as water use patterns (Uchimura, 1994; Komatsu et al., 2010). Some bamboo stands were reported to evaporate more water than tree-dominated forests (Komatsu et al., 2010; Ichihashi et al., 2015), but studies focusing on water use patterns of bamboos are still rare thus far (Pereira and Hosegood, 1962; Dierick et al., 2010; Komatsu et al., 2010; Kume et al., 2010; Ichihashi et al., 2015).

Water use patterns of bamboos and trees differ in several aspects. In contrast to trees, bamboos are monocotyledonous species and lack secondary growth (Zimmermann and Tomlinson, 1972). Therefore, vascular conduits of bamboo xylem have to remain functional throughout the ontogeny of a bamboo culm. Bamboos consequently have great ability to avoid cavitation (Cochard et al., 1994; Cao et al., 2012; Petit et al., 2014); root pressure mechanisms may contribute to repairing embolized conduits at night (Cao et al., 2012). Such features and structural traits of bamboos may also lead to particular water use patterns.

In general, plant water use is driven by micrometeorological factors and can be limited by soil water availability (O'Brien et al., 2004; Bovard et al., 2005; Kume et al., 2007); it is regulated by stomata opening and closing (Jarvis, 1989) and can be influenced by internal water storage mechanisms (Waring and Running, 1978; Goldstein et al., 1998; Carrasco et al., 2014). Xylem sap flux reflects these multiple factors. For some tree species, for example, hysteresis in the diurnal sap flux response to radiation and vapor pressure deficit (VPD) of the air have been reported (Goldstein et al., 1998; O'Brien et al., 2004). Sap flux measurements thus appear suitable to study the water use patterns of bamboos as well as their controlling environmental factors.

Thermal dissipation probes (TDP) are widely used to measure sap flux density ($J_s$) in trees (Granier, 1985). Several studies suggest calibrating the method before studying new species (Lu et al., 2004; Wullschleger et al., 2011; Vandegehuchte and Steppe, 2013). To our knowledge, only two studies have applied the TDP method on bamboos so far. Both reported an underestimation of bamboo sap flux compared to stem heat balance (SHB) and reference gravimetric measurements (GM) when the TDP method was not calibrated (Dierick et al., 2010; Kume et al., 2010). In contrast, the SHB method (Sakuratani, 1981) was suggested to be well suited for sap flux measurements on bamboos (Dierick et al., 2010). Bamboo culms are hollow; hence heat loss in the form of heat storage inside culms is marginal, so that steady thermal conditions as a main assumption of the method are met (Baker and Van Bavel, 1987).

The aim of this study was to analyze water use patterns of tropical bamboo species and particularly the response of $J_s$ to the principal environmental drivers. First, we calibrated the SHB and the TDP method with reference GM in an experiment on potted culms of *Bambusa vulgaris*. We then measured $J_s$ in the field in four bamboo species including *B. vulgaris* with both the TDP and SHB method, and calibrated the TDP method with the SHB method. Herein, three factors which may influence the quality of the calibration were tested: time step of the data, formula specificity and calibration formula type. After calibration of the TDP method, we applied it to monitor $J_s$ in four bamboo and three tree species in a common garden in Bogor, Indonesia. Differences in the response of $J_s$ to fluctuations in environmental conditions were assessed. The study intends to contribute to expanding the yet limited knowledge on the eco-hydrological functioning of bamboos.

## MATERIALS AND METHODS

### Study Sites and Species Selection

The pot calibration experiment was conducted in Guangzhou, China (23°26′13″ N, 113°12′33″E, 13 m asl). The field calibration experiment and monitoring campaign were carried out in a common garden in Bogor, Indonesia (6°33′40″ S, 106°43′27″ E, 182 m asl). Average annual temperature in Bogor is 25.6°C and annual precipitation is 3978 mm. Relatively dry conditions with consecutive rainless days can occur between June and September. During this dry period, monthly precipitation is on average 40% lower than during the wet period (230 vs. 383 mm), and the number of consecutive dry days (rainfall < 1 mm) is twice that of the wet period (8 vs. 4 days, 1989–2008, Van Den Besselaar et al., 2014). During our study period (July 2012–January 2013), differences between dry and wet period were more pronounced, i.e., 155 vs. 489 mm monthly precipitation, 14 vs. 2 consecutive dry days, and 0.29 vs. 0.39 $m^{-3}$ $m^{-3}$ daily soil water content. In Bogor, four bamboo species (*B. vulgaris*, *Dendrocalamus asper*, *Gigantochloa atroviolacea*, *G. apus*) with five culms per species and three tree species (*Gmelina arborea*, *Shorea leprosula* and *Hevea brasiliensis*, Table 1) with five stems per species were selected and their $J_s$ were monitored with the TDP method for 7 months.

**Table 1.** Structural characteristics of the studied bamboo and tree species ($n$ = 5 per species; mean ± SD)

| | Species | DBH (cm) | Bamboo culm wall thickness (cm) | Height (m) |
|---|---|---|---|---|
| Bamboo | *B. vulgaris* | 7.0 ± 0.3 | 1.3 ± 0.1 | 17.9 ± 0.8 |
| | *G. apus* | 8.6 ± 0.4 | 1.2 ± 0.2 | 16.2 ± 2.7 |
| | *D. asper* | 11.9 ± 1.9 | 2.4 ± 0.2 | 21.1 ± 0.9 |
| | *G. atroviolacea* | 8.9 ± 0.6 | 1.6 ± 0.1 | 17.0 ± 1.0 |
| Tree | *H. brasiliensis* | 27.4 ± 2.3 | — | 25.2 ± 3.0 |
| | *G. arborea* | 26.3 ± 7.7 | — | 26.5 ± 2.3 |
| | *S. leprosula* | 20.7 ± 4.8 | — | 19.2 ± 2.5 |

*Culm wall thickness (derived from five culms per species) and culm height (derived from three cut culms per species) of the studied bamboos.*

## TDP Construction and Installation

To measure $J_s$ in trees and bamboos, we used self-made TDP (1 and 2 cm length, respectively). In sensor design and construction, we followed Wang et al. (2012). Each TDP sensor was comprised of a heating (downstream) and a reference (upstream) probe made of steel hypodermic needles. The probes were placed 10 cm apart (vertically). For bamboos and trees, TDP installation depths in culms and stems were 1 and 2 cm, respectively. After installation, each TDP was supplied with a constant current of 120 mA; the respective power outputs of 1 and 2 cm length TDP were ~0.1 and ~0.2 W. TDP signals were sampled every 30 s and stored as 10-min averages for the pot calibration experiment and as 1-min averages for all other experiments by data loggers and multiplexers (CR1000, AM16/32, Campbell Scientific Inc., USA).

## Calibration of the TDP Method

### *Pot Calibration Experiment: TDP, SHB, and GM*

Five culms of *B. vulgaris* (diameter 5.3–7.3 cm, height 2.2–3.2 m) with trimmed canopies were transplanted into plastic bags (diameter 30 cm, height 25 cm) 6 months before the calibration experiment. One day before the experiment, the five bamboos were transplanted into bigger plastic pots (diameter 50 cm, height 65 cm). The pots were filled with cobblestones and water and were then fully sealed with plastic cover and aluminum foil to prevent evaporation of water from the pots (Figure 1A). A scaled syringe tube was attached to each pot and connected into the pot through a U-type tube. At the beginning of the experiment, the water was added into the pot through the syringe tube to a fixed level (5 cm below the pot cover). Subsequently, water was added manually every 30 min to reach the pre-defined level. The weight of the added water was determined gravimetrically (GM). To measure $J_s$, each bamboo culm was equipped with three pairs of 1 cm length TDP which were evenly installed circumferentially, about 15 cm above the plastic cover. To minimize potential measurement errors induced by circumferential variations of $J_s$, the thermocouple wires of the three TDP were connected in parallel to get an average voltage signal for each bamboo culm (Lu et al., 2004). For a second $J_s$ estimate, a SHB gage (SGB50 or SGA70, Dynagage Inc., USA) was installed about 1.5 m above the TDP. Both sensor types were protected by foil and the sensor signals were subsequently recorded as described in Section "TDP Construction and Installation." For the comparison to reference GM, 10-min TDP and SHB derived values were aggregated to half-hourly values.

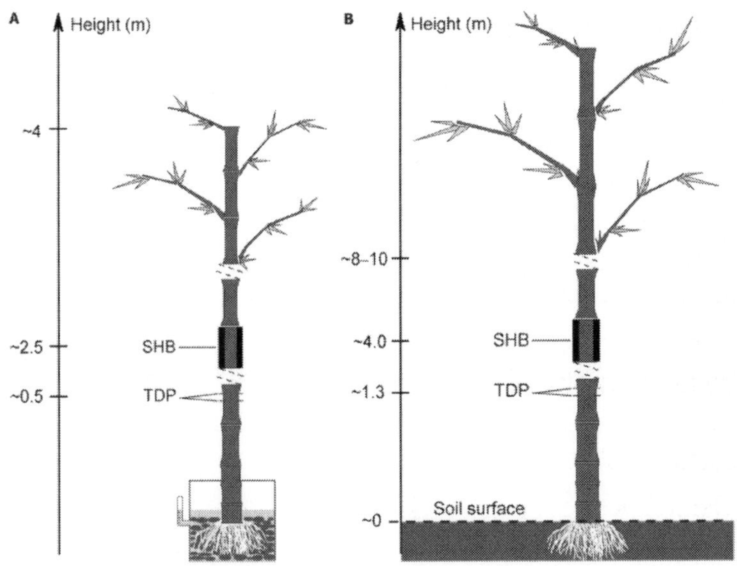


**Figure 1.** Installation of thermal dissipation probe (TDP) and stem heat balance (SHB) sensors on bamboo culms for the calibration experiments on potted plants (A) and for field calibration (B).

To assess the performance of TDP and SHB in the pot experiments, $J_s$ derived from TDP and SHB ($J_{s\_TDP}$ and $J_{s\_SHB}$, respectively) on daily and 30-min scales were compared to GM derived $J_s$ ($J_{s\_GM}$) with paired *t*-tests. Additionally, the slopes of the respective linear fits between $J_{s\_TDP}$, $J_{s\_SHB}$, and $J_{s\_GM}$ were tested for significant differences from one with the test of homogeneity of slopes. The same statistical analyses were applied again later when testing for significant differences between $J_{s\_TDP}$ and $J_{s\_SHB}$ in the field calibration experiments.

### *Field Calibration Experiment: TDP and SHB*

Five culms per bamboo species (*B. vulgaris*, *D. asper*, *G. atroviolacea*, *G. apus*) were selected for TDP measurements (Table 1), three to four of which were additionally measured with SHB for a field calibration of the TDP method. TDP sensors were installed at 1.3 m height, and SHB gages (SGB50, SGA70, Dynagage Inc., USA) were installed about 2.5 m above the TDP. Simultaneous TDP-SHB measurements were conducted for a minimum of 5 days per culm (Figure 1B). Heat storage inside bamboo culms is assumed to be negligible, which was confirmed by installing thermocouple wires inside the measured segments of the respective bamboo culms to detect fluctuations in culm temperature (Dierick et al., 2010). The observed fluctuations were marginal, which meant stable thermal conditions as a requirement of the SHB method were met.

### *Parametrization for TDP Calibration*

We derived cross-sectional water conductive areas ($A_{TDP}$) from the culm wall thickness at the location of TDP sensor installation. In the pot calibration experiment, reference $J_s$ were calculated by dividing water flow rates (g h$^{-1}$, GM-derived) by $A_{TDP}$. In the field calibration experiment, reference $J_s$ were taken from the SHB measurements. The reference $J_s$ could subsequently be used to calibrate $J_{s\_TDP}$. Nighttime sap flux values were excluded in both calibration experiments.

In the field calibration, three factors were considered for obtaining a TDP calibration formula from reference (SHB) measurements: time step of the data, formula specificity and calibration formula type. To examine effects of varying time steps, the formulas were built and tested on data at varying intervals (1-, 10-, 30-, and 60-min averages, respectively). The effects of formula specificity were examined by using common (i.e., all bamboo species pooled), species-specific and culm-specific formulas, respectively. Regarding the calibration formula type, two formulas were compared: one was non-linear ($J_s = aK^b$) and generated by deriving new *a* and *b* parameters for the original Granier (1985) formula. The second was a linear formula ($J_{s\_SHB} = c \times J_{s\_TDP}$) which was calculated from the linear relationship between $J_{s\_TDP}$ and $J_{s\_SHB}$.

To obtain stable calibration formulas, pooled data sets were randomly split in half for calibration and independent validation, respectively (Niu et al., 2015). First, for each time step (1-, 10-, 30-, and 60-min, respectively), a data pool was built. Three culms of each bamboo species were randomly chosen, and for each, 3 days of data were randomly chosen from an initial common dataset. With these data pools, formula specificity was examined. For the common calibration, culms of all four species were selected for calibration. For species-specific and culm-specific calibration, only the data of the respective species or culms was selected. Next, the selected data was randomly split in half, for building the calibration formula and testing it, respectively. When testing the formula, the differences between $J_{s\_SHB}$ and calibrated $J_{s\_TDP}$ ($J_{s\_TDP\_cali}$, abnormal distribution, $P > 0.05$) were examined with the Wilcoxon Signed-Rank Test (no significant differences at $P > 0.05$). The process of randomly building and testing the formula was iterated 10,000 times. Final calibration formula parameters were derived by averaging the parameters of those iterations which passed the Wilcoxon Signed-Rank Test ($P > 0.05$).

For an evaluation of the performance of the different formulas and the influence of the three factors (time scale, formula specificity and calibration formula type), differences in normalized Root-Mean-Square Errors (nRMSE) were assessed for each culm, species and formula factor, respectively. First, the RMSE for each day was derived with the $J_{s\_SHB}$ and $J_{s\_TDP\_cali}$ values, and the nRMSE was calculated by normalizing the RMSE with the observed daily range of $J_{s\text{-}SHB}$ (difference between maximum and minimum $J_{s\_SHB}$). Then, the nRMSE were analyzed regarding the three formula factors (data time scale, formula specificity and calibration formula type) by ANOVA (Analysis of variance). Additionally, for each day, $J_{s\_TDP\_cali}$ with each formula type was tested for significant differences from $J_{s\_SHB}$ with the Wilcoxon Signed-Rank Test. The rates of passing the Wilcoxon Signed-Rank Test ($P > 0.05$ when no significant difference between TDP and SHB derived values) were assessed for each formula.

## Field Study

### *Monitoring Bamboo and Tree Sap Flux*

Four calibrated bamboo species as well as three tree species (*G. arborea*, *S. leprosula*, and *H. brasiliensis*) were monitored with the TDP method for 7 months (July, 2012–January, 2013). Five bamboo culms and five tree trunks per species were selected for the measurements. On bamboos, three pairs of TDP (10 mm in length) were installed evenly around each culm at 1.3 m height and connected in parallel (see TDP Construction and Installation). On trees, two pairs of 20 mm TDP were installed in the trunk 1.3 m above the ground, in the North and South, respectively. $J_s$ for the two sensors were separately derived with the original calibration formula (Granier, 1985) and subsequently averaged to obtain values for each tree. For bamboos, $J_s$ derived with the original formula were calibrated with species-specific calibration parameters (from reference SHB field measurements) to obtain final $J_s$ values.

### *Environmental Measurements and Analyses*

A micrometeorological station was set up in an open area. It was about 100 and 600 m away from the closer measurement sites (*D. asper*, *G. arborea*, *G. atroviolacea*, *G. apus*, *S. leprosula*) and farthest sites (*B. vulgaris*, *H. brasiliensis*), respectively. Air temperature (Ta, °C) and air relative humidity (RH, %) were measured with a temperature and relative humidity probe (CS215, Campbell) installed in a radiation shield. VPD (kPa) was calculated from Ta and RH. Radiation (J $m^{-2}$ $s^{-1}$) was measured with a pyranometer (CS300, Campbell). Data were recorded with the previously described data loggers every minute.

In addition to the mentioned micrometeorological variables, soil moisture (SM, $m^{-3}$ $m^{-3}$) was measured with time domain reflectometry sensors (TDR, CS616, Campbell) at 0–20 cm depth. As the clump of *D. asper* and the stand of *G. arborea* were next to each other, one TDR was positioned between them to measure soil moisture. Likewise, one sensor was used for measurements of *G. atroviolacea* and *G. apus*. One TDR each were used for the remaining species (*S. leprosula*, *B. vulgaris*, *H. brasiliensis*). TDR measurements ran in parallel to the sap flux field campaign and data were recorded with the described data loggers every minute.

For the day-to-day analysis of influences of fluctuations in environmental conditions (VPD, radiation, SM) on $J_s$ in the studied bamboo and tree species, daily accumulated $J_s$ (kg cm$^{-2}$ d$^{-1}$) were normalized by setting the highest daily observation of each species to one and the lowest to zero. For a more isolated analysis of potentially limiting influences of soil moisture on $J_s$, we focused on 'dry period conditions' with consecutive rainless days, which occurred between June and September in the study area. During this period, monthly precipitation was only 32% of monthly wet period precipitation (155 vs. 489 mm), and the number of consecutive dry days (rainfall < 1 mm) was seven times higher than during the wet period (14 vs. 2 days). Dry period conditions are also characterized by higher VPD (average daily VPD > 0.74 kPa on 92% of the days). 0.74 kPa was chosen as the threshold to distinguish between dry and wet period because it constituted the mean maximum ('turning point') in the fitted $J_s$ response functions to VPD in three of the four studied bamboo species (except *D. asper*, see Figure 4B).

For the diurnal analysis of influences of fluctuations in environmental conditions on $J_s$, time lags between $J_s$ and micrometeorological drivers (radiation and VPD) were calculated as the time difference between the respective occurrences of maximal $J_s$ ($J_{s\_max}$) and maximal radiation and VPD. *T*-tests were used to test time lags for significant differences from 0 min. 30-min $J_s$ values (average values of three selected sunny days) of each species were plotted against radiation and VPD to examine occurrences of hysteresis. The respective areas of hysteresis were compared between bamboos and trees with *t*-tests.

All data analyses were performed with SAS 9.3 (SAS Institute Inc., Cary, NC, USA, 2013).

# RESULTS

## Calibration of the TDP Method for Bamboos

### *Pot Calibration Experiment: TDP, SHB, and GM*

In the pot calibration experiment with *B. vulgaris*, SHB yielded similar absolute values of $J_s$ as GM on daily and 30-min scales ($P > 0.05$). The slope of the linear fit between SHB and GM on the 30-min scale was 0.98 ($R^2 = 0.93$, $P < 0.01$). It did not significantly differ from 1 ($P > 0.05$, Figure 2A). In contrast to this, TDP estimates, with the original parameters of the calibration formula (Granier, 1985), differed substantially from GM values at both the daily (60% underestimation of accumulated $J_s$, $P < 0.01$) and 30-min scale (56% underestimation, $P < 0.01$). The slope of the linear fit between TDP and GM on the 30-min scale was 0.44 ($R^2 = 0.84$, $P < 0.01$). It was significantly different from 1 ($P < 0.01$, Figure 2A).

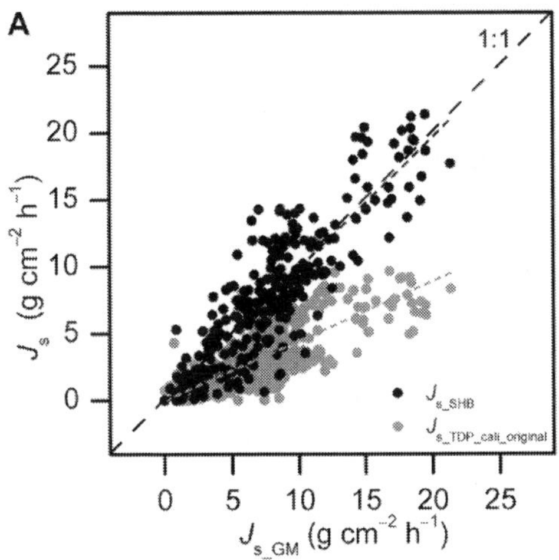

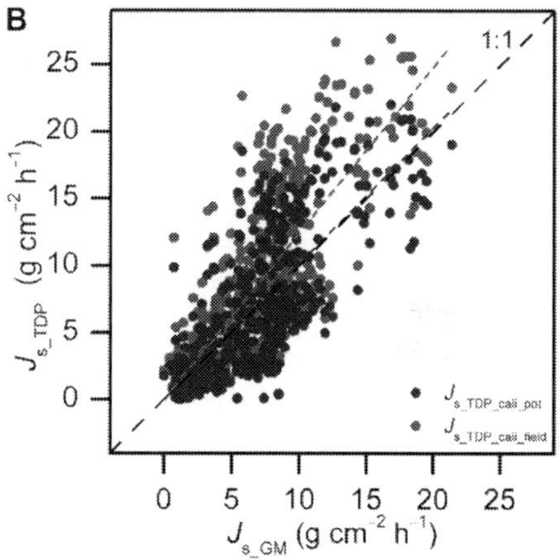


**Figure 2.** Half-hourly sap flux density ($J_s$) measured with thermal dissipation probes (TDP) and stem heat balance (SHB) sensors on five potted *Bambusa vulgaris* **culms plotted against GM-derived reference sap flux densities ($J_{s\_GM}$)** before (A; $J_{s\_TDP\_cali\_original}$: $Y = 0.44X$, $R^2 = 0.84$, $P < 0.01$; $J_{s\_SHB}$: $Y = 0.98X$, $R^2 = 0.93$, $P < 0.01$) **and after (B;** $J_{s\_TDP\_cali\_field}$: $Y = 1.24X$, $R^2 = 0.84$, $P < 0.01$; $J_{s\_TDP\_cali\_pot}$: $Y = 1.01X$, $R^2 = 0.84$, $P < 0.01$) **species-specific calibration and field calibrations of the TDP method.** Pooled data from 2 to 5 days of simultaneous TDP, SHB, and gravimetric measurements (GM).

After applying the TDP calibration parameter for *B. vulgaris* derived from the pot experiment ($c = 2.28$), the 30-min $J_{s\_TDP}$ were in line with those from GM. The slope was not significantly different from 1 ($P > 0.05$, Figure 2B). When applying the calibration parameters derived for *B. vulgaris* from the SHB field calibration experiment ($c = 2.79$), $J_{s\_TDP}$ was 19% higher than $J_{s\text{-}GM}$ ($P < 0.01$, Figure 2B).

## *Field Calibration Experiment: TDP and SHB*

Formula type and data time step had no significant influence on the performance of the calibration formula, but it mattered whether culm- or species-specific or a common calibration formula was used (Appendix Tables A1 and A2 Supplementary Material). Based on the nRMSE and the passing rate of the Wilcoxon test ($P > 0.05$) between calibrated $J_{s\_TDP}$ and $J_{s\_SHB}$, culm-specific formulas performed better than species-specific and common formulas. In our study, there was no statistically significant difference between the species-specific and the common calibration parameters (Table 2, $P > 0.05$). For two of the four studied bamboo species (*G. apus* and *B. vulgaris*), however, using species-specific formulas slightly improved the quality of predictions as compared to applying the common formula ($P = 0.06$ and 0.07, respectively, Table 2). These two bamboo species had lower nRMSE and higher passing rates than *D. asper* and *G. atroviolacea* (Appendix Table A2 in Supplementary Material). The linear calibration parameters of the four bamboo species were significantly different from each other ($P < 0.01$). The linear calibration parameters, the slopes of $J_{s\_TDP}$ vs. $J_{s\_SHB}$, were examined with the test of homogeneity of slopes and were found to differ significantly from each other ($t > 0.01$).

**Table 2.** Values of the parameter *c* **of different bamboo calibrations (species-specific/com**mon) for TDP sap flux estimates

| Formula specificity | Species | c | nRMSE | | |
|---|---|---|---|---|---|
| | | | Species-specific formula | Common formula | *P*-value |
| Species | *B. vulgaris* | $2.79 \pm 0.13^{a}$ | 0.10 | 0.11 | 0.07 |
| | *G. apus* | $3.32 \pm 0.08^{b}$ | 0.10 | 0.12 | 0.06 |
| | *D. asper* | $2.42 \pm 0.06^{c}$ | 0.18 | 0.18 | 0.97 |
| | *G. atroviolacea* | $2.53 \pm 0.11^{d}$ | 0.12 | 0.13 | 0.81 |
| Common | | $2.74 \pm 0.07^{e}$ | | | |

*Significant differences between species-specific and common c estimates (Tukey's test, P < 0.01) are indicated by superscripted letters. P-values < 0.05 indicate significant differences between Normalized Root-Mean-Square Errors (nRMSE) of species-specific and common formula.*

Before calibration, $J_{s\_TDP}$ was on average 66 and 63% lower than SHB-derived reference values on the daily and 30-min scales, respectively ($P < 0.01$). This deviation was reduced to 10 and 8% underestimations ($P < 0.01$) when using species-specific calibration parameters (Table 2). On average, for 77 ± 6% of the days that were included in the analysis, the species-specific post-calibration 30-min $J_{s\_TDP}$ values were not significantly different from the respective reference $J_{s\_SHB}$ (Wilcoxon Signed-Rank test, $P > 0.05$).

## Field Study

### *Monitoring Bamboo and Tree Sap Flux*

$J_{s\_max}$ in the studied bamboo species (averages from five individuals per species) were 70.5, 21.6, 49.7, and 56.2 g cm$^{-2}$ h$^{-1}$ for *B. vulgaris*, *D. asper*, *G. apus*, and *G. atroviolacea*, respectively. In trees, corresponding values were 17.7, 10.5, and 23.3 g cm$^{-2}$ h$^{-1}$ for *H. brasiliensis*, *G. arborea*, and *S. leprosula*, respectively. Across bamboo species, $J_{s\_max}$ decreased with increasing culm diameter ($R^2 = 0.97$, $P = 0.02$, Figure 3).

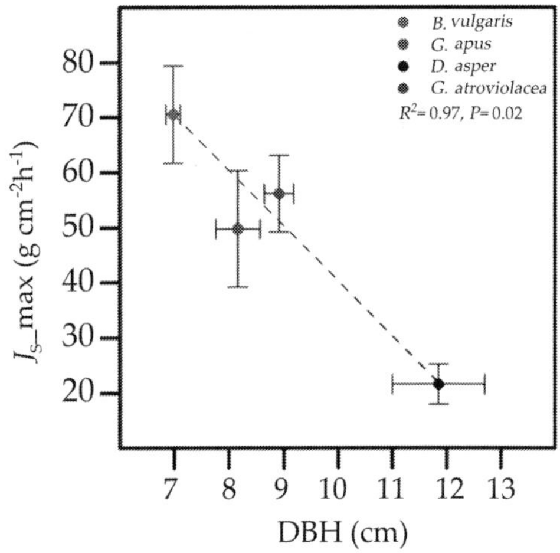


**Figure 3.** Relationship between diameter at breast height (DBH) of bamboo culms and maximum observed sap flux density ($J_{s\_max}$) in four bamboo species. Horizontal error bars indicate DBH standard errors, vertical bars standard errors of $J_{s\_max}$. Data of five culms pooled per species, average of the highest 10% of daily $J_{s\_max}$ values of each culm used for the analysis.

## *Environmental Measurements and Analyses*

The normalized daily accumulated $J_s$ of all studied species increased with increasing daily integrated radiation. This relationship did not fully hold up for accumulated $J_s$ and average daily VPD. In several species, daily $J_s$ increased with increasing VPD only to a certain VPD threshold (approximately 0.74 kPa, Figure 4); after this threshold, accumulated $J_s$ decreased with further increasing VPD. Such conditions of high VPD were characteristic of the dry period. For days with VPD > 0.74 kPa, daily accumulated $J_s$ of most studied species (except in *D. asper* and *G. arborea*) declined with decreasing soil moisture content ($R^2 =$ 0.39, 0.44, 0.4,0.52, and 0.55 for *B. vulgaris*, *G. apus*, *G. atroviolacea*, *S. leprosula*, and *H. brasiliensis*, respectively; $P < 0.05$, Figures 5A,B).

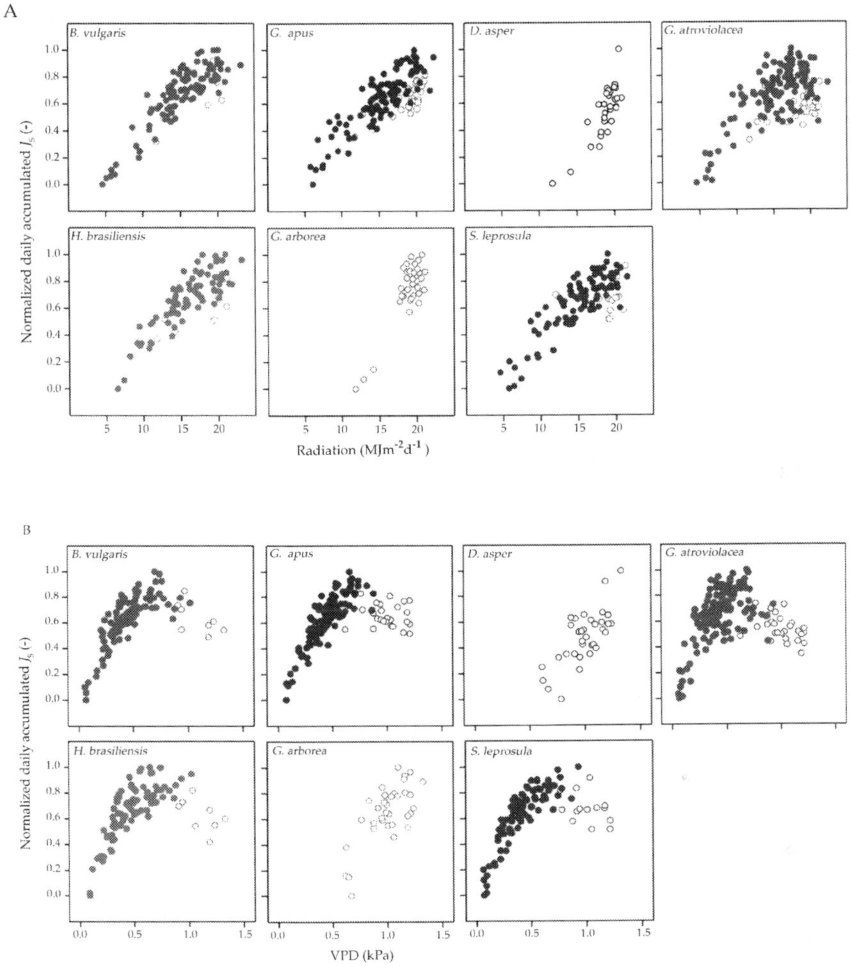


**Figure 4**. Normalized daily accumulated sap flux density ($J_s$) plotted against absolute values of (A) integrated daily radiation and (B) average daily vapor pressure deficit (VPD). Daily values of four bamboo (upper row) and three tree species (lower row); data from 7 months of measurements (July 2012–January 2013) encompassing both wet (filled circles) and dry (open circles) periods (except for *Dendrocalamus asper* and *Gmelina arborea*, mainly dry period). Daily averages derived from measurements of five culms per species.

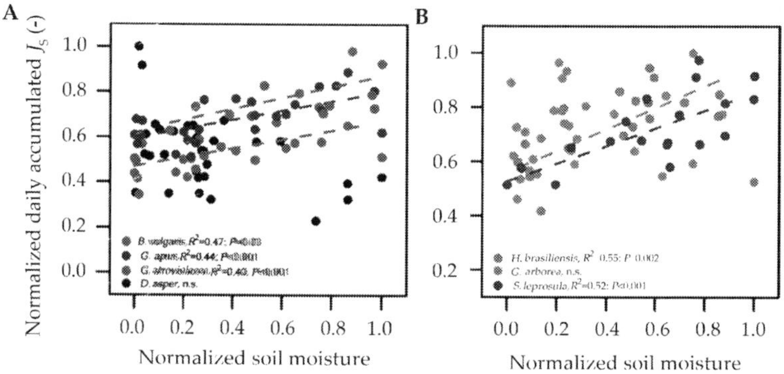

**Figure 5.** Normalized daily accumulated sap flux density ($J_s$) of four bamboo species (A) and three tree species (B) in the 'dry period' (characterized with mean daily VPD > 0.74 kPa) plotted against normalized mean daily soil moisture content (SM). There was a significant linear relationship between $J_s$ and SM ($P < 0.05$) for all species except *D. asper* and *G. arborea*. Normalized values do not reach 1.0 for all species in the figure as the normalization was performed by setting the maximum value of the full measurement period of each species (including wet period) to one, while the figure displays only values in dry period. Daily averages derived from measurements on five culms per species, data of at least 10 dry period days per species.

Diurnal peaks in $J_s$ in the studied bamboo species occurred relatively early (on average at about 11 am), which was significantly earlier than the peaks of radiation and VPD (20–82 and 131–206 min, respectively). In the studied tree species, maximal hourly $J_s$ values were observed after the peak of radiation (3–97 min), but still before (51–108 min) VPD peaked. All time lags were significantly different from 0 min ($P < 0.01$; Table 3), except for the time lag to radiation for the tree species *S. leprosula* ($P > 0.05$).

Positive values indicate a peak of radiation/VPD after the peak of Js, negative values indicate a peak before Js; N, culms/trunks per species averaged (mean ± SD). Significant differences in bamboo/tree mean time lags are indicated by different superscripted letters (Tukey's test, P < 0.01). Significant differences between species are indicated by capital letters (P < 0.01).

**Table 3.** Time lags between diurnal peaks of radiation and VPD and peaks of $J_s$ **in studied bamboos and trees**

| Species | *N* | Time lag with radiation (min) | Time lag with VPD (min) |
|---|---|---|---|
| *B. vulgaris* | 5 | 82 ± 62 | 171 ± 63 |
| *D. asper* | 5 | 41 ± 57 | 206 ± 57 |
| *G. apus* | 4 | 20 ± 61 | 131 ± 53 |
| *G. atroviolacea* | 5 | 64 ± 30 | 170 ± 35 |
| **Bamboo_mean** | **19** | **$51^A$** | **$169^A$** |
| *H. brasiliensis* | 5 | $-37 \pm 12^a$ | 51 ± 9 |
| *G. aborea* | 5 | $-97 \pm 87^b$ | 67 ± 87 |
| *S. leprosula* | 5 | $-3 \pm 25^a$ | 108 ± 20 |
| **Tree_mean** | **15** | **$-46^B$** | **$75^B$** |

Diurnally, some of the studied species showed pronounced hysteresis of hourly $J_s$ to radiation and VPD. Direction of rotation (i.e., order of observations) was counter-clockwise for radiation (Figure 6A) and clockwise for VPD (Figure 6B). The area of the hysteresis to VPD was on average 32% larger in bamboos than in trees, while the area of hysteresis to radiation was on average 50% smaller in bamboos ($P < 0.01$).

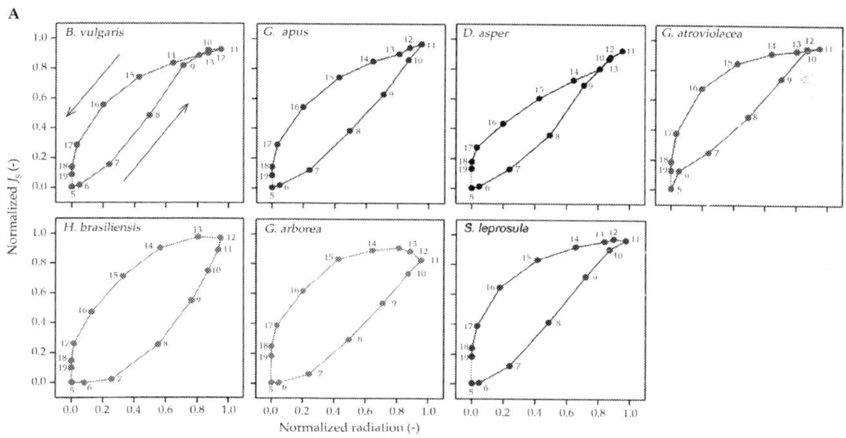

B

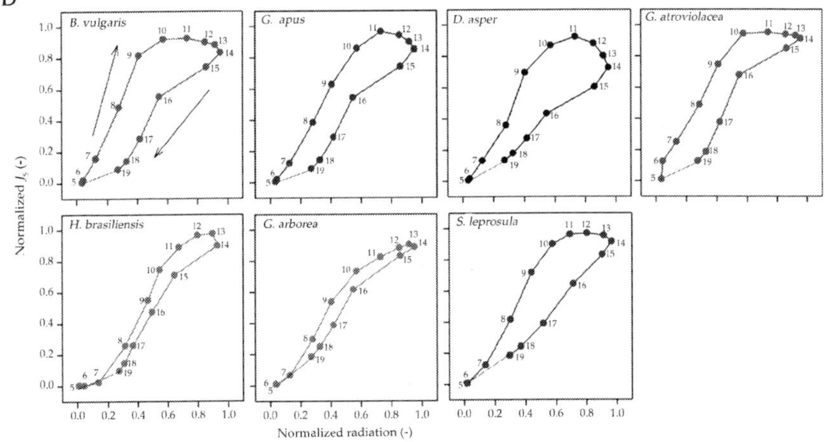


**Figure 6.** Normalized hourly sap flux density ($J_s$) plotted against (A) normalized hourly radiation and (B) VPD. Data of four bamboo (upper row) and three tree species (lower row). Hourly averages derived from simultaneous measurements on five culms per species and by averaging the values of three sunny days to minimize influences of weather. The numbers in the sub-figures indicate the respective time of the day.

# DISCUSSION

## Calibration Experiments

In the pot calibration experiment, SHB yielded similar results as reference GM measurements. Bamboos seem well suited for the SHB method (Dierick et al., 2010) due to their round shape and smooth and barkless surface, which allows for tight contact with the gages. Additionally, the hollow center and thin culm walls result in relatively low energy losses to heat storage so that the heat balance conditions required for the SHB method are met. 'Zero sap flux' conditions to obtain the heat conductivity of the sheath ($K_{sh}$, Sakuratani, 1981) as a further requirement of the SHB method are difficult to determine *in situ* due to potential root pressure induced night time sap flux in bamboos (Cao et al., 2012); however, using $K_{sh}$ derived from field conditions of very low night-time sap flux likely introduced only negligible errors into the calculation of daytime sap flux (Grime and Sinclair, 1999). As we observed very low sap flux over several hours during our experiments (e.g., about 1 g $cm^{-2}$ $h^{-1}$ during the pot experiment), our obtained $K_{sh}$ were likely reliable.

In contrast to SHB, the TDP method was found to substantially underestimate $J_s$ of bamboos in the pot and field calibration experiments. Underestimations by TDP were also reported in two other bamboo species: respective average underestimations of 13% for *B. blumeana* (Dierick et al., 2010) and 31% for *Phyllostachys pubescens* (Moso bamboo, Kume et al., 2010) were reported. Reasons for the observed underestimations could lie in the distinct hydraulic and physiological features of bamboos. Diurnal variations of stem water storage, for example, could affect the accuracy of TDP measurements (Vergeynst et al., 2014). Bamboos have approximately 50% parenchyma in culm walls (Dransfield and Widjaja, 1995), which potentially provides large water reservoirs. The depletion and refilling of the stem during the day and night, respectively, could cause diurnal fluctuations in culm thermal diffusivity. Higher water content during the night could lead to a lower maximum temperature difference ($\Delta T_{max}$) between heated and reference probe under "zero sap flux" conditions. Likewise, lower water content during the day could lead to higher observed $\Delta T$ values. As $\Delta T_{max}/\Delta T$ constitutes the basis for calculations of daytime $J_s$, substantial underestimations of $J_s$ could be introduced when using the original calibration parameters (Granier, 1985; Vergeynst et al., 2014). This hypothesis was assessed further by comparing the linear calibration parameters of *B. vulgaris* from the pot and the field calibration experiment ($c$ = 2.28 and 2.79, respectively). In the pot experiment, the bamboos were always supplied with plenty of water, so that the variability of the culm water content was likely smaller than under field conditions. Effects of varying stem water content on $\Delta T_{max}/\Delta T$ are thus likely much smaller in the pot experiment, which may explain why pot and field calibration experiment yield different parameters for the linear calibration of the same species (*B. vulgaris*). Another potential factor for the divergence could be that the maximum observed $J_s$ in the field (about 70 g $cm^{-2}$ $h^{-1}$) was much larger than in the pot experiment (about 20 g $cm^{-2}$ $h^{-1}$). Higher daytime sap flux (and thus transpiration) may cause a quicker depletion of the potential culm water storage, which consequently leads to a higher variability of culm water content between night and day.

We expected the calibration formula type (linear vs. non-linear) and data time step to have an impact on the performance of TDP predictions. However, both were not as important as the factor formula specificity. Even though species-specific calibration formulas generally did not perform significantly better than the common formula, species-specific formulas tended to show slightly better performance (Table 1) for two of the studied species (*G. apus* and *B. vulgaris*). Also, the calibration parameters were significantly different among the four studied bamboo species (Table 2). Confronting this insight with results from sap flux studies on other bamboo species (Dierick et al., 2010; Kume et al., 2010), differences among species become even more apparent. We thus used the derived species-specific formulas for further analysis. The observed differences among species may be indicative of highly heterogeneous wood anatomical properties among bamboo species. For example, size and shape of vascular bundles and parenchyma of 15 bamboo species were reported to be highly variable (Rúgolo de Agrasar and Rodríguez, 2003). For two further bamboo species (*Chusquea ramosissima* and *Merostachys claussenii*), it was suggested that differences in number of vascular bundles per unit area (1000 vs. 225 per $cm^2$) and vessel length (~1 m vs. 20 cm) could lead to differences in xylem hydraulic conductivity (Saha et al., 2009). Differences in wood anatomical properties may also lead to heterogeneous heat conductive properties, which potentially affects applicability and accuracy of sap flux measurements and particularly of the TDP method (Wullschleger et al., 2011).

In our study, culm-specific formulas performed better at predicting $J_s$ than species-specific and common calibration formulas (Appendix Tables A1 and A2 in Supplementary Material). This result indicates heterogeneity in conductive properties among culms of the same species. Potential reasons could lie in the age and the ontogeny of individual culms. Even though we carefully selected culms of similar age (approximately 2 years old), the exact age of individual bamboo culms within a given clump is difficult to assess. As all monocot species, bamboos lack secondary growth (Zimmermann and Tomlinson, 1972), so culm diameters are not related to culm age. Additionally, over the ontogeny of a certain culm, events and processes such as conductive circuit failure (drought- or metabolism-related; Cochard et al., 1994; Liese and Weiner, 1996), lignification (Lin et al., 2002) or increasing hydraulic limitations with height (Renninger and Phillips, 2010; Cao et al., 2012) could result in overall reduced hydraulic conductivity and thus lower sap flux densities with increasing culm age. However, these processes remain difficult to assess from the outside of the culm; further studies linking the age and ontogeny of bamboos to (TDP-derived) sap flux and water use patterns are suggested.

## Water Use Patterns of Bamboos and Trees

Half-hourly $J_{s\_max}$ in the four studied bamboo species ranged from 21.6 to 70.5 g cm$^{-2}$h$^{-1}$ and were (on average) almost two times greater than in the studied tree species. The observed range for both bamboos and trees falls into the range of $J_{s\_max}$ values reported for tropical tree species in a variety of sap flux studies (Meinzer et al., 2001; O'Brien et al., 2004). For *D. asper*, the $J_{s\_max}$ (21.6 g cm$^{-2}$ h$^{-1}$) was similar to values reported for *B. blumeana* culms (25.7 g cm$^{-2}$ h$^{-1}$, Dierick et al., 2010) and Moso bamboos (approximately 20 g cm$^{-2}$ h$^{-1}$, Kume et al., 2010) of similar size. Our four studied bamboo species showed significant differences in $J_{s\_max}$, which were negatively correlated with species-specific differences in DBH (Figure 3). Consistent with this, in a study on 27 tropical tree species, the negative correlation between $J_{s\_max}$ and DBH was also observed (Meinzer et al., 2001). It was assumed to be related to a decline of the leaf area to sapwood area ratio with increasing DBH. This was also observed in a study on *Eucalyptus grandis* trees (Dye and Olbrich, 1993). In our study, we harvested leaves of three bamboo species (*B. vulgaris*, *D. asper*, and *G. apus*) and found that the leaf weight to sapwood area ratio was positively correlated with $J_{s\_max}$ ($R^2$ = 0.45, $P < 0.05$). However, studies connecting such anatomical and eco-hydrological properties of bamboos are yet scarce (Saha et al., 2009).

On the day-to-day level, accumulated $J_s$ of both the studied bamboo and tree species were significantly correlated with radiation and VPD (Figure 4). During the long wet period, accumulated $J_s$ linearly increased with higher integrated radiation and average daily VPD. Likewise, linear relationships in the day-to-day behavior of $J_s$ to micrometeorological drivers have been reported for some tropical bamboo and several dicot tree species (Dierick and Hölscher, 2009; Köhler et al., 2009). During the dry period characterized by higher radiation and VPD (13 and 100% higher, respectively) than during the wet period, however, the observed linear relationship to VPD did not hold. Higher average daily VPD ('dry period conditions') led to decreases in accumulated $J_s$ of several studied species (Figure 4B). Similar decreases after a certain peak value have been reported for some previously studied tree species (Kubota et al., 2005; Jung et al., 2011), but in most species studied so far, higher average daily VPD leads to increases in accumulated $J_s$ or water use (Wullschleger and Norby, 2001; Tang et al., 2006; Kume et al., 2007; Hernández-Santana et al., 2008; Peters et al., 2010; Horna et al., 2011). This was also reported for Moso bamboo (Komatsu et al., 2010). The observed decreasing accumulated $J_s$ in bamboos under high VPD in our study were related to a reduction of soil moisture in the dry period (for three of the four bamboo and two of the three studied tree species). During the dry period, VPD was generally much higher than during the wet period. Soil moisture may become a limiting factor after several days without rainfall in the dry period. Accumulated $J_s$ decreased strongly and linearly with decreasing soil moisture under 'dry period conditions' (i.e., VPD > 0.74 kPa) for all studied bamboo (except *D. asper*) and tree species (except *G. arborea*, Figure 5). Similarly, in a throughfall reduction experiment in Indonesia, declines of monthly $J_s$ of Cacao and *Gliricidia sepium* were found to linearly correlate with reduced soil moisture (Köhler et al., 2010). Such sensitivity of daily $J_s$ to fluctuating soil moisture may be related to a relatively shallow rooting depth (Kume et al., 2007).

Regarding the diurnal course of $J_s$, the studied bamboo species showed earlier peaks than radiation and VPD, and also earlier than the respective peaks of the studied tree species. In contrast to this, previous studies on tropical trees reported rather small time-lags between peaks of $J_s$ and radiation and VPD, respectively (Dierick and Hölscher, 2009; Köhler et al., 2009; Horna et al., 2011). Pre-noon peaks of $J_s$ have only been described for few species thus far, for example, *Acer rubrum* (Johnson et al., 2011) and oil palms (Niu et al., 2015). The early diurnal peaks of $J_s$ result in substantial hysteresis of $J_s$ particularly to VPD. For another monocot species, oil palm, it has been suggested that such pre-noon peaks of $J_s$ and the resulting large hysteresis to VPD could be indicative of internal trunk water storage and/or root pressure mechanisms (Niu et al., 2015; Röll et al., 2015). Early peaks of $J_s$ could be due to a pre-noon contribution of internal water storage to bamboo transpiration. Likewise, the decoupling of hourly $J_s$ particularly from VPD in the afternoon, i.e., the drop in bamboo $J_s$ (after an early peak) despite further rising VPD, could be connected to the reduced water availability for leaves after the depletion of internal water storage at a certain time of the day. The depletion of stored stem water may be compensated for during the night by root pressure mechanisms (Cao et al., 2012; Yang et al., 2012). Other potential reasons for the diurnally relatively early decline of bamboo $J_s$ and the consequent decoupling of the sap flux response from micrometeorological drivers could be a decline in leaf hydraulic conductance in the afternoon hours, which could contribute to prevent stem water potential loss and subsequent xylem cavitation (Saha et al., 2009; Yang et al., 2012).

## CONCLUSION

Adjusting and applying the TDP method for sap flux measurements on four bamboo species pointed to substantial differences in water use patterns between the studied bamboos and three tree species studied. Bamboos had higher $J_s$, and respective hourly maxima were reached earlier in the day than in tree species. This resulted in strong diurnal hysteresis, particularly to VPD, and in significant time lags between the peaks of $J_s$ in bamboos and the respective peaks of radiation and VPD. Both may point to a strong contribution of internal water storage mechanisms to bamboo transpiration. We found substantial differences in the day-to-day $J_s$ response of most studied bamboo and tree species to fluctuations in environmental conditions between the dry and the wet period. Reduced $J_s$ under conditions of high VPD in the dry period could largely be explained by limiting soil moisture content. The regulation of bamboo water use thus seems to involve mechanisms at the leaf-, culm-, and root- level. However, these mechanisms yet remain to be inter-connected convincingly.

## CONFLICT OF INTEREST STATEMENT

The authors declare that the research was conducted in the absence of any commercial or financial relationships that could be construed as a potential conflict of interest.

## ACKNOWLEDGMENTS

The authors acknowledge the funding from the German Research Foundation (DFG, code number: HO 2119 and CRC 990). DF and FN received scholarships from the China Scholarship Council (CSC). We thank the Departments of Facility and Property, the Ecology Lab and the Agriculture Faculty of Bogor Agriculture University (IPB) for supporting our field work. We especially thank the IPB Faculty of Forestry for close cooperation in our project and our assistants Wahyu Iskandar and Popi Puspita for their work with great enthusiasm in our field experiments. We are very grateful to Diego Dierick for suggestions on experimental methods and to Michael Köhler for suggestions on data processing.


## SUPPLEMENTARY MATERIAL

**Appendix** Table1. The influence of the three factors formula type, time step and formula specificity on the performance of the linear calibration model. Results of multi-ANOVA of the three factors against normalized Root-Mean-Square Error (nRMSE, calculated by normalizing the RMSE with the observed range of sap flux densities from stem heat balance (SHB) measurements; RMSE derived from SHB measurements vs. model-predicted values of each day). Data of 63 days were used. $P<0.01$ indicates significant difference.

| **Source** | **DF** | **Type III SS** | **Mean Square** | **F Value** | **P** |
|---|---|---|---|---|---|
| Formula type | 1 | 0.0046 | 0.0046 | 0.68 | 0.4093 |
| Time step | 3 | 0.0046 | 0.0015 | 0.23 | 0.8777 |
| Formula specificity | 2 | 7.7675 | 3.8838 | 577.54 | <.0001 |

**Appendix** Table2. Performance of culm-specific vs. species-specific vs. common linear formulas for a simple linear field calibration of the thermal dissipation probe (TDP) method with the stem heat balance (SHB) method on four bamboo species Normalized Root-Mean-Square Errors (nRMSE) and passing rates of the Wilcoxon Signed-Rank test for each species and formula specificity type. Superscripted letters indicate significant differences between nRMSEs and passing rates, respectively, within each species (Tukey's test, $P<0.05$).

| Species | Formula specificity | nRMSE | Passing rate (%) |
|---|---|---|---|
| *B. vulgaris* | common | 0.11[a] | 83[a] |
| | culm | 0.04[b] | 94[b] |
| | species | 0.10[a] | 84[a] |
| *G. apus* | common | 0.12[a] | 77[a] |
| | culm | 0.06[b] | 90[b] |
| | species | 0.10[a] | 81[a] |
| *D. asper* | common | 0.18[a] | 74[a] |
| | culm | 0.04[b] | 94[b] |
| | species | 0.18[a] | 70[a] |
| *G. atroviolacea* | common | 0.13[a] | 74[a] |
| | culm | 0.06[b] | 89[b] |
| | species | 0.12[a] | 74[a] |

## REFERENCES

1. Baker, J. M., and Van Bavel, C. H. M. (1987). Measurement of mass flow of water in the stems of herbaceous plants. *Plant Cell Environ.* 10, 777–782. doi: 10.1111/1365-3040.ep11604765
2. Bovard, B. D., Curtis, P. S., Vogel, C. S., Su, H.-B., and Schmid, H. P. (2005). Environmental controls on sap flow in a northern hardwood forest. *Tree Physiol.* 25, 31–38. doi: 10.1093/treephys/25.1.31
3. Cao, K.-F., Yang, S.-J., Zhang, Y.-J., and Brodribb, T. J. (2012). The maximum height of grasses is determined by roots. *Ecol. Lett.* 15, 666–672. doi: 10.1111/j.1461-0248.2012.01783.x
4. Carrasco, L. O., Bucci, S. J., Francescantonio, D. D., Lezcano, O. A., Campanello, P. I., Scholz, F. G., et al. (2014). Water storage dynamics in the main stem of subtropical tree species differing in wood density, growth rate and life history traits. *Tree Physiol.* 35, 354–365. doi: 10.1093/treephys/tpu087
5. Chen, X., Zhang, X., Zhang, Y., Booth, T., and He, X. (2009). Changes of carbon stocks in bamboo stands in China during 100 years. *For. Ecol. Manag.* 258, 1489–1496. doi: 10.1016/j.foreco.2009.06.051

6. Cochard, H., Ewers, F. W., and Tyree, M. T. (1994). Water relations of a tropical vine-like bamboo (*Rhipidocladum racemiflorum*): root pressures, vulnerability to cavitation and seasonal changes in embolism. *J. Exp. Bot.* 45, 1085–1089. doi: 10.1093/jxb/45.8.1085
7. Dierick, D., and Hölscher, D. (2009). Species-specific tree water use characteristics in reforestation stands in the Philippines. *Agric. For. Meteorol.* 149, 1317–1326. doi: 10.1016/j.agrformet.2009.03.003
8. Dierick, D., Hölscher, D., and Schwendenmann, L. (2010). Water use characteristics of a bamboo species (*Bambusa blumeana*) in the Philippines. *Agric. For. Meteorol.* 150, 1568–1578. doi: 10.1016/j.agrformet.2010.08.006
9. Dransfield, S., and Widjaja, E. (1995). *Plant Resources of South-East Asia No. 7: Bamboos.* Leiden: Backhuys Publishers.
10. Dye, P. J., and Olbrich, B. W. (1993). Estimating transpiration from 6-year-old Eucalyptus grandis trees: development of a canopy conductance model and comparison with independent sap flux measurements. *Plant Cell Environ.* 16, 45–53. doi: 10.1111/j.1365-3040.1993.tb00843.x
11. Food Agriculture Organization [FAO] (2010). *Global Forest Resources Assessment 2010?: Main Report, FAO Forestry Paper, 0258-6150; 163.* Rome: Food and Agriculture Organization of the United Nations.
12. Goldstein, G., Andrade, J. L., Meinzer, F. C., Holbrook, N. M., Cavelier, J., Jackson, P., et al. (1998). Stem water storage and diurnal patterns of water use in tropical forest canopy trees. *Plant Cell Environ.* 21, 397–406. doi: 10.1046/j.1365-3040.1998.00273.x
13. Granier, A. (1985). Une nouvelle méthode pour la mesure du flux de sève brute dans le tronc des arbres. *Ann. For. Sci.* 42:8. doi: 10.1051/forest:19850204
14. Grime, V. L., and Sinclair, F. L. (1999). Sources of error in stem heat balance sap flow measurements. *Agric. For. Meteorol.* 94, 103–121. doi: 10.1016/S0168-1923(99)00011-8
15. Hernández-Santana, V., David, T. S., and Martínez-Fernández, J. (2008). Environmental and plant-based controls of water use in a Mediterranean oak stand. *For. Ecol. Manag.* 255, 3707–3715. doi: 10.1016/j.foreco.2008.03.004

16. Horna, V., Schuldt, B., Brix, S., and Leuschner, C. (2011). Environment and tree size controlling stem sap flux in a perhumid tropical forest of Central Sulawesi. *Indonesia. Ann. For. Sci.*68, 1027–1038. doi: 10.1007/s13595-011-0110-2
17. Ichihashi, R., Komatsu, H., Kume, T., Onozawa, Y., Shinohara, Y., Tsuruta, K., et al. (2015). Stand-scale transpiration of two *Moso bamboo* stands with different culm densities. *Ecohydrology* 8, 450–459. doi: 10.1002/eco.1515
18. International Network for Bamboo and Rattan [INBAR] (2014). *International Trade in Bamboo and Rattan 2012.* Beijing: International Network for Bamboo and Rattan. Available at: http://www.inbar.int/publications/?did=292 [accessed June 19, 2015].
19. Jarvis, N. J. (1989). A simple empirical model of root water uptake. *J. Hydrol.* 107, 57–72. doi: 10.1016/0022-1694(89)90050-4
20. Johnson, D. M., McCulloh, K. A., Meinzer, F. C., Woodruff, D. R., and Eissenstat, D. M. (2011). Hydraulic patterns and safety margins, from stem to stomata, in three eastern US tree species. *Tree Physiol.* 31, 659–668. doi: 10.1093/treephys/tpr050
21. Jung, E. Y., Otieno, D., Lee, B., Lim, J. H., Kang, S. K., Schmidt, M. W. T., et al. (2011). Up-scaling to stand transpiration of an Asian temperate mixed-deciduous forest from single tree sapflow measurements. *Plant Ecol.* 212, 383–395. doi: 10.1007/s11258-010-9829-3
22. Köhler, M., Dierick, D., Schwendenmann, L., and Hölscher, D. (2009). Water use characteristics of cacao and Gliricidia trees in an agroforest in Central Sulawesi. *Indonesia. Ecohydrol.* 2, 520–529. doi: 10.1002/eco.67
23. Köhler, M., Schwendenmann, L., and Hölscher, D. (2010). Throughfall reduction in a cacao agroforest: tree water use and soil water budgeting. *Agric. For. Meteorol.* 150, 1079–1089. doi: 10.1016/j.agrformet.2010.04.005
24. Komatsu, H., Onozawa, Y., Kume, T., Tsuruta, K., Kumagai, T., Shinohara, Y., et al. (2010). Stand-scale transpiration estimates in a *Moso bamboo* forest: II. Comparison with coniferous forests. *For. Ecol. Manag.* 260, 1295–1302. doi: 10.1016/j.foreco.2010.06.040
25. Kubota, M., Tenhunen, J., Zimmermann, R., Schmidt, M., Adiku,

S., and Kakubari, Y. (2005). Influences of environmental factors on the radial profile of sap flux density in Fagus crenata growing at different elevations in the Naeba Mountains, Japan. *Tree Physiol.* 25, 545–556. doi: 10.1093/treephys/25.5.545

26. Kume, T., Onozawa, Y., Komatsu, H., Tsuruta, K., Shinohara, Y., Umebayashi, T., et al. (2010). Stand-scale transpiration estimates in a *Moso bamboo* forest: (I) Applicability of sap flux measurements. *For. Ecol. Manag.* 260, 1287–1294. doi: 10.1016/j.foreco.2010.07.012
27. Kume, T., Takizawa, H., Yoshifuji, N., Tanaka, K., Tantasirin, C., Tanaka, N., et al. (2007). Impact of soil drought on sap flow and water status of evergreen trees in a tropical monsoon forest in northern Thailand. *For. Ecol. Manag.* 238, 220–230. doi: 10.1016/j.foreco.2006.10.019
28. Liese, W., and Weiner, G. (1996). Ageing of bamboo culms. A review. *Wood Sci. Technol.* 30, 77–89. doi: 10.1007/BF00224958
29. Lin, J., He, X., Hu, Y., Kuang, T., and Ceulemans, R. (2002). Lignification and lignin heterogeneity for various age classes of bamboo (*Phyllostachys pubescens*) stems. *Physiol. Plant.* 114, 296–302. doi: 10.1034/j.1399-3054.2002.1140216.x
30. Lu, P., Urban, L., and Zhao, P. (2004). Granier's thermal dissipation probe (TDP) method for measuring sap flow in trees: theory and practice. *ACTA Bot. Sin. Engl. Ed.* 46, 631–646.
31. Meinzer, F. C., Goldstein, G., and Andrade, J. L. (2001). Regulation of water flux through tropical forest canopy trees: do universal rules apply? *Tree Physiol.* 21, 19–26. doi: 10.1093/treephys/21.1.19
32. Niu, F., Röll, A., Hardanto, A., Meijide, A., Köhler, M., Hendrayanto, et al. (2015). Oil palm water use: calibration of a sap flux method and a field measurement scheme. *Tree Physiol.* 35, 563–573. doi: 10.1093/treephys/tpv013
33. O'Brien, J. J., Oberbauer, S. F., and Clark, D. B. (2004). Whole tree xylem sap flow responses to multiple environmental variables in a wet tropical forest. *Plant Cell Environ.* 27, 551–567. doi: 10.1111/j.1365-3040.2003.01160.x
34. Pereira, H. C., and Hosegood, P. H. (1962). Comparative water-use

of softwood plantations and bamboo forest. *J. Soil Sci.* 13, 299–313. doi: 10.1111/j.1365-2389.1962.tb00709.x

35. Peters, E. B., McFadden, J. P., and Montgomery, R. A. (2010). Biological and environmental controls on tree transpiration in a suburban landscape. *J. Geophys. Res. Biogeosci.* 115:G04006. doi: 10.1029/2009JG001266
36. Petit, G., DeClerck, F. A. J., Carrer, M., and Anfodillo, T. (2014). Axial vessel widening in arborescent monocots. *Tree Physiol.* 34, 137–145. doi: 10.1093/treephys/tpt118
37. Renninger, H. J., and Phillips, N. (2010). Intrinsic and extrinsic hydraulic factors in varying sizes of two Amazonian palm species (*Iriartea deltoidea* and *Mauritia flexuosa*) differing in development and growing environment. *Am. J. Bot.* 97, 1926–1936. doi: 10.3732/ajb.1000015
38. Röll, A., Niu, F., Meijide, A., Hardanto, A., Hendrayanto, Knohl, A., et al. (2015). Transpiration in an oil palm landscape: effects of palm age. *Biogeosciences* 12, 5619–5633. doi: 10.5194/bg-12-5619-2015
39. Rúgolo de Agrasar, Z., and Rodríguez, M. F. (2003). Culm anatomy of native woody bamboos in argentina and neighboring areas: cross section. *Bamboo Sci. Cult. J. Am. Bamboo Soc.* 17, 28–43.
40. Saha, S., Holbrook, N. M., Montti, L., Goldstein, G., and Cardinot, G. K. (2009). Water Relations of *Chusquea ramosissima* and *Merostachys claussenii* in Iguazu National Park, Argentina. *Plant Physiol.* 149, 1992–1999. doi: 10.1104/pp.108.129015
41. Sakuratani, T. (1981). A heat balance method for measuring water flux in the stem of intact plants. *J. Agric. Meteorol.* 37, 9–17. doi: 10.2480/agrmet.37.9
42. Tang, J., Bolstad, P. V., Ewers, B. E., Desai, A. R., Davis, K. J., and Carey, E. V. (2006). Sap flux–upscaled canopy transpiration, stomatal conductance, and water use efficiency in an old growth forest in the Great Lakes region of the United States. *J. Geophys. Res. Biogeosci.* 111:G02009. doi: 10.1029/2005JG000083
43. Uchimura, E. (1994). *Invitation to Bamboos: Wonder of Biology.* Tokyo: Kenseisha.
44. Van Den Besselaar, E. J. M., Klein Tank, A. M. G., Van Der Schrier, G., Abass, M. S., Baddour, O., Van Engelen, A. F. V., et al. (2014).

International climate assessment & dataset: climate services across borders. *Bull. Am. Meteorol. Soc.* 96, 16–21. doi: 10.1175/BAMS-D-13-00249.1

45. Vandegehuchte, M. W., and Steppe, K. (2013). Sap-flux density measurement methods: working principles and applicability. *Funct. Plant Biol.* 40, 213–223. doi: 10.1071/FP12233_CO
46. Vergeynst, L. L., Vandegehuchte, M. W., McGuire, M. A., Teskey, R. O., and Steppe, K. (2014). Changes in stem water content influence sap flux density measurements with thermal dissipation probes. *Trees* 28, 949–955. doi: 10.1007/s00468-014-0989-y
47. Wang, H., Zhao, P., Hölscher, D., Wang, Q., Lu, P., Cai, X. A., et al. (2012). Nighttime sap flow of Acacia mangium and its implications for nighttime transpiration and stem water storage. *J. Plant Ecol.* 5, 294–304. doi: 10.1093/jpe/rtr025
48. Waring, R. H., and Running, S. W. (1978). Sapwood water storage: its contribution to transpiration and effect upon water conductance through the stems of old-growth Douglas-fir. *Plant Cell Environ.* 1, 131–140. doi: 10.1111/j.1365-3040.1978.tb00754.x
49. Wullschleger, S. D., Childs, K. W., King, A. W., and Hanson, P. J. (2011). A model of heat transfer in sapwood and implications for sap flux density measurements using thermal dissipation probes. *Tree Physiol.* 31, 669–679. doi: 10.1093/treephys/tpr051
50. Wullschleger, S. D., and Norby, R. J. (2001). Sap velocity and canopy transpiration in a sweetgum stand exposed to free-air $CO_2$ enrichment (FACE). *New Phytol.* 150, 489–498. doi: 10.1046/j.1469-8137.2001.00094.x
51. Yang, S.-J., Zhang, Y.-J., Sun, M., Goldstein, G., and Cao, K.-F. (2012). Recovery of diurnal depression of leaf hydraulic conductance in a subtropical woody bamboo species: embolism refilling by nocturnal root pressure. *Tree Physiol.* 32, 414–422. doi: 10.1093/treephys/tps028
52. Zimmermann, M. H., and Tomlinson, P. B. (1972). The vascular system of monocotyledonous stems. *Bot. Gaz.* 133, 141–155. doi: 10.1086/336628

CHAPTER

# 5

# WATER RELATIONS AND FOLIAR ISOTOPIC COMPOSITION OF PROSOPIS TAMARUGO PHIL., AN ENDEMIC TREE OF THE ATACAMA DESERT GROWING AT THREE LEVELS OF WATER TABLE DEPTH

Marco Garrido[1], Paola Silva[2] and Edmundo Acevedo[2]

[1] PhD Program in Silvicultural and Veterinary Sciences, University of Chile, Santiago, Chile

[2]Soil-Plant-Water Relations Laboratory, Agricultural Production Department, Faculty of Agronomical Sciences, University of Chile, Santiago, Chile

*Prosopis tamarugo* Phil. is a strict phreatophyte tree species endemic to the "Pampa del Tamarugal", Atacama Desert. The extraction of water for various uses has increased the depth of the water table in the Pampa aquifers threatening its conservation. This study aimed to determine the effect of the groundwater table depth on the water relations of *P. tamarugo* and to present thresholds of groundwater depth (GWD) that can be used in the groundwater management of the *P. tamarugo* ecosystem. Three levels of GWD, 11.2 ± 0.3 m, 10.3 ± 0.3 m, and 7.1 ± 0.1 m, (the last GWD being our reference) were selected and groups of four individuals per GWD were studied in the months of January and July of the years 2011 through 2014. When the water table depth exceeded 10 m, *P. tamarugo* had lower pre-dawn and mid-day water potential but no differences were observed in minimum leaf stomatal resistance when compared to the condition of 7.1 m GWD; the leaf tissue increased its $\delta^{13}C$ and $\delta^{18}O$ composition. Furthermore, a smaller green canopy fraction of the trees and increased foliage loss in winter with increasing water table depth was observed. The differences observed in the physiological behavior of *P. tamarugo* trees, attributable to the ground water depth; show that increasing the depth of the water table from 7 to 11 m significantly affects the water status of *P. tamarugo*. The results indicate that *P. tamarugo* has an anisohydric stomatal behavior and that given a reduction in water supply it regulates the water demand via foliage loss. The growth and leaf physiological activities are highly sensitive to GWD. The foliage loss appears to prevent the trees from reaching water potentials leading to complete loss of hydraulic functionality by cavitation. The balance achieved between water supply and demand was reflected in the low variation of the water potential and of the variables related to gas exchange over time for a given GWD. This acclimation capacity of *P. tamarugo* after experiencing increases in GWD has great value for the implementation of conservation strategies. The thresholds presented in this paper should prove useful for conservation purposes of this unique species.

**Keywords:** $^{13}C$ isotopic composition, $^{18}O$ isotopic composition, phreatophyte, stomatal behavior, water table depth

## INTRODUCTION

The mortality rate of forest communities in different biomes has experienced substantial increases due to higher temperatures and incidence of more frequent and severe droughts (Peng et al., 2011; Liu et al., 2013; Williams et al., 2013). In addition, the disturbance of the environment by human activity, such as pollution or mining, exerts a substantial control over the species composition and structure of the forests, either by itself or by enhancing the effects of a stressor (Frelich, 2002). Whatever the case, the impact of natural stress or disturbance on plant species generates damage at various levels, from the photosynthetic activity to the survival capacity of the individual, the extent of the damage being dependent on the intensity of the phenomenon and the plant resistance to the stress (Thomas and Packham, 2007).

Human economic advance has caused high pressure on the use of natural resources. An example of this is the removal of water from the deserts for its use in urban centers and industrial activities (Pringle, 2001) endangering the species that live there. Such is the case of the Pampa del Tamarugal and Llamara aquifer, ecosystem having a hyper-arid climate (Pliscoff and Leubert, 2006), located in northern Chile in the Atacama Desert.

The vegetation of the Llamara Salt Flat is a tropical thorn forest composed by tamarugo (*Prosopis tamarugo*) and retama (*Tessaria absinthioides*). The shrub layer is composed mainly by retama (*T. absinthioides*; *Caesalpinia aphylla*), cachiyuyo (*Atriplex atacamentis*), algarrobilla (*Prosopis strombulifera*), and mesquite (*Prosopis burkartii*), while the herbaceous layer is almost completely dominated by salt grass (*Distichlis spicata*; Tréllez et al., 2011). The vegetation in the study site is composed almost entirely of *P. tamarugo* and some individuals of *T. absithioides*.

In this environment, the extraction of groundwater for the supply of human communities and mining (Rojas and Dassargues, 2007) has led to a drop in the water table out of which plant species live. The most abundant species in this ecosystem is *P. tamarugo* Phil., an endemic, strict phreatophyte legume tree (Altamirano, 2006), highly adapted to the conditions of high temperature and radiation of the Pampa del Tamarugal (Lehner et al., 2001) and described as a water stress resistant species (Acevedo et al., 1985). Water stress strongly increases the rate of defoliation of *P. tamarugo* under controlled conditions, decreases its paraheliotropic capacity (Chávez et al., 2013a), and its photosynthetic rate (Delatorre et al., 2008). Field studies in mature trees have shown that the effect of lowering the water table, increases the variability among individual trees of *P. tamarugo* (Altamirano, 1994), decreases its water potential, and causes loss of foliage (Chávez et al., 2013b), decreases NDVI (Normalized Difference Vegetation Index) as canopy activity indicator (Ortiz et al., 2012; Chávez et al., 2014) and eventually there is loss of paraheliotropic ability (Chávez et al., 2014).

A theoretical framework has been proposed in which the mortality of woody species to water stress would be governed by three factors

(McDowell et al., 2008): (i) biotic agents, which are a constant factor that can intensify by the incidence of stress, (ii) hydraulic failure, which plays an important role in the mortality of leaves, shoots, and peripheral roots under water stress (Rice et al., 2004; Hoffmann et al., 2011), and (iii) carbon starvation as the depletion of the plant reserves, reducing its resilience and increasing susceptibility to other stresses (Bigler et al., 2006) such as oxidative stress (Sage and Kubien, 2007; Delatorre et al., 2008). The predominant mechanism depends on the nature of drought, its intensity, and duration (McDowell et al., 2008; McDowell, 2011) and stomatal behavior of each species (Tardieu and Simonneau, 1988). It is expected that species having isohydric stomatal regulation possess an hydraulic system more succeptible to cavitation when compared to anisohydric species (Vogt, 2001; McDowell et al., 2008; Mitchell et al., 2013), the former being associated to regions of moderate humidity, and the latter to arid zone communities (Maherali et al., 2004).

Stomatal regulation and carbon assimilation can be studied through the leaf carbon and oxygen isotopic composition; the carbon isotopic composition ($\delta^{13}C$) of the leaf tissue can be used as a proxy for carbon assimilation (the more negative its value, the higher is the C assimilation in a given period of time), and the oxygen isotopic composition ($\delta^{18}O$) of the leaf tissue can be used as an indicator of stomatal resistance integrated over a period of time, when the leaf to air vapor pressure deficit is similar across the study area (the higher the oxygen isotopic composition, the higher is the stomatal resistance; Farquhar et al., 2007); further, the ratio of these two values may be used as a proxy for the intrinsic water use efficiency (assimilation per unit of stomatal conductance; Farquhar et al., 1989). Combined analyses of the carbon and oxygen isotopic composition of leaf biomass provide a means to distinguish the separate effects of stomatal conductance and net photosynthesis on water use efficiency (Scheidegger et al., 2000). The conceptual model of Scheidegger et al. (2000) has been successfully applied in the field (Keitel et al., 2003; Sullivan and Welker, 2007), so the dual-isotope approach has proven a valuable concept for ecological applications.

The objective of this study is to provide physiological information about how tamarugo trees behave as the GWD is increasing. This information should prove useful for the management of the aquifers where *P. tamarugo* grows and for the conservation of this unique species.

In the process, we use integrative variables such as carbon and oxygen isotopic composition of the leaf tissue as indicators of gas exchange and the fraction of green canopy as indicator of defoliation. The advantage of using this type of variables has already been established (Farquhar and Lloyd, 1993; Brugnoli and Farquhar, 2000; Chávez et al., 2014).

# MATERIALS AND METHODS

## Site Characterization and Monitoring

The study site is located south of the Pampa del Tamarugal, Atacama Desert, northern Chile, in the Llamara Salt Flat (19K 434222.0 m E, 7658495.0 m S) dominated by the Llamara aquifer, located in the mid depression of the Tarapaca Region. This confined aquifer has 30 to 40 Km width at a mean altitude of 1000 m (DICTUC, 2007). The Pampa del Tamarugal has a hyper-desertic tropical bio-climate (Pliscoff and Leubert, 2006) characterized by the almost complete absence of rainfall (long term mean of 5 mm year$^{-1}$), low relative humidity, and wide temperature variation (Campillo and Hojas, 1975). Table 1 shows meteorological information obtained at the Canchones Experiment Station, Arturo Prat University, commune of Pozo al Monte, Tamarugal Province (19K 444490.5 m E; 7739128.5 m S). Rainfall during the study period was near zero.

**Table 1.** Meteorological data from a meteorological station for the time of measurements (see text), mean monthly values for daily mean temperature, relative humidity, and wind speed; maximum solar radiation in the month and total rainfall

| Meteorological variable | Year | | | | Mean ± SD |
|---|---|---|---|---|---|
| | 2011 | 2012 | 2013 | 2014 | |
| **JANUARY** | | | | | |
| Temperature (°C) | 20.5 | 21.1 | 21.4 | 21.5 | 21.1 ± 0.4 |
| Relative Humidity (%) | 44.0 | 48.3 | 43.0 | 50.6 | 46.5 ± 3.6 |
| Wind speed (ms$^{-1}$) | 1.1 | 1.2 | 1.1 | 1.1 | 1.1 ± 0 |
| Maximum solar radiation (Wm$^{-2}$) | 1059.2 | 1014.5 | 1067.8 | 1050.3 | 1047.9 ± 23.4 |
| Rainfall (mm)* | 3.4 | 0.8 | 0.0 | 0.0 | 1.1 ± 1.6 |
| **JULY** | | | | | |
| Temperature (°C) | 13.7 | 16.1 | 14.4 | 12.9 | 14.3 ± 1.4 |
| Relative Humidity (%) | 45.4 | 27.3 | 32.2 | 32.5 | 34.3 ± 7.8 |
| Wind speed (ms$^{-1}$) | 0.6 | 0.3 | 0.5 | 0.3 | 0.4 ± 0.1 |
| Maximum solar radiation (Wm$^{-2}$) | 704.9 | 685.6 | 745.2 | 795.6 | 732.8 ± 48.6 |
| Rainfall (mm)** | 0.4 | 0.4 | 0.2 | 0.0 | 0.25 ± 0.2 |

*Canchones Experiment Station, commune of Pozo al Monte, Tamarugal Province. (19K 444490.5 m E; 7739128.5 m S). Rainfall during this period was 5.2 mm, practically nil.*
**Amount of rainfall between January and June. **Amount of rainfall between July and December.*

In the Llamara Salt Flat site there is a gradient of groundwater depth (GWD) caused by the presence of water extraction wells that generate a depression cone with a radius of about 4 km. After 5 years of the onset of water extraction (2006–2010), the water table depth was stabilized, and three zones of GWD were established in a linear transect from the extraction wells: (i) High-GWD (11.2 ± 0.3 m), (ii) Mid-GWD (10.3 ± 0.3 m), and (iii) Low-GWD (7.1 ± 0.1 m). The 7.1 m was used as reference GWD since previous studies covering the period 2005–2011 and GWDs from 4.0 m to 6.0–7.0 m (Calderón et al., 2015) had shown that except for a small decrease in twig growth rate, the physiological parameters studied were not affected at 7.1 m GWD. Four *P. tamarugo* trees per GWD zone were selected for the study (total of 12 trees). The depth of the water table at each tree was estimated through spacial interpolation, using actual GWDs obtained from observation wells located in the area. The depth of the water table in the observation wells was measured with a water level indicator (Waterline, USA). Table 2 shows the mean and standard deviation of the GWD for the four *P. tamarugo* trees of each group over the 4 years of this study (January 2011 through July 2014). Notably, the groundwater depth at each GWD zone has been stable over the 4 years of study. Water extraction began in February 2006 and the GWD prior to water extraction is indicated by the value of December 2005 in Table 2.

**Table 2.** Estimated mean water table depths ± SD in a transect away from water extraction wells at the Llamara Salt Flat, Pampa del Tamarugal, Atacama Desert

| | Mean depth of the water table (m) | | |
|---|---|---|---|
| **Measuring date** | **High-GWD** | **Mid-GWD** | **Low-GWD** |
| Dec-05* | 4.0 ± 0.1 | 4.3 ± 0.1 | 5.9 ± 0.5 |
| Jan-11 | 11.1 ± 0.0 | 10.4 ± 0.3 | 7.0 ± 0.2 |
| Jul-11 | 10.5 ± 0.2 | 9.8 ± 0.2 | 7.0 ± 0.2 |
| Jan-12 | 11.4 ± 0.0 | 10.5 ± 0.5 | 7.1 ± 0.3 |
| Jul-12 | 11.1 ± 0.2 | 10.2 ± 0.3 | 7.0 ± 0.2 |
| Jan-13 | 11.3 ± 0.2 | 10.2 ± 0.4 | 7.1 ± 0.2 |
| Jul-13 | 11.6 ± 0.1 | 10.4 ± 0.6 | 7.2 ± 0.2 |
| Jan-14 | 11.6 ± 0.1 | 10.6 ± 0.5 | 7.3 ± 0.2 |
| Jul-14 | 11.3 ± 0.2 | 10.5 ± 0.3 | 7.3 ± 0.2 |
| Mean between Jan-11 and Jul-14 | 11.2 ± 0.3 | 10.3 ± 0.3 | 7.1 ± 0.1 |
| Δ between Dec-05 and Jul-14 | 7.3 | 6.2 | 1.4 |

Pumping started in December 2005. High GWD (11.2 m deep, 150 m away from the water extraction well); Mid-GWD (10.3 m deep, 300 m away from the water extraction well); and Low-GWD (7.1 m deep, reference GWD, 3000 m away from the extraction well). Data collected from January 2011 through July 2014.

*Water table depth prior to water extraction.

The data were collected in the months of January (summer) and July (winter). The group of trees associated with the High-GWD zone was at approximately 150 m away from the water extraction wells, those associated with the Mid-GWD zone were around 300 m away and those in the Low-GWD zone were at about 3000 m away from the water extraction wells. All the trees of this study are on the Llamara aquifer of the Llamara Salt Flat, the most distant two trees are 3500 m away from the water extraction wells.

## Plant Water Status: Water Potential and Stomatal Resistance

The water potential and the stomatal resistance were measured in each field campaign. The water potential was measured at pre-dawn ($\Psi_{pd}$; around 6:00 h in summer and around 7:00 h in winter, dawn in the area is 6:44 h in January and 7:50 h in July) and around solar noon (mid-day water potential; $\Psi_{md}$) between 12:30 and 14:30 h using a Scholander pressure chamber (Scholander et al., 1965). The twigs used for water potential determination were lignified, one season twigs which were cut at the base. Two twigs per tree which had been covered with a plastic bag and foil the previous evening were used. The mid-day water potential was measured only in the trees of the High-GWD and Low-GWD zones.

The stomatal resistance was measured between 8:30 and 10:30 h, time during which the stomata have their maximum aperture in these trees (Calderón et al., 2015). The stomatal resistance ($r_s$) was measured in four fully expanded, nutritionally, and sanitary healthy leaves per tree, fully exposed to the sun (facing east). A steady-state diffusion porometer (DECAGON Devices, Steady State Diffusion Leaf Porometer Model SC-1) was used. The measurements of water potential and stomatal resistance were done the same day in the same trees.

## Soil Water Content

In each campaign soil samples at 20, 30, 40, 60, and 100 cm depth were taken to determine the soil water content gravimetrically. The samples were taken under the canopy of one selected tree in each GWD zone. The samples were placed in aluminum tins, weighed to determine wet weight, and dried at 105°C to determine the weight of dry soil. The soil bulk density for each depth was determined using the cylinder method (Allen, 1989), and the soil water content was expressed as a volume fraction ($cm^3$ $cm^{-3}$). Finally, weighted mean soil water (weighted per soil depth, WWC) was calculated for each 1 m soil profile.

## Leaf Sampling and Measurement of $\delta^{13}C$ and $\delta^{18}O$ Composition

In the campaigns of January of each year a composite sample of leaves per each tree was taken. Sampling was done by taking leaves of the present growing season in the entire perimeter of each tree at about 1.5 m height. The size of each sample was ~50 g. Each sample was dried at 60°C in a forced air oven for 48 h. The samples were then crushed to a homogeneous powder with a power mill. Two sub-samples were taken from each sample and put in tin and silver capsules to measure isotopic composition of $^{13}C$ ($\delta^{13}C$) and $^{18}O$ ($\delta^{18}O$), respectively. Finally, the samples were sent to the Stable Isotope Facility at the University of California, Davis, for analysis (data available in Supplementary Material; Table 1). The samples for $\delta^{13}C$ were measured using a PDZ Europa ANCA-GSL elemental analyzer interfaced to a PDZ Europa 20–20 isotope ratio mass spectrometer (Sercon Ltd., Cheshire, UK) with a precision of 0.2‰. The samples for $\delta^{18}O$ were measured with an Elementar PyroCube (Elementar Analysensysteme GmbH, Hanau, Germany) interfaced to a PDZ Europa 20–20 isotope ratio mass spectrometer (Sercon Ltd., Cheshire, UK), with a precision of 0.5‰.The standards used to measure the $\delta^{13}C$ isotopic composition were USGS-41 (Glutamic Acid) and Nylon 5 as international standards and Glutamic Acid and Peach leaves as the internal check. For the measurement of the $\delta^{18}O$ isotopic composition the international standards were IAEA-600 (Caffeine) and USGS-35 (Sodium Nitrate) and Cellulose as an internal check.

## Green Canopy Fraction

The green canopy fraction ($f$GCC), used to quantitatively assess defoliation, was defined as the green surface of a side of a *P. tamarugo* tree respect to the total area of the same side and was determined according to the methodology described in Chávez et al. (2013b). In each campaign a 10 mega pixels picture of each tree was taken with a digital camera (FUJIFILM Model FINEPIX S2000HD) in automatic mode and from the same point each time. The photographs were processed with the sub-routine for image analysis based on objects of the program eCognition®Developer. Objects were identified within each image with a scale parameter of 3.0, a shape factor of 0.1, and compactness of 0.5. Then, the objects associated with "leaves" and "branches" were identified through an analysis of the green, red, and blue bands of each pixel that made up the picture. The area of "leaves" and "branches" was determined according to the number of pixels. The $f$GCC was calculated as follows,

$$fGCC = \frac{\text{Leaf area}}{\text{Leaf area} + \text{Branch area}} \quad (1)$$

In order to assess the magnitude of change in $f$GCC, each year between January and July the relative difference of green coverage from January to July (R$f$GCC$_{Jan-Jul}$) was calculated as follows,

$$RfGCC_{Jan-Jul} = \frac{fGCC_{Jan} - fGCC_{Jul}}{fGCC_{Jan}} \quad (2)$$

## Data Analysis

Statistical analyses were performed using the statistical software InfoStat (Di Rienzo et al., 2015) and/or R version 3.2.1 (R Core Team, 2015). The significance level was set at 0.05 for all comparisons. The variables $f$GCC, $\Psi_{pd}$, $\Psi_{md}$, $r_s$, $\delta^{13}C$, $\delta^{18}O$, and R$f$GCC$_{Jan-Jul}$ were analyzed using general linear and mixed models for repeated measures, with the measuring date, GWD zone and the interaction between factors as fixed factors, and we modeled the correlation with an autocorrelation structure of order 1 (corAR1) for the measurements over the same individual (tree) between different times (measurement date), followed by LSD -Fisher *post-hoc* analyzes when appropriate (Supplementary Material; Table

2). The model selection in each case was based on the fulfillment of the assumptions of the statistical analysis and under AIC index values (Akaike information criterion; Bozdogan, 2000). The only variable that had to be transformed to meet the statistical assumptions for analysis was stomatal resistance.

# RESULTS

## Effect of the Depth of the Water Table on the Water Potential and Gas Exchange of *P. tamarugo*

The groups of trees of the GWD zones were significantly different in pre-dawn ($\Psi_{pd}$; $P = 0.007$) and noon water potential ($\Psi_{md}$; $P = 0.0176$; Figures 1A,B, respectively). The measuring date factor and interaction between GWD × measuring date were not significant ($P > 0.05$). The mean Low-GWD $\Psi_{pd}$ was higher than the mean Mid-GWD $\Psi_{pd}$ and mean High-GWD $\Psi_{pd}$ (Figure 1A). This difference was also observed among the water potential values at mid-day (Figure 1B). The GWD had no significant effect ($P > 0.05$) on the stomatal resistance ($r_s$) but the measuring date was significant ($P = 0.001$) with high $r_s$-values in January 2011 which decreased from July 2011 onwards (Figure 1C). The mean soil water content (WWC, mean weighted by soil depth) of the Low-GWD zone was higher than the mean water content of the Mid-GWD and High-GWD zones (Figure 1D).

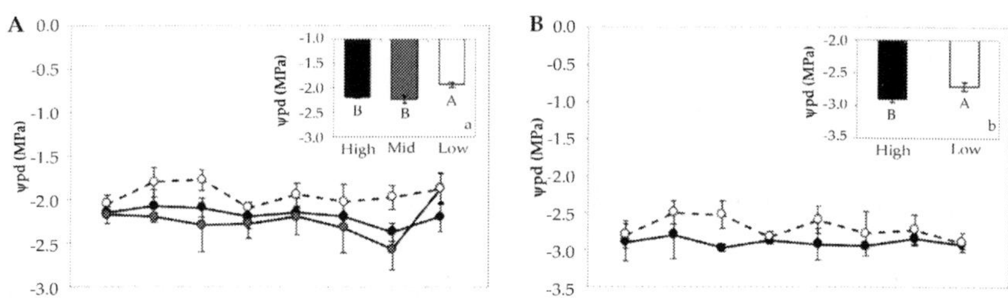

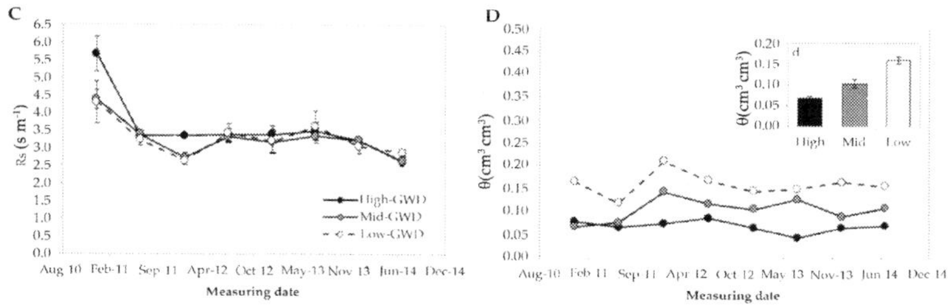


**Figure 1.** Means ± 1 S.E. of (A) predawn water potential, (B) mid-day water potential, (C) stomatal resistance, and (D) weighted mean of soil water content. Each curve within a graph represents a groundwater depth (GWD, High = 11.2 m; Mid = 10.3 m; Low = 7.1 m) between January 2011 and July 2014. The insertions (**a**; predawn water potential), (**b**; mid-day water potential), and (**d**; weighted mean soil water content) represent the mean ± 1 S.E. for each variable at each GWD ($n = 32$) when there was no significant GWD × measuring date interaction ($p > 0.05$). Different letters indicate significant differences according LSD-Fisher *post-hoc* analysis.

The leaf $\delta^{13}C$ (Figure 2) differed between GWDs at 10% level ($P = 0.10$), with lower values in the Low-GWD zone respect to the Mid-GWD and High-GWD zones (Figure 2a) and between measuring dates ($P = 0.0053$; Figure 2b). No significant interaction between GWD and measuring date ($P > 0.05$) was observed for $\delta^{13}C$.

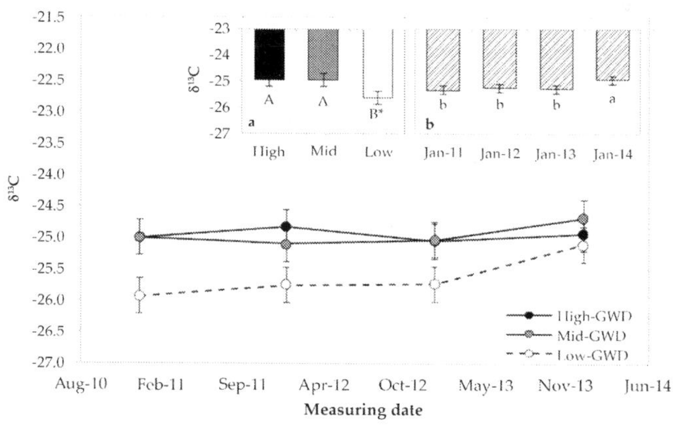


**Figure 2.** Means ± 1 S.E. of isotopic composition of $^{13}C$ ($\delta^{13}C$) measured in January from 2011 through 2014 in foliar samples of *P. tamarugo* trees growing at three GWDs (High, Mid, and Low as in Figure 1). The insertions (a)

and (b) show the mean values ± 1 S.E. of δ$^{13}$C at each GWD ($n$ = 16) and at each measuring date ($n$ = 12), respectively. Different letters indicate significant differences according LSD–Fisher *post-hoc* analysis. *Significant difference at 10% level.

The leaf δ$^{18}$O (Figure 3) was significantly different among GWDs ($P$ = 0.0206) and between measuring dates ($P$ < 0.0001). The interaction GWD x measuring date was not significant ($P$ > 0.05). Figure 3a shows that the High-GWD trees had a higher value of leaf δ$^{18}$O than the Low-GWD trees, while the Mid-GWD trees had an intermediate value equal to the leaf δ$^{18}$O of the trees of the other two depths of the water table. Regarding the measuring date (Figure 3b), a marked seasonality was observed such that leaf δ$^{18}$O decreased significantly from January 2011 to January 2012 repeating the cycle in 2013 and 2014 (a similar oscilation was observed in the mean relative humidity for January from 2011 through 2014, Table 1, what may indicate changes in the atmosfetric demand in January across these years). The high value of leaf δ$^{18}$O measured in January 2011 was consistent with the high $r_s$ values measured on the same date (Figure 1C).

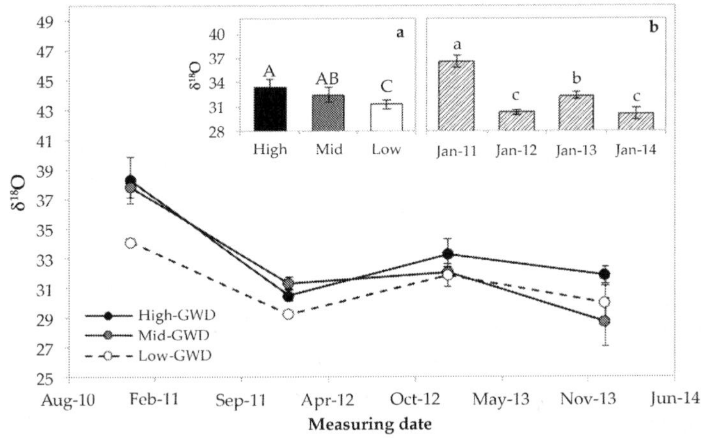


**Figure 3.** Mean of leaf isotopic $^{18}$O composition ± 1 S.E. (δ$^{18}$O) measured between January 2011 and January 2014 in *P. tamarugo* groups ($n$ = 4) located at three GWDs (High, Mid, and Low as in Figure 1). The insertions (a) and (b) show the mean ± 1 S.E. of δ$^{18}$O at each GWD ($n$ = 16) and at each measuring date ($n$ = 12), respectively. Different letters indicate significant differences according LSD–Fisher *post-hoc* analysis.

## Effect of the Water Table Level on the Green Cover and Defoliation of *P. tamarugo*

The green cover fraction (*f*GCC; Figure 4) was significantly different between the trees at the various GWDs ($P < 0.001$) and between measuring dates ($P < 0.001$). The interaction GWD × measuring date was not significant ($P > 0.05$). Figure 4a shows that the average *f*GCC of the trees in the Low-GWD zone was significantly higher than in the trees of the Mid-GWD and High-GWD zones, which were equal among them. The mean *f*GCC of *P. tamarugo* in January of each year was significantly higher than in July in all years (Figure 4b).

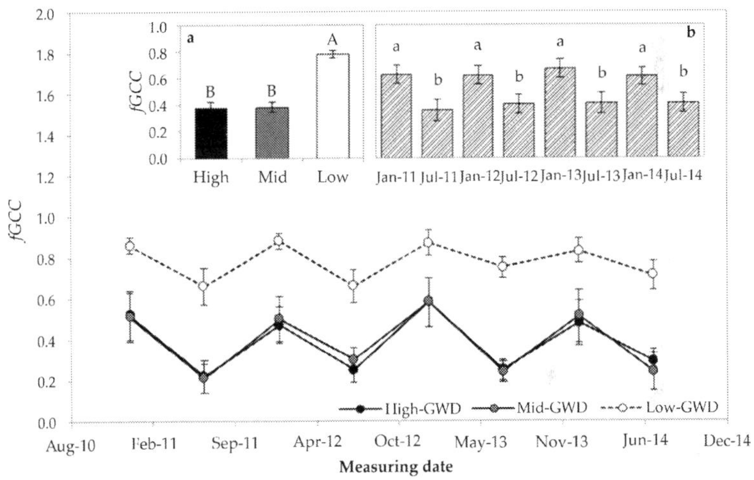


**Figure 4.** Means ± 1 S.E. of green canopy fraction (***f*****GCC**) for three GWDs. Data obtained in January and July from 2011 through 2014 ($n = 4$). The insertions (a) and (b) show the mean *f*GCC ± 1 S.E. at each GWD ($n = 32$) (High, Mid, and Low as in Figure 1) and measuring date ($n = 12$), respectively. Different letters indicate significant differences according LSD–Fisher *post-hoc* analysis.

The relative difference of *f*GCC between January and July ($Rf\text{GCC}_{Jan-Jul}$) was significantly different between GWDs ($P < 0.0001$) and between measuring dates ($P = 0.0227$). The interaction GWD x measuring date was not significant ($P > 0.05$). The Mid-GWD and High-GWD trees had equal *f*GCC and had $Rf\text{GCC}_{Jan-Jul}$ values significantly higher than the trees of the Low-GWD zone (Table 3), therefore, *P. tamarugo* trees subject to greater depths of water table, besides having a lower green coverage over time, lost a higher proportion of their foliage during the winter, however they recovered the green canopy fraction in summer.

**Table 3.** Mean values of the relative difference between January and July green canopy fraction ($RfGCC_{Jan-Jul}$) at each GWD and date, significance of the factors and their interaction

| | $RfGCC_{Jan-Jul}$ | |
|---|---|---|
| **GWD ZONE** | | |
| High-GWD | 0.51 | A |
| Mid-GWD | 0.53 | A |
| Low-GWD | 0.20 | B |
| **MEASURING DATE** | | |
| 2011 | 0.50 | A |
| 2012 | 0.38 | B |
| 2013 | 0.42 | AB |
| 2014 | 0.35 | B |
| **Factors** | ***p*-values** | |
| GWD zone | <0.0001 | |
| Measuring date | 0.0405 | |
| Interaction | 0.1509 | |

*Different letters indicate significant differences according LSD-Fisher post-hoc analysis.*

# DISCUSSION

## Water Relations of *P. tamarugo* in Different Scenarios of Water Table Depth

The water table depth values given in Table 2 as well as the values of the physiological variables measured in this study indicate that the trees were acclimated at each GWD (there was practically no change of the variables from 2011 through 2014 at each GWD). Furthermore, the results of this study demonstrate that the GWD affects the water status of *P. tamarugo*. When the GWD was 7.1 m the $\Psi_{pd}$ and $\Psi_{md}$ were higher compared to a GWD of 10.3 or 11.2 m (Figures 1A,B). The differences are consistent with those reported by Chávez et al. (2013b), where at a water table depth of 11 m (Pintados Salt Flat) the trees had significantly lower $\Psi_{pd}$ compared to a water table depth of 6 m (Llamara Salt Flat).

The minimum water potential values observed were −2.7 and −3.0 MPa for GWDs of 7.1 and 11.2 m, respectively. The maximum pre-dawn water potentials measured were −2.0 and −2.2 MPa, respectively, therefore $\Psi$ decreased throughout the day in a parallel way across GWD zones. *P. tamarugo* in the High-GWD zone decreased its water potential in 0.7 MPa and in the Low-GWD the decrease in water potential was 0.8 MPa. The observation of a similar twig water potential change along the day for the two GWDs plus the fact that the trees are apparently not increasing the minimum $r_s$ at the leaf level (Figure 1C), point toward a relatively anisohydric stomatal behavior of *P. tamarugo* (Tardieu and Simonneau, 1988; McDowell et al., 2008, 2011). The species was able to maintain low values of $r_s$ under a condition of reduced water availability. This behavior is a characteristic of species that have evolved in high aridity (Maherali et al., 2004).

While instant measurements of $r_s$ indicated that there were no differences between trees across the various GWDs, the foliar isotopic composition (an integrated value over time) indicated otherwise. The trees growing in the High-GWD zone had a higher mean value of $\delta^{13}C$ (Figure 2a) and a higher mean $\delta^{18}O$ (Figure 3a) compared to the trees growing in the Low-GWD zone. These results indicate that under increased GWD, *P. tamarugo* has lower assimilation (higher $\delta^{13}C$ isotopic composition) and a higher integrated $r_s$, indicated by higher values of $^{18}O$ isotopic composition at the leaf level (Hasselquist et al., 2010).

The high value of leaf $\delta^{18}O$ measured in January 2011 was consistent with the high $r_s$-values measured on the same date (Figure 1C). We attribute the values of $^{18}O$ isotopic composition of the leaf tissue to leaf transpiration, controlled by the stomatal resistance as the $^{18}O$ isotopic composition of the groundwater of the Llamara Salt Flat is homogeneous at the GWDs considered in this study. Indeed, the $^{18}O$ composition is not correlated with GWD ($r = 0.072$, $P = 0.81$), with mean $^{18}O$ composition values of −6.0‰, $SD = 0.7$, $n = 13$ (Dirección General de Aguas DGA, 1985). In 2012, water samples were taken from two observation wells to measure $\delta^{18}O$, one close to the High, and the other to the Low GWD zone; the values obtained were −6.10 and −6.09‰, respectively (unpublished data, personal communication, Geohidrologia Consultores, ARCADIS, Chile).

The results suggest that *P. tamarugo* exerts control over its leaf resistance when the GWD is beyond 10.0 m and it is probable that twig Ψ values of −3.0 MPa measured under this condition could be near the threshold of loss of hydraulic functionality. It is necessary, however, to consider that these thresholds tend to be variable throughout the growing season (Kolb and Sperry, 1999), and further research is required to fully assess this point.

## Water Use Strategy of *P. tamarugo* Growing at Increased Water Table Depths

Mooney et al. (1980) proposed that during the night, when stomatal resistance of *P. tamarugo* is maximum, the water flow in the soil-plant-atmosphere occurs from the water table through the deep roots into the plant, and from the shallow root mass into the upper part of the soil, following a strong water potential gradient generated by soil salinity and tree transpiration of the previous day. During the day, the water flow would occur from the water table and the soil surrounding the shallow root mass, to the atmosphere through the plant. Aravena and Acevedo (1985) demonstrated that the water surrounding the shallow root mass and the water present in the tree had the same isotopic composition ($^{18}O$ and $^{2}H$) than the water of the ground water. Figure 1D shows that the soil water content in the soil root mass zone decreased with increasing GWD. The water depletion in the soil root mass zone at higher GWDs indicates that the water flow through the deep roots of *P. tamarugo* has decreased affecting the tree water balance via decreased water supply.

At the same time *P. tamarugo* regulates its water demand via partial stomatal closure ($r_s$ at the time of maximum stomatal opening did not decrease but $^{18}O$ isotopic composition increased) and essentially through foliage loss, as has been described in other phreatophytes (Cooper et al., 2003, 2006). Studies done in young *P. tamarugo* plants subject to water stress showed an accelerated senescence (Chávez et al., 2013a), and adult plants growing in the Pintados Salt Flat (close to our study area) experienced decreased activity measured by NDVI, reaching minimum values at 10 m of GWD (Ortiz et al., 2012). A similar response has been observed in *Populus deltoides* (Cooper et al., 2003), *Populus fremontii, Salix gooddingii*, and *Tamarix chinensis* (Horton et al., 2001) when these trees face a water table drop. In this study, GWDs from 7.1 to 10.3 m decreased the mean fraction of green canopy cover from 78 to 38% (Figure 4a). Furthermore, the foliage loss in winter increased from 20 to 50% when the GWD increased from 7.1 to 10.3 m. Besides the green cover loss in winter, increased under water stress, *P. tamarugo* recovered its green canopy fraction during the summer period (Table 3), supporting an acclimation hypothesis. The decrease in green cover during winter had been already described (Acevedo et al., 2007) but the increase in seasonal foliage loss due to water stress had not been reported. The lowest values of twig water potential measured in this study were around −3.0 MPa in the *P. tamarugo* trees of the High-GWD zone. It is probable that, with this value of water potential, the trees lose a significant fraction of their hydraulic functionality, what would be consistent with the observations of *f*GCC since the foliage loss is due to a loss of hydraulic conductivity in anisohydric species experiencing low water potentials as a result of high stomatal conductance under stress conditions (Hoffmann et al., 2011). In this study we observed that in the summer periods the *f*GCC of the trees was 0.86 under Low-GWD, 0.53 under Mid-GWD, and 0.51 in the High-GWD zone, while in the winter months the values were 0.69 for the trees of the Low-GWD, 0.24 for those of the the Mid-GWD and 0.25 in High-GWD zone. While this strategy may be useful to tolerate prolonged periods of water stress under the conditions of the Atacama Desert, it also could have a disadvantage given the high intensities of radiation, temperature, and wind speed prevailing in this ecosystem. Low coverage implies that a larger fraction of the tree leaf area is exposed to direct radiation, increasing the likelihood of photoinhibition, and oxidative stress (Delatorre et al., 2008). Chávez et al. (2013a) and Chávez et al. (2014) described the mechanism of paraheliotropism in *P. tamarugo* as an avoiding mechanism of high radiation stress. They also found that paraheliotropism decreased in summer, when water stress was higher. On the other hand, lower leaf coverage increases the turbulence inside of the cup of *P. tamarugo*, decreasing the aerodynamic resistance, what would increase the rate of transpiration, inducing stomatal closure. This situation is reflected in the observed values of $\delta^{13}C$ and $\delta^{18}O$ composition when GWD was beyond 10 m.

The fact that each GWD was constant from 2011 through 2014 (Table 2) allow us to infer what will happen to *P. tamarugo* growing in the Llamara Salt Flat as GWD increases beyond 10 to 11 m. There will be a decrease in twig water potential, an increased integrated stomatal resistance over time (increased $\delta^{18}O$ composition), a lower photosynthetic rate (increased $\delta^{13}C$) and a decreased green canopy fraction (*f* GCC). Notably, these variables changed with mean GWD but did not interact with measuring date during the 4 years of this study.

The increasing scarcity of water as a resource, particularly in arid and desert zones, will force an increased water extraction of water which will exacerbate the effect of the stressful conditions of the Atacama Desert. This will have a negative impact on the sustainability of the *P.tamarugo*ecosystem, therefore research is needed to better characterize the phenotypic plasticity of this phreatophyte in response to changes in the groundwater regime.

The balance achieved between water supply and demand was reflected in the low variation of the water potential and of the variables related to gas exchange over time for a given GWD. This acclimation capacity of *P. tamarugo* after experiencing increases in GWD has great value for the implementation of conservation strategies. Further research is needed, however, to determine the limits of the acclimation.

Even though, we are not yet in a position to definitely establish a GWD beyond which water extraction from the Tamarugal basin should be completely stopped, the physiological information reported here clearly indicates that special care should be taken when the aquifer reaches 11 m deep.

An additional observation pointing toward a *P. tamarugo* conservation strategy relates to the soil water content in the upper 1.0 m of soil, where the *P. tamarugo* root mass lies. If the *P. tamarugo* tap roots are providing water in appropropriate amount to the tree, the soil water content of this zone should not change significantly over periods of several days; on the contrary, if the *P. tamarugo* tap roots are not providing enough water to fulfill the water demand of the tree, a decrease in the soil water content of this zone, as the one reported in this article should be observed. A word of caution should be expressed here, to remind the reader that the water flux from the ground water to the soil root mass zone through the tap roots of the tree occurs along a water potential gradient, therefore, the soil salinity in the soil root mass zone will have an important influence in the water flux from the ground water to the soil of the root mass zone via tap roots.

## CONCLUSIONS

The differences observed in the physiological behavior of *P. tamarugo* trees, attributable to the ground water depth; show that increasing the depth of the water table from 7 to 11 m significantly affects the water status of *P. tamarugo*.

*P. tamarugo* has a strategy of foliage loss to balance the water demand against a diminished water supply, a characteristic of a relative anisohydric species.

At present and for conservation purposes we advise to monitor the green canopy fraction of the trees as an indicator of their status. This is an easy to measure, very sensitive variable whose maximum and minimum values were maintained during the summer and winter across several years (2011 through 2014) at 11.2 m GWD. A disruption of the equilibrium, detected by increased defoliation beyond the values reported in this paper, would indicate the need to decrease or even stop water extraction to avoid irreversible damage to the trees. In the meantime, additional information should be gathered to tune-up this recommendation.

## AUTHOR CONTRIBUTIONS

MG collected data and drafted this manuscript. PS provided logistic support, participated in designing the experiment and edited the manuscript. EA designed the experiment, followed up on data collection, provided support through the Soil-Water Relations Laboratory and edited the manuscript.

## CONFLICT OF INTEREST STATEMENT

The authors declare that the research was conducted in the absence of any commercial or financial relationships that could be construed as a potential conflict of interest.

The Soil-Plant-Water Relations Laboratory of the University of Chile has an agreement with SQM, a Chilean Mining Company under which it monitors the water relations of *Prosopis tamarugo* Phil. This is a requirement that the Chilean Environmental Authority imposes on SQM to allow ground water pumping from the Llamara Salt Flat. This research has been partially financed by this agreement.

## ACKNOWLEDGMENTS

The authors thank Dr. Nicolás Franck, Dr. Carlos Prado, and Mr. Cristian Ortiz for critical reading of this manuscript and acknowledge the contribution to data acquisition of Ing. Agr. Marcelo Becerra (UCH-Soil-Plant-Water Relations Laboratory) and the MS student Mr. Manuel Cáceres. This study was financed by a joint venture between the Agronomy Faculty of the University of Chile and SQM as well as FONDECYT Project N°1150799 and Advanced Human Capital Project CONICYT-PCHA/National Doctorate/2015-21150807.

## SUPPLEMENTARY MATERIAL

**Supplementary Table 1:** Isotopic composition of leaf tissue $\delta^{13}C$ and δ18O of *Prosopis tamarugo* growing at three groundwater depths (Low=7.1m;Midium=10.0m ; High= 11.7m). The Low GWD is a reference depth at which physiological processes are not affected. Data are provided for each year and Measuring Date (January).

| Measuring date | GWD | Tree | $\delta^{13}C$ (‰) | $\delta^{18}O$ (‰) |
|---|---|---|---|---|
| Jan-11 | High | A01 | -25.1 | 36.5 |
| Jan-11 | High | A02 | -24.2 | 42.6 |
| Jan-11 | High | A17 | -25.5 | 38.6 |
| Jan-11 | High | A19 | -25.2 | 35.4 |
| Jan-12 | High | A01 | -25.3 | 29.7 |
| Jan-12 | High | A02 | -23.7 | 30.4 |
| Jan-12 | High | A17 | -25.2 | 31.5 |
| Jan-12 | High | A19 | -25.1 | 30.4 |
| Jan-13 | High | A01 | -25.5 | 31.5 |
| Jan-13 | High | A02 | -23.9 | 31.9 |

| Jan-13 | High | A17 | -25.2 | 36 |
|---|---|---|---|---|
| Jan-13 | High | A19 | -25.6 | 33.7 |
| Jan-14 | High | A01 | -25.1 | 30.1 |
| Jan-14 | High | A02 | -24.1 | 32.4 |
| Jan-14 | High | A17 | -24.9 | 32.4 |
| Jan-14 | High | A19 | -25.6 | 32.5 |
| Jan-11 | Mid | A05 | -24.6 | 39.5 |
| Jan-11 | Mid | A15 | -25.2 | 38.4 |
| Jan-11 | Mid | A20 | -25.0 | 36.3 |
| Jan-11 | Mid | A50 | -25.2 | 37.0 |
| Jan-12 | Mid | A05 | -24.8 | 31.2 |
| Jan-12 | Mid | A15 | -25.4 | 32.2 |
| Jan-12 | Mid | A20 | -24.6 | 30.4 |
| Jan-12 | Mid | A50 | -25.6 | 31.6 |
| Jan-13 | Mid | A05 | -24.7 | 32.2 |
| Jan-13 | Mid | A15 | -25.4 | 33 |
| Jan-13 | Mid | A20 | -24.8 | 31.2 |
| Jan-13 | Mid | A50 | -25.2 | 31.9 |
| Jan-14 | Mid | A05 | -24.7 | 24.0 |
| Jan-14 | Mid | A15 | -25.3 | 29.8 |
| Jan-14 | Mid | A20 | -24.1 | 31.4 |
| Jan-14 | Mid | A50 | -24.6 | 29.5 |
| Jan-11 | Low | A38 | -25.8 | 33.8 |
| Jan-11 | Low | A41 | -26.8 | 33.8 |
| Jan-11 | Low | A43 | -25.0 | 34.8 |
| Jan-11 | Low | A44 | -26.1 | 34.0 |
| Jan-12 | Low | A38 | -26.0 | 29.2 |
| Jan-12 | Low | A41 | -26.2 | 29.3 |
| Jan-12 | Low | A43 | -25.4 | 29.4 |
| Jan-12 | Low | A44 | -25.4 | 29.1 |
| Jan-13 | Low | A38 | -25.7 | 29.9 |
| Jan-13 | Low | A41 | -26.4 | 33.6 |
| Jan-13 | Low | A43 | -25.0 | 32 |
| Jan-13 | Low | A44 | -25.8 | 31.9 |
| Jan-14 | Low | A38 | -24.7 | 31.5 |

| | | | | |
|---|---|---|---|---|
| Jan-14 | Low | A41 | -25.9 | 31.9 |
| Jan-14 | Low | A43 | -24.7 | 29.8 |
| Jan-14 | Low | A44 | -25.1 | 26.8 |

**Supplementary Table 2:** ANOVA Tables for physiological variables (Water relations and foliar isotopic composition of Prosopis tamarugo Phil. a native legume tree of the Atacama Desert growing at three levels of water table depth)

| **Predawn water potential** | | | |
|---|---|---|---|
| **Factor** | **numDF** | **F-value** | **p-value** |
| **(Intercept)** | 1 | 2913.55 | <0.0001 |
| **Measuring date** | 7 | 1.4 | 0.2187 |
| **GWD** | 2 | 5.32 | 0.007 |
| **Measuring date:GWD** | 14 | 0.59 | 0.8679 |
| | | | |
| | | | |
| **Midday water potential** | | | |
| **Factor** | **numDF** | **F-value** | **p-value** |
| **(Intercept)** | 1 | 5059 | <0.0001 |
| **Measuring date** | 7 | 0.44 | 0.874 |
| **GWD** | 1 | 6.04 | 0.0176 |
| **Measuring date:GWD** | 7 | 0.36 | 0.9221 |
| | | | |
| | | | |
| **Stomatal resistance** | | | |
| **Factor** | **numDF** | **F-value** | **p-value** |
| **(Intercept)** | 1 | 3111.48 | <0.0001 |
| **Measuring date** | 7 | 18.5 | <0.0001 |
| **GWD** | 2 | 2.3 | 0.1078 |
| **Measuring date:GWD** | 14 | 1.53 | 0.1223 |
| | | | |
| | | | |
| **Leaf $\delta^{13}C$** | | | |

| **Factor** | **numDF** | **F-value** | **p-value** |
|---|---|---|---|
| **(Intercept)** | 1 | 29849.3 | <0.0001 |
| **Measuring date** | 3 | 5 | 0.0053 |
| **GWD** | 2 | 2.38 | 0.1006 |
| **Measuring date:GWD** | 6 | 1.35 | 0.2623 |
| | | | |
| | | | |
| **Leaf $\delta^{18}O$** | | | |
| **Factor** | **numDF** | **F-value** | **p-value** |
| **(Intercept)** | 1 | 11530.14 | <0.0001 |
| **Measuring date** | 3 | 41.76 | <0.0001 |
| **GWD** | 2 | 4.33 | 0.0206 |
| **Measuring date:GWD** | 6 | 1.97 | 0.0959 |
| | | | |
| | | | |
| ***f*GCC** | | | |
| **Factor** | **numDF** | **F-value** | **p-value** |
| **(Intercept)** | 1 | 224.42 | <0.0001 |
| **Measuring date** | 7 | 34.59 | <0.0001 |
| **GWD** | 2 | 14.56 | <0.0001 |
| **Measuring date:GWD** | 14 | 1.12 | 0.3526 |
| | | | |
| | | | |
| **Relative difference of *f*GCC** | | | |
| **Factor** | **numDF** | **F-value** | **p-value** |
| **(Intercept)** | 1 | 364.43 | <0.0001 |
| **Measuring date** | 3 | 3.6 | 0.0227 |
| **GWD** | 2 | 25.54 | <0.0001 |
| **Measuring date:GWD** | 6 | 2.06 | 0.0822 |

## REFERENCES

1. Acevedo, E., Ortiz, M., Franck, N., and Sanguineti, P. (2007). *Relaciones Hídricas de Prosopis Tamarugo Phil. Uso de Isótopos Estables.* Santiago: Universidad de Chile.
2. Acevedo, E., Sotomayor, D., and Zenteno, V. (1985). "Water uptake is affected by the environment in *Prosopis tamarugo* (Phil.)," in *The Current State of Knowledge of Prosopis tamarugo*, ed M. Habit (New York, NY: Food and Agriculture Organization of the United Nations), 273–281.
3. Allen, S. E. (1989). *Chemical Analysis of Ecological Materials. 2nd Edn.* Oxford: Blackwell Scientific Publications.
4. Altamirano, H. (1994). *Plan de Operaciones Demostrativas en el Sector Pillalla.* Informe No. 1. Chile: Corporación Nacional Forestal.
5. Altamirano, H. (2006). "*Prosopis tamarugo* Phil. Tamarugo," in *Las Especies arbóreas de los Bosques Templados de Chile y Argentina. Autoecología*, ed C. Donoso (Valdivia: Marisa Cuneo Ediciones), 534–540.
6. Aravena, R., and Acevedo, E. (1985). "The use of environmental isotopes oxigen-18 and deuterium in the study of water relations of *Prosopis tamarugo* Phil.," in *The Current State of Knowledge of Prosopis tamarugo*, ed M. Habit (New York, NY: Food and Agriculture Organization of The United Nations), 251–256.
7. Bigler, C., Braker, O., Bugmann, H., Dobbertin, M., and Rigling, A. (2006). Drought as an inciting mortality factor in Scots pine stands of the Valais of Switzerland. *Ecosystems* 9, 330–343. doi: 10.1007/s10021-005-0126-2
8. Bozdogan, H. (2000), Akaike's information criterion recent, developments in information complexity. *J. Math. Psychol.* 44, 62–91. doi: 10.1006/jmps.1999.1277.
9. Brugnoli, E., and Farquhar, G. (2000). "Photosynthetic fractionation of carbon isotopes," in *Photosynthesis: Physiology and Metabolism*, eds R. Leegood, T. Sharkey, and S. von Caemmerer (Dordrecht: Kluwer Akademic Publisher), 399–434.
10. Calderón, G., Garrido, M., and Acevedo, E. (2015). *Prosopis*

*tamarugo* Phil.: a native tree from the Atacama Desert. Groundwater table depth thresholds for conservation. *Rev. Chil. Hist. Nat.* 88, 18. doi: 10.1186/s40693-015-0048-0

11. Campillo, U., and Hojas, A. (1975). *Hidrogeología de la Pampa del Tamarugal*. Santiago: Instituto de Investigación de Recursos Natrales; Corporación de Fomento de la Producción.
12. Chávez, R., Clevers, J., Herold, M., Ortiz, M., and Acevedo, E. (2013a). Modelling the spectral response of the desert tree *Prosopis tamarugo* to water stress. *Int. J. Appl. Earth Obs. Geoinformation* 21, 53–61. doi: 10.1016/j.jag.2012.08.013
13. Chávez, R., Clevers, J., Verbesselt, J., Naulin, P., and Herold, M. (2014). Detecting leaf pulvinar movements on NDVI time series of desert trees: a new approach for water stress detection. *PLoS ONE* 9:e106613. doi: 10.1371/journal.pone.0106613
14. Chávez, R., Jan, G., Clevers, W., Herold, M., Acevedo, E., and Ortiz, M. (2013b). Assessing water stress of desert tamarugo trees using *in situ* data and very high spatial resolution remote sensing. *Remote Sens.* 5, 5064–5088. doi: 10.3390/rs5105064
15. Cooper, D., D'Amico, D. R., and Scott, M. (2003). Physiological and morphological response of *Populus deltoides* to alluvial groundwater pumping. *Environ. Manage.* 31, 215–226. doi: 10.1007/s00267-002-2808-2
16. Cooper, D., Sanderson, J., Stannard, D., and Groeneveld, D. (2006). Effects of long-term water table drawdown on evapotranspiration and vegetation in an arid region phreatophyte community. *J. Hydrol.* 325, 21–36. doi: 10.1016/j.jhydrol.2005.09.035
17. Delatorre, J., Pinto, M., and Cardemil, L. (2008). Effects of water stress and high temperature on photosynthetic rates of two species of Prosopis. *J. Photochem. Photobiol. B* 92, 67–76. doi: 10.1016/j.jphotobiol.2008.04.004
18. DICTUC. (2007). "Anexo IX.2. Geología del Salar de LLamara," in *EIA proyecto Pampa Hermosa* (Santiago), 154.
19. Di Rienzo, J., Casanoves, F., Balzarini, M., Gonzalez, L., Tablada, M., Robledo, C., and InfoStat versión (2015). *Grupo InfoStat, FCA, Universidad Nacional de Córdoba*, Argentina. Available online: http://www.infostat.com.ar

20. Dirección General de Aguas (DGA) (1985). *Estudio de Hidrología Isotópica Area Salar de Llamara Desierto de Atacama-Chile.* Santiago: Proyecto de investigación OIEA 301-F3-CHI-3715/GS, 43.

21. Farquhar, G., Ehleringer, J., and Hubick, K. (1989). Carbon isotope discrimination and photosynthesis. *Annu. Rev. Plant Physiol. Plant Mol. Biol.* 40, 503–537. doi: 10.1146/annurev.pp.40.060189.002443

22. Farquhar, G., Cernusak, L. A., and Barnes, B. (2007). Heavy Water Fractionation during Transpiration. *Plant Physiol.* 143, 11–18. doi: 10.1104/pp.106.093278

23. Farquhar, G., and Lloyd, J. (1993). "Carbon and oxygen isotope effects in the exchange of carbon dioxide between terrestrial plants and the atmosphere," in *Stable Isotopes and Plant Carbon-Water Relations*, eds J. Ehleringer, A. Hall, and G. Farquhar (San Diego, CA: Academic Press), 47–70.

24. Frelich, L. (2002). *Forest Dynamics and Disturbance Regimes—Studies from Temperate Evergreen-Deciduous Forests.* Cambridge: Cambridge University Press.

25. Hasselquist, N. J., Michael, M. F., and Louis, L. S. (2010). Water relations of evergreen and drought-deciduous trees along a seasonally dry tropical forest chronosequence. *Oecologia* 164, 881–890. doi: 10.1007/s00442-010-1725-y

26. Hoffmann, W., Marchin, R., Abit, P., and Lau, O. (2011). Hydraulic failure and tree dieback are associated with high wood density in a temperate forest under extreme drought. *Glob. Change Biol.* 17, 2731–2742. doi: 10.1111/j.1365-2486.2011.02401.x

27. Horton, J., Kolb, T., and Hart, S. (2001). Physiological response to groundwater depth avies among species and with rivers flow regulation. *Ecol. Appl.* 11, 1046–1059. doi: 10.1890/1051-0761(2001)011[1046:PRTGDV]2.0.CO;2

28. Keitel, C., Adams, M., Holst, T., Matzarakis, A., Mayer, H., Rennenberg, H., et al. (2003). Carbon and oxygen isotope composition of organic compounds in the phloem sap provides a shortterm measure for stomatal conductance of European beech (*Fagus sylvatica* L.). *Plant Cell Environ.* 26, 1157–1168. doi: 10.1046/j.1365-3040.2003.01040.x

29. Kolb, K., and Sperry, J. (1999). Transport constraints on water use by the Great Basin shrub, *Artemisia tridentata. Plant Cell Environ.* 22, 925–935. doi: 10.1046/j.1365-3040.1999.00458.x
30. Lehner, G., Delatorre, J., Lütz, C., and Cardemil, L. (2001). Field studies on the photosynthesis of two desert Chilean plants: *Prosopis chilensis* and *Prosopis tamarugo. J. Photochem. Photobiol. B Biol.* 64, 36–44. doi: 10.1016/S1011-1344(01)00187-7
31. Liu, H., Park Williams, A., Allen, C. D., Guo, D., Wu, X., Anenkhonov, O. A., et al. (2013). Rapid warming accelerates tree growth decline in semi-arid forests in Inner Asia. *Glob. Change Biol.* 19, 2500–2510. doi: 10.1111/gcb.12217
32. Maherali, H., Pockman, W., and Jackson, R. (2004). Adaptive variation in the vulnerability of woody plants to xylem cavitation. *Ecology* 85, 2184–2199. doi: 10.1890/02-0538
33. McDowell, N. (2011). Mechanisms linking drought, hydraulics, carbon metabolism, and vegetation mortality. *Plant Physiol.* 155, 1051–1059. doi: 10.1104/pp.110.170704
34. McDowell, N. G., Beerling, D. J., Breshears, D. D., Fisher, R., Raffa, K., and Stitt, M. (2011). The interdependence of mechanisms underlying climate-driven vegetation mortality. *Trends Ecol. Evol.* 26, 523–532. doi: 10.1016/j.tree.2011.06.003
35. McDowell, N., Pockman, W., Allen, C. D., Breshears, D., Cobb, N., Kolb, T., et al. (2008). Mechanisms of plant survival and mortality during drought: why do some plants survive while others succumb to drought? *New Phytol.* 178, 719–739. doi: 10.1111/j.1469-8137.2008.02436.x
36. Mitchell, P. J., O'Grady, A. P., Tissue, D. T., White, D. T., Ottenschlaeger, M. L., and Pinkard, E. (2013). Drought response strategies define the relative contributions of hydraulic dysfunction and carbohydrate depletion during tree mortality. *New Phytol.* 197, 862–872. doi: 10.1111/nph.12064
37. Mooney, H., Gulmon, S., Rudel, P., and Ehleringer, J. (1980). Further observations on the water relations of *Prosopis tamarugo* of the northern Atacama Desert. *Oecologia* 44, 177–180. doi: 10.1007/BF00572676
38. Ortiz, M., Morales, L., Silva, P., and Acevedo, E. (2012). Estimación

del nivel freático a partir del NDVI Landsat en la Pampa del Tamarugal (Chile). *Revista de Teledetección* 37, 42–50.

39. Peng, C., Ma, Z., Lei, X., Zhu, Q., Chen, H., Wang, W., et al. (2011). A drought-induced pervasive increase in tree mortality across Canada's boreal forests. *Nat. Clim. Change* 1, 467–471. doi: 10.1038/nclimate1293
40. Pliscoff, P., and Leubert, F. (2006). *Sinopsis Bioclimática y Vegetacional de Chile*. Santiago de Chile: Editorial Universitaria.
41. Pringle, C. (2001). Hydrologic connectivity and the management of biological reserves: a global perspective. *Ecol. Appl.* 11, 981–998. doi: 10.1890/1051-0761(2001)011[0981:H.C.A.T.M.O.]2.0.C.O.;2
42. R Core Team (2015). *R: A Language and Environment for Statistical Computing.* R Foundation for Statistical Computing, Vienna, Austria. Available online: http://www.R-project.org/
43. Rice, K. J., Matzner, S., Byer, W., and Brown, J. (2004). Patterns of tree dieback in Queensland, Australia: the importance of drought stress and the role of resistance to cavitation. *Oecologia* 139, 190–198. doi: 10.1007/s00442-004-1503-9
44. Rojas, R., and Dassargues, A. (2007). Groundwater flow modelling of the regional aquifer of the Pampa del Tamarugal, northern Chile. *Hydrogeol. J.* 15, 537–551. doi: 10.1007/s10040-006-0084-6
45. Sage, R., and Kubien, D. (2007). The temperature response of C3 and C4 photosynthesis. *Plant Cell Environ.* 30, 1086–1106. doi: 10.1111/j.1365-3040.2007.01682.x
46. Scheidegger, Y., Saurer, M., Bahn, M., and Siegwolf, R. (2000). Linking stable oxygen and carbon isotopes with stomatal conductance and photosynthetic capacity: a conceptual model. *Oecologia* 125, 350–357. doi: 10.1007/s004420000466
47. Scholander, P. F., Bradstreet, E. D., Hemmingsen, E. A., and Hammel, H. T. (1965). Sap pressure in vascular plants: negative hydrostatic pressure can be measured in plants. *Science* 148, 339–346. doi: 10.1126/science.148.3668.339
48. Sullivan, P. F., and Welker, J. (2007). Variation in leaf physiology of *Salix arctica* within and across ecosystems in the High Arctic: test of a dual $\delta^{13}C$ and $\delta^{18}O$ conceptual model. *Oecologia* 151, 372–386. doi: 10.1007/s00442-006-0602-1

49. Tardieu, F., and Simonneau, T. (1988). Variability among species of stomatal control under fluctuating soil water status and evaporative demand: modeling isohydric and anisohydric behaviours. *J. Exp. Bot.* 49, 419–432. doi: 10.1093/jxb/49.Special_Issue.419

50. Thomas, P., and Packham, J. (2007). *Ecology of Woodlands and Forests. Description, Dynamics and Diversity.* Cambridge: Cambridge University Press.

51. Tréllez, E., Mamani, M., Valenzuela, F., and Vera, C. (2011). *Guía: Educación y Sensibilización Ciudadana para la Conservación y uso Sustentable de los Humedales de la Región de Tarapacá*, ed C. Burgos (Santiago: Andros Impresores), 103.

52. Vogt, U. (2001). Hydraulic vulnerability, vessel refilling, and seasonal courses of stem water potential of *Sorbus aucuparia* L. *and Sambucus nigra* L. *J. Exp. Bot.* 52, 1527–1536. doi: 10.1093/jexbot/52.360.1527

53. Williams, A., Allen, C., Macalady, A., Griffin, D., Woodhouse, C., Meko, D., et al. (2013). Temperature as a potent driver of regional forest drought stress and tree mortality. *Nat. Clim. Change* 3, 292–297. doi: 10.1038/nclimate1693

CHAPTER

# 6

# VARIATION IN ECOPHYSIOLOGICAL TRAITS AND DROUGHT TOLERANCE OF BEECH (FAGUS SYLVATICA L.) SEEDLINGS FROM DIFFERENT POPULATIONS

Claudia Cocozza[1], Marina de Miguel[2], Eva Pšidová[3], L'ubica Ditmarová[3], Stefano Marino[4], Lucia Maiuro[4], Arturo Alvino[4], Tomasz Czajkowski[5], Andreas Bolte[5] and Roberto Tognetti[6,7]

[1] Institute for Sustainable Plant Protection, National Research Council, Sesto Fiorentino, Italy

[2]BIOGECO, INRA, Univ. Bordeaux, Cestas, France

[3]Institute of Forest Ecology, Slovak Academy of Science, Zvolen, Slovak Republic

[4] Department of Agriculture, Environment and Food, University of Molise, Campobasso, Italy

[5]Johann Heinrich von Thünen Institute, Institute of Forest Ecosystems, Eberswalde, Germany

[6] Department of Biosciences and Territory, University of Molise, Pesche, Italy

[7]The EFI Project Centre on Mountain Forests (MOUNTFOR), Edmund Mach Foundation, San Michele all'Adige, Italy

Frequency and intensity of heat waves and drought events are expected to increase in Europe due to climate change. European beech (*Fagus sylvatica* L.) is one of the most important native tree species in Europe. Beech populations originating throughout its native range were selected for common-garden experiments with the aim to determine whether there are functional variations in drought stress responses among different populations. One-year old seedlings from four to seven beech populations were grown and drought-treated in a greenhouse, replicating the experiment at two contrasting sites, in Italy (Mediterranean mountains) and Germany (Central Europe). Experimental findings indicated that: (1) drought (water stress) mainly affected gas exchange describing a critical threshold of drought response between 30 and 26% SWA for photosynthetic rate and $C_i/C_a$, respectively; (2) the $C_i$ to $C_a$ ratio increased substantially with severe water stress suggesting a stable instantaneous water use efficiency and an efficient regulation capacity of water balance achieved by a tight stomatal control; (3) there was a different response to water stress among the considered beech populations, differently combining traits, although there was not a well-defined variability in drought tolerance. A combined analysis of functional and structural traits for detecting stress signals in beech seedlings is suggested to assess plant performance under limiting moisture conditions and, consequently, to estimate evolutionary potential of beech under a changing environmental scenario.

**Keywords: Water stress,** ecophysiology, gas exchange, chlorophyll *a* **fluorescence, European beech**

# INTRODUCTION

Warming-induced drought is threatening forest ecosystems worldwide, increasing water stress and mortality risk for trees (Allen et al., 2010). The vulnerability of plants to drought varies in dependence of stress severity, its duration, and the combination with other stresses (Niinemets, 2010). Intraspecific variation of tree response to drought, recently, has received increasing attention in the case of important forest species, such as *Fagus sylvatica* L. (beech; e.g., Borghetti et al., 1993; Tognetti et al., 1995; García-Plazaola and Becerril, 2000; Peuke et al., 2002; Aranda et al., 2015; Knutzen et al., 2015; Pšidová et al., 2015), in order to inform forest managers on adaptive capacities of populations for stress tolerance and decrease tree vulnerability to climate change.

Acclimation of trees to water deficit is the result of adaptive changes in plant development and ecophysiological processes, such as gas exchange, growth rate, and water relations (Sala et al., 2010). Drought-induced hydraulic limitation on carbohydrate use may prolong survival in plants under stress. However, if drought persists, reduced photosynthetic carbon assimilation due to stomatal closure (isohydric behavior) may promote carbon starvation, as carbohydrate demand continues for maintenance of osmoregulation, and plants fail to maintain hydraulic integrity (McDowell, 2011). If plants maintain their stomata open during drought (anisohydric behavior), hydraulic failure may occur, thus leading to mortality. Tree mortality may occur when drought has caused >50% loss of stem hydraulic conductivity, corresponding to −4.5 MPa in beech (Barigah et al., 2013). The capacity for adaptive changes to the environment may ultimately be critical in determining tree species survival under climate change (Aitken et al., 2008). Physiological responses, including adaptation and evolution to environmental changes, define phenotypic plasticity that can be assumed as the dominant underlying process with consequences on ecosystem functions (Hovenden and Vander Schoor, 2003; Thomas, 2011). A better understanding of geographic pattern and genetic variation in functional and structural traits of important tree species is essential for implementing adaptive forest management strategies to mitigate anticipated impacts of climate change on plant growth and drought tolerance.

Beech is a naturally dominant tree species in many European forests and sensitive to water deficit (Tognetti et al., 1995; Backes and Leuschner, 2000; Czajkowski et al., 2005; Bolte et al., 2007; Rose et al., 2009). The distribution of beech in Europe is characterized by high genetic diversity, resulting in high potential to adapt to changing environmental conditions (Dounavi et al., 2016). Acclimation to drought and heat stress in beech may occur after increasing levels of proline amino acid that plays as osmo-protectants to raise the osmotic pressure and thus maintain membrane integrity and stabilize proteins (Rennenberg et al., 2006). Beech can also respond to water stress through decrease in photosynthetic efficiency and light sensitivity of the photosynthetic apparatus (Tognetti et al., 1995, 1997; Peuke et al., 2002). In southern Europe, the recent decline in basal area increment of beech has been linked to decreasing water availability (Jump et al., 2006; Piovesan et al., 2008), which can affect carbon dynamics and sequestration potentials at the southern limit of this species distribution (Tognetti et al., 2014). However, this is not a general response and positive growth (tree-ring width) in beech at Mediterranean latitudes has been observed (Tegel et al., 2013). In central Europe, the extreme 2003 drought has not been found to induce dramatic growth reduction in beech (Leuzinger et al., 2005; Van der Werf et al., 2007), while recording large reductions in gross primary production (GPP) in temperate forests (Ciais et al., 2005; Granier et al., 2007). Nevertheless, the radial growth of beech has been reported to be reduced in the context of more frequent and intensive drought extremes (Beck and Heußner, 2012). Although beech is considered vulnerable to climate change (Ohlemuller et al., 2006; Geßler et al., 2007; Scherrer et al., 2011), its distribution has been modulated, by forest conversion from coniferous forests to mixed stands (Fischer et al., 2013). Beech populations have been able to adapt to environmental changes depending on the level and distribution of genetic variation within and between these populations and their phenotypic plasticity (Bresson et al., 2011; Vitasse et al., 2014).

Plasticity of functional traits plays a determinant role in plant response to different environments, providing relevant information in resistance and resilience of beech to global warming (Stojnić et al., 2015). Köcher et al. (2009) found that beech adjusted the vitality and productivity as function of the drought sensitivity of stem growth and leaf and root production, and the success of rejuvenation under a drier climate. Furthermore, changes in photosynthetic characteristics induced by water stress were similar between beech seedlings originated from different sites, but the intensity of change was dependent on the origin (Pšidová et al., 2015). However, phenotypic differences in populations are related to the different directional selection defined by environmental conditions and evolutionary processes (Poorter et al., 2010). Disturbances may also lead to higher genotypic variability at the population level (Borghetti et al., 1993). Beech may exhibit intraspecific variation in drought resistance strategies characterized by varying degree of anisohydricity or isohydricity. A prompt stomatal reaction in beech populations from drier sites may be expected, vs. low stomatal regulation in the plants from mesic conditions (Aranda et al., 2015). Differences in the sensitivity of photosystem II (PSII) to drought among populations may also occur, because those with a drought-avoiding strategy limit carbon assimilation more than drought-tolerant seed sources.

Within this framework, provenance trials represent a valuable tool for assessing adaptive potential to a changing environment (Stojnić et al., 2015). Common garden investigations are carried out to test hypotheses on the geographic variation of adaptive traits in tree species, and study the relationships between trait variability and seed source environments (i.e., Sork et al., 2013). In the present study, beech from a wide range of European populations was selected throughout its native range, considering the role of local adaptation and phenotypic plasticity of this species (according to Bolte et al., 2016). We hypothesized that divergence in the plasticity of the response to environmental conditions should have occurred in beech populations originating from different localities. Specific hypothesis were: (1) population differentiation of traits will be moderate to high based on beech distribution; (2) population traits will match seed source climate and will reflect local adaptation (higher water use efficiency in populations from drier sites); (3) populations will distribute in European sites based on morphological and physiological characteristics. Two common garden experiments were carried out in Germany and Italy to quantify geographic variation of adaptive traits in beech populations and investigate relationships between trait variability and seed source, determining the effects of water stress on assimilation rate, predawn water potential, chlorophyll *a* fluorescence and chlorophyll content, (Experiment 1) and the assimilation potential (e.g., gas exchanges and leaf traits) (Experiment 2).

# MATERIALS AND METHODS

## Plant Material

Beech populations were collected through the pan-European EU Cost STReESS network, from seven sites in six European countries. Seeds were collected in the fall of 2013 from numerous maternal families in forest stands throughout the native range of the species, representing a gradient in mean air temperature and rainfall (Table 1). Uniformly sized seeds of each population were surface-sterilized by soaking in 3% sodium hypochlorite for 5 min and rinsing with deionized water. Therefore, a stratification procedure was performed: (1) the seed moisture content was reduced to about 8% of the fresh seeds' moisture content (e.g., by storing them about 1 week at a dry and cool place); (2) seeds were preserved in plastic bags in a freezer at −5°C until the mid of February 2014 (stratification by frost); (3) the seed moisture was increased (using a water sprayer) at a temperature of 3–5°C; (4) as soon as the first little sprout was visible, the seedlings were transplanted in pots. Seedlings obtained from seeds were planted in pots.

**Table 1**. Location, average of mean annual air temperature (°C), and total annual rainfall (mm) at the seven sites where the seedlings originated

| Altitude order | PV1 | PV2 | PV3 | PV4 | PV5 | PV6 | PV7 |
|---|---|---|---|---|---|---|---|
| Country | Denmark | Germany | France | France | Romania | Bosnia | Spain |
| Provenance | *Stenderup Midtskov* | *Sellhorn* | *Crecy* | *Montagne Noir* | *Valea Baromului* | *Nevesinje* | *Erro* |
| Longitude | 9.65 | 9.93 | 1.88 | 2.22 | 21.68 | 18.13 | -1.47 |
| Latitude | 55.47 | 53.35 | 50.225 | 43.50 | 41.77 | 43.27 | 43.00 |
| Altitude (GIS Interpol.) | 18 | 86 | 30 | 341 | 145 | 862 | 931 |
| Mean annual air temperature (°C) | 7.7 | 8.2 | 10.5 | 12.4 | 9.3 | 9.6 | 9.1 |
| Total annual rainfall (mm) | 720 | 748 | 637 | 791 | 722 | 1199 | 1166 |

A different level of success of seed germination was observed, thus different populations were considered in each experiment. Seven populations ("Denmark" PV1, "Germany" PV2, "France-Crecy" PV3, "France-Montagne Noir" PV4, "Romania" PV5, "Bosnia" PV 6, and "Spain" PV 7) were tested in Experiment 1 in the North-East of Germany. Four populations (PV1, PV3, PV4, and PV5) were used in Experiment 2 in Central Italy.

## Experimental Set-Up

Two common garden studies were carried out accurately to estimate differentiation under diverse environmental conditions determined by different latitude and elevation between sites (Central Europe, Germany; Mediterranean mountains, Italy). The two experiments took place in greenhouses; plants were subjected to the ambient light without additional illumination; plants were grown in cylindrical PVC pots (1.4 L) filled with a mixture of 70% silty-sand soil (grain size 0–2 mm) and 30% peat-based substrate amended with 2 kg m$^{-3}$ Osmocote (NPK 14:13:13+7$SO_3$, plus micro elements).

Twenty replicate seedlings for each population per treatment were designated, by selecting homogenous "matched pairs" based on visual comparison. Well-watered (control) seedlings were regularly watered with tap water to field capacity throughout the experiment. Seedlings subjected to drought treatment were submitted to progressive drought by withdrawing irrigation, when full leaf unfolding had occurred. Initial soil water content was estimated by weighing the samples before and after oven drying at 105°C for 48 h. Pots were initially watered to saturation. After excess water has drained away field capacity (FC, Blume et al., 2016) was reached at −0.06 MPa soil water potential (pF 1.8). Subsequent changes in pot weight were attributed to changes in soil water content.

A group of 20 seedlings (control) was maintained near to FC by daily watering; whereas, water supply was suspended for those seedlings (20 individuals) subjected to drought treatments. Since watering interruption, each pot was weighed three times per week, in order to estimate the plant available water content ($\theta_{AWC}$), using the following Equation (1, cf. Veihmeyer and Hendrickson, 1927):

$$\theta_{AWC} = \theta_{FC} - \theta_{PWP} \quad (1)$$

where θ is the soil water content (g) at FC (pF 1.8 ≈ −0.06 MPa soil water potential) and at the visual loss of leaf turgor (commonly defined permanent wilting point in agricultural plants; PWP, pF 4.2 ≈ −1.5 MPa soil water potential). The permanent wilting point (PWP = −1.5 MPa) is used as a conventional threshold for determining the soil water availability (SWA) within the entire effective rooting depth according to the concept of Reid et al. (1984). The PWP was derived for agricultural plants, and young trees like European beech may deplete water resources near to the fine roots at lower soil water potentials than the PWP.

The residual soil water availability (SWA; %) is defined as the proportion of available soil water content ($\theta_{AWC}$) during drought treatment referred to the initial value under condition of field capacity.

During the drought experiment the pots were placed at random position and re-randomized after every watering event. Each treatment group was ideally represented by 20 individuals corresponding to replicate seedlings for each population per drought treatment. The pots were allowed to reach different thresholds of drought (50, 40, and 30% SWA) at different time.

Plant responses to extreme drought were studied focusing on gas exchange and water relations, and anatomical traits. We put specific emphasis on the assimilation ability and leaf anatomical traits determining gas exchange and transpiration processes.

Leaf gas exchange was measured on one leaf of five seedlings per treatment once per week, using a portable gas exchange system (GFS-3000, Heinz Walz GmbH Germany) in Germany, and using a LI-COR photosynthesis system (LI-6400; LI-COR Inc., Lincoln, NE, USA) in Italy. The adjusted values were: leaf/cuvettes temperature 25°C, (Air-to-leaf), Vapor Pressure Deficit 13.8 −16.1 (Pa/kPa), $CO_2$ concentrations 360 ppm, flow 750 μmol $s^{-1}$, and the light intensity (LED Light Source 3040-L) 1500 μmol $m^{-2}$ $s^{-1}$. Measurements (assimilation rates and water vapor conductance) were taken between 09:30 and 14:30 (UT) on fully expanded un-shaded leaves and calculated according to von Caemmerer and Farquhar (1981). Instantaneous photosynthetic rate and corresponding stomatal conductance, ratio of intercellular ($C_i$) to ambient ($C_a$) $CO_2$concentration ($C_i/C_a$), and ratio of photosynthetic rate (A) to stomatal conductance ($A/g_s$), namely leaf intrinsic water use efficiency (WUE), were measured at light intensity of 1000 μmol $m^{-2}$ $s^{-1}$ (LED source, red blue 6400-02B) and with a controlled flow of 400 μmol $s^{-1}$ of ambient air with $CO_2$ concentration fixed at 400 μmol $mol^{-1}$. Air temperature and relative humidity were maintained close to ambient values.

For specific insights in plant water status variation during treatment predawn (1 h before the start of the daily light regime) leaf water potential ($\Psi_{pd}$; MPa) was measured in three seedlings per treatment per provenance to obtain the mean soil water potential next to the roots, closely correlated to the relative transpiration rate (Améglio et al., 1999). Predawn leaf water potential was measured between 1:00 and 5:00 (UT). Exact measurement times varied over the course of the experiment, as the predawn time changed during the experiment. One leaf per five seedlings were detached and rapidly enclosed in a Scholander-type pressure chamber (SKPM1400, Skye Instruments, Llandrindod Wells, UK).

## Experiment 1

The Experiment 1 was carried out in late summer 2014 by the Thünen Institute of Forest Ecosystems—University of Sustainable Development (HNE) in Eberswalde (52°49′28″ N 13°47′29″ E, 30 m a.s.l.). Within the treatment period relative air humidity averaged 69%, with a minimum of 30% and a maximum of 88%. Air temperature ranged between 11°C (minimum during night) and 31°C (maximum during day), and attained a mean of 19.0°C.

Seedlings reached the ontogenic stage of four fully expanded leaves in 06 August 2014, when the water irrigation was interrupted in mid-summer (06 August 2014) until desiccation of seedlings (06 October 2014), which defined the end of experiment (the duration of experiment was 61 days). Root collar diameter, plant height, leaf area and leaf number were measured at the beginning of the experiment in control seedlings.

Fast chlorophyll *a* fluorescence kinetics were determined using the chlorophyll fluorimeter Handy PEA (Hansatech, Instruments Ltd, UK). The following parameters were measured: minimal ($F_0$), maximal ($F_m$), and variable ($F_v = F_m - F_0$) fluorescence of dark-adapted leaves; the dark-adapted leaf is illuminated with weak modulated measuring light to give the background fluorescence level in the dark ($F_0$); application of a saturation pulse allows measurement of the maximum fluorescence level in the dark ($F_m$). The maximal efficiency of PSII photochemistry was calculated as $F_v/F_m$. The performance index, PI, indicator of sample vitality reflecting the energy cascades, was also determined. Relevant fluorescence parameters were measured using the portable chlorophyll fluorometer MINI-PAM (Walz, Heinz Walz GmbH, Effeltrich, Germany). Rapid light curves (RLCs) were recorded after each chlorophyll fluorescence kinetics measurement: yield ($\Phi_{PSII}$), the photochemical yield of photosystem II; ETR, the relative rate of electron transport; qP, the coefficient of photochemical quenching; qN, the coefficient of non-photochemical quenching; NPQ, the non-photochemical quenching. Relative chlorophyll content of the leaves was estimated by portable chlorophyll meter (CL-01, Hansatech, Instruments Ltd., Kings Lynn, UK). The results were expressed as chlorophyll index (Chl index; Cassol et al., 2008). Values of Chl index were estimated by device on the basis of the absorbance at 620 and 940 nm.

## Experiment 2

The Experiment 2 took place during mid-summer 2014 at the University of Molise in Pesche (41°37′00″ N 14°17′00″ E, 732 m a.s.l.). During the experiment, relative humidity ranged between 60 and 70% and air temperature between 15°C (night) and 30°C (day), with a mean of 22.5°C. The light intensity at the plant level never exceeded 1000 μmol photons $m^{-2} s^{-1}$ in sunny days (>80% of the days during which the experiment was carried out), whereas the vapor pressure deficit (VPD) rarely exceeded 2 kPa. In both experiments, the plants were subjected to the ambient light without additional illumination.

Due to the different climatic conditions, the seasonal drought is stronger, occurs earlier and lasts longer in Italy than in Germany. Therefore, seedlings reached the ontogenic stage of four fully expanded leaves in 26 June 2014 in Italy, when water irrigation was interrupted in early summer until desiccation of seedlings in 26 July 2014 (the duration of experiment was 30 days).

Light response curves (A/Q) and intercellular $CO_2$ response curves (A/$C_i$) were measured on full expanded leaves from five seedlings grown in control conditions at the end of experiment using a portable photosynthesis system (LI-6400, Li-Cor, Lincoln, NE; according to Tognetti et al., 2004). Light response curves were obtained by measurements at Q-values of 2000, 1500, 1000, 500, 200, 100, 50, 20, and 0 μmol $m^{-2} s^{-1}$ (LED source, red blue 6400-02B). Measurements were recorded automatically at each set point when A had equilibrated; irradiance was changed at intervals of 120–200 s. The $CO_2$ entering the cuvette was adjusted to maintain a chamber $CO_2$ concentration ([$CO_2$]) of 400 μmol $mol^{-1}$. The response of leaf A to Q was modeled by a non-rectangular hyperbola where the initial slope was apparent quantum efficiency (φ), light compensation point ($\Gamma_l$), and apparent dark respiration ($R_d$) were estimated from axis intercepts, and the light-saturated maximum photosynthetic rate ($A_{max}$) was the upper asymptote. All parameters were determined by fitting data to the model function (Prioul and Chartier, 1977). See Tognetti et al. (2004) for further details on curve fitting and equations.

The $CO_2$ response curves were obtained by changing the [$CO_2$] entering the cuvette from 50 to 800 μmol $mol^{-1}$ with an external $CO_2$ cartridge mounted on the LI-6400 console and automatically controlled by a $CO_2$ injector. The $CO_2$ assimilation rate was first measured by setting the reference [$CO_2$] near ambient (400 μmol $mol^{-1}$) and then at 300, 200, 100, 50, 400, 400, 600, and 800 μmol $mol^{-1}$. Gas exchange was determined at each step after exposure of the leaf to the new [$CO_2$], waiting for A to reach equilibrium, which was typically < 3 min; Q was maintained at 1000 μmol $m^{-2}$ $s^{-1}$. The response of leaf A to Ci was analyzed according to the mechanistic model of $CO_2$ assimilation proposed by Farquhar et al. (1980) and subsequently modified by Sharkey (1985). A non-linear regression technique was used to estimate $R_{day}$, the maximum rate of carboxylation ($V_{cmax}$), the light-saturated rate of electron transport ($J_{max}$), and the rate of triose phosphate utilization for sucrose and starch synthesis (TPU; Sharkey, 1985; Wullschleger, 1993). Description of parameter estimates can be found in Tognetti et al. (2004). The entire leaf area entered the cuvette, limiting errors associated with area determination and the occurrence of a patchy distribution of stomata.

Leaf thickness was measured (lower and upper epidermis, palisade, and spongy parenchyma) on transversal sections (four measurements in each of five plants were averaged). Stomatal density (number of stomata per $mm^2$ of leaf area), and the average of polar (length) and equatorial (width) stomatal size, and distance between guard cells were determined. Three regions were measured in the median leaf blade, summarizing 30 observations per population. Structural leaf traits were measured on three leaves of three plants per population on all plants at the beginning of the experiment using the uppermost fully expanded leaves. Leaf portions (1–2 cm in diameter) were fixed with 3% glutaraldehyde (v/v) in 0.1-M phosphate buffer (pH 7.2) for 6–8 h under 4°C, post-fixed in 1% osmium tetroxide for 1 h, and immersed in 0.1-M phosphate buffer (pH 7.2) for 1–2 h. The samples were then dehydrated in a graded ethanol series (50, 60, 70, 80, 90, 95, and 100%) with a last wash in acetone for a better $CO_2$ substitution during the dehydration procedure at a pressure of 1200 bars. Dry tissue samples were coated with gold in a sputter coater and observed in a scanning electron microscope (SEM Zeiss DSM 940A, Oberkochen, Germany) operated at 10 keV (according to Hultine and Marshall, 2001).

## Data Analysis

Because all seedlings in each experiment shared the same substrate conditions (nutrient availability, soil texture, etc.), we assumed a significant main effect of imposed dry-down conditions on phenotypic plasticity of physiological traits. Data were examined for assumptions of the homogeneity of variance and normality and were found to conform to model requirements. Variations in leaf gas exchange parameters during the experiment were evaluated by two-way repeated measures ANOVA with "population," as random factor, and "soil water availability," as fixed factor. One-way ANOVA was performed to analyze the effect of population on A/Ci and A/Q curve parameters, root collar diameter, plant height, leaf area and number, and microscopic leaf traits. Statistical analysis was conducted with OriginPro version 8.5.1 (OriginLab, Northampton, MA). Statistical analyses were separately performed considering the different experimental period between Germany and Italy.

# RESULTS

## Experiment 1

Beech seedlings differed in the root collar diameter, plant height, and leaf number ($p < 0.0001$) between populations in control conditions (Table 2). Root collar diameter was higher in German (PV2), French (PV3 and PV4), and Bosnian (PV6) populations; plant height in French (PV3 and PV4), Romanian (PV5), and Bosnian (PV6) populations; leaf number in German (PV2), French (PV3), Romanian (PV5), and Bosnian (PV6) populations (Table 2).

**Table 2.** Root collar diameter, plant height, leaf area, and number for beech seedlings at the beginning of the water-stress experiment

| | PV1 | PV2 | PV3 | PV4 | PV5 | PV6 | PV7 | ANOVA | $F$-value | $P$-level |
|---|---|---|---|---|---|---|---|---|---|---|
| Root collar diameter (mm) | 2.00 (0.00) | 2.44 (0.09) | 2.25 (0.10) | 2.45 (0.11) | 2.00 (0.00) | 2.75 (0.10) | 2.00 (0.00) | Population | 4.696 | 0.000 |
| Plant height (cm) | 11.67 (0.32) | 12.44 (0.62) | 17.50 (0.40) | 11.92 (0.76) | 15.42 (0.47) | 16.77 (0.79) | 13.80 (0.45) | | 7.568 | 0.000 |
| Leaf area (cm$^2$) | 77.13 (5.47) | 99.62 (12.70) | 99.44 (12.19) | 69.99 (11.50) | 111.25 (16.12) | 116.13 (7.47) | 90.12 (11.61) | | 1.918 | 0.098 |
| Leaf number ($n$) | 9.50 (0.90) | 13.13 (2.38) | 12.75 (1.50) | 4.40 (0.57) | 15.00 (1.64) | 15.63 (0.86) | 9.63 (2.27) | | 5.171 | 0.000 |

The parameters derived from rapid light curves decreased with decreasing SWA (Figure 1). Yield and qP showed the same decreasing pattern in all populations. After 48 days of drought treatment, when 20% of SWA was reached, ETR, NPQ, and qN were reduced in comparison with control; however, PV2 showed higher values and PV1 and PV5 than other populations (Figure 1). The reduction of SWA showed an increasing trend for $F_0$, and decreasing pattern for PI and Chl index; whereas, $F_v/F_m$ was not substantial affected by water stress (Figure 2).

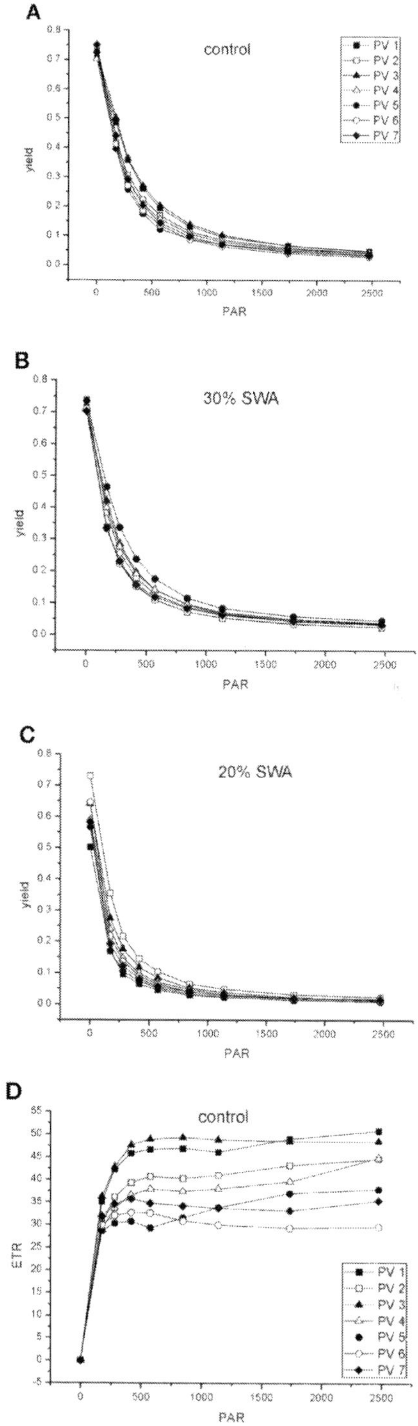
A
control
PV 1
PV 2
PV 3
PV 4
PV 5
PV 6
PV 7
yield
PAR
B
30% SWA
yield
PAR
C
20% SWA
yield
PAR
D
control
ETR
PV 1
PV 2
PV 3
PV 4
PV 5
PV 6
PV 7
PAR

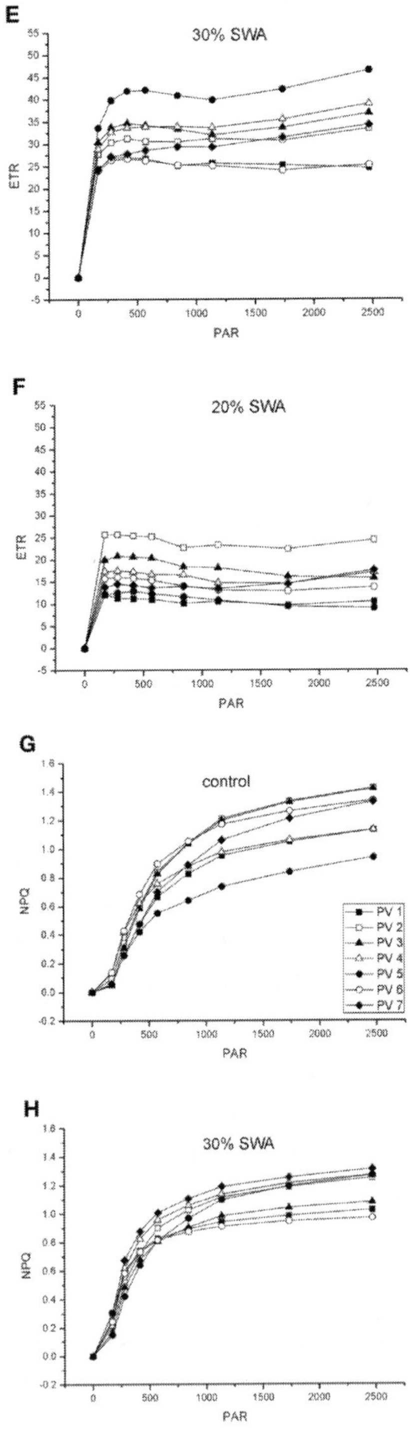
E
30% SWA
ETR
PAR
F
20% SWA
ETR
PAR
G
control
NPQ
PAR
PV 1
PV 2
PV 3
PV 4
PV 5
PV 6
PV 7
H
30% SWA
NPQ
PAR

I

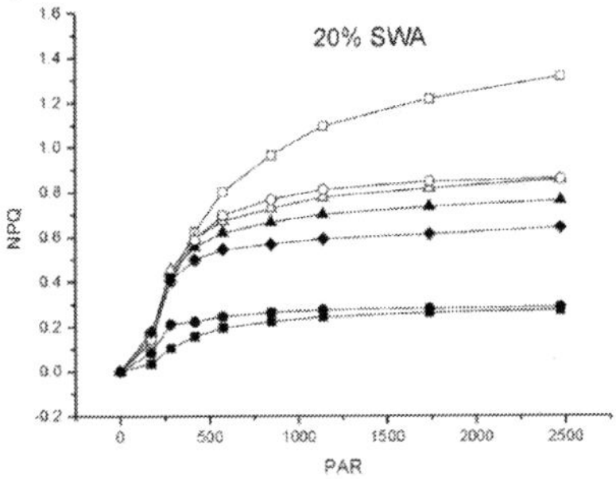
20% SWA
NPQ
PAR

J

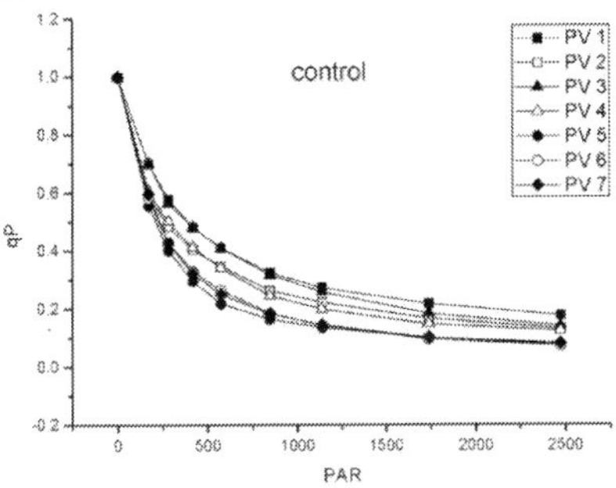
control
PV 1
PV 2
PV 3
PV 4
PV 5
PV 6
PV 7
qP
PAR

K

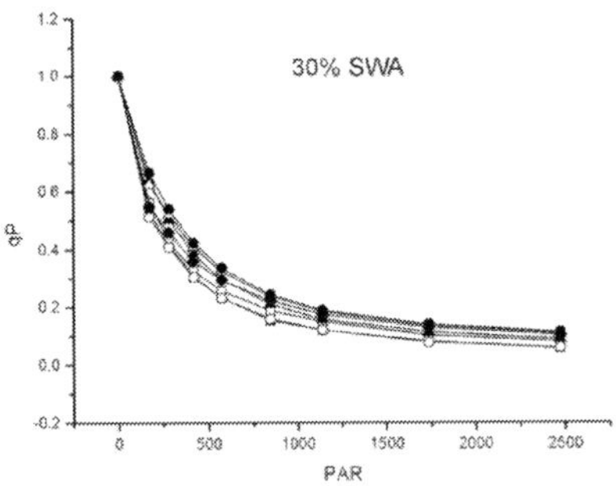
30% SWA
qP
PAR

L

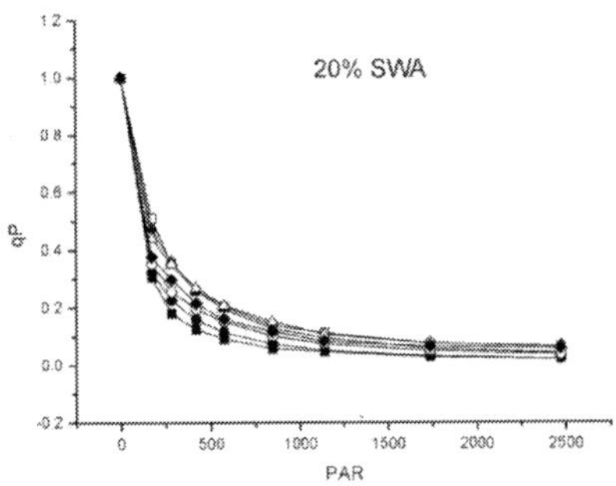
20% SWA
qP
PAR

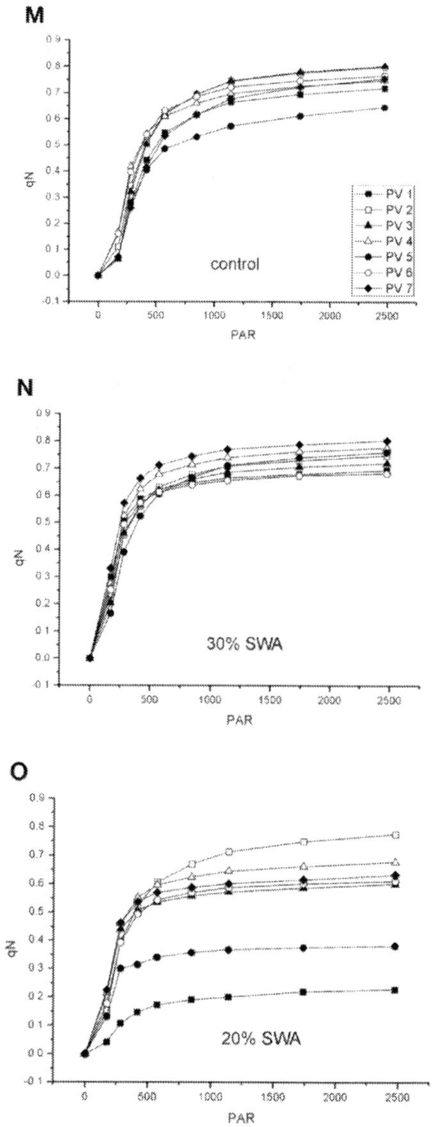


**Figure 1.** Effects of drought treatment on rapid light response curves of the maximal yield of photochemical energy conversion (yield; A-C), the relative photosynthetic electron transport rate (ETR; D-F), the quantum yield of regulated energy dissipation of PSII (NPQ; G-I), the coefficient for photochemical quenching (qP; J-L), the coefficient for non-photochemical quenching (qN; M-O) in beech seedlings in days of drought treatment (experiment 1).

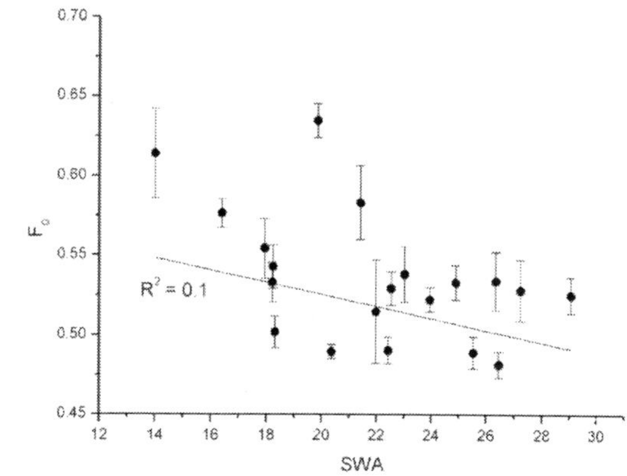

$F_o$
$R^2 = 0.1$
0.70
0.65
0.60
0.55
0.50
0.45
12 14 16 18 20 22 24 26 28 30
SWA


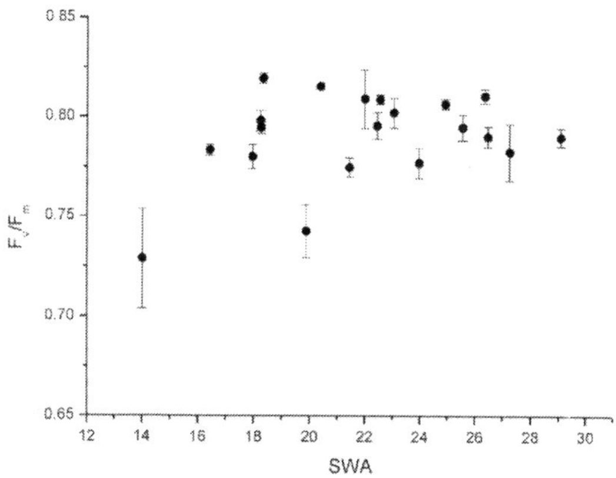

$F_v/F_m$
0.85
0.80
0.75
0.70
0.65
12 14 16 18 20 22 24 26 28 30
SWA


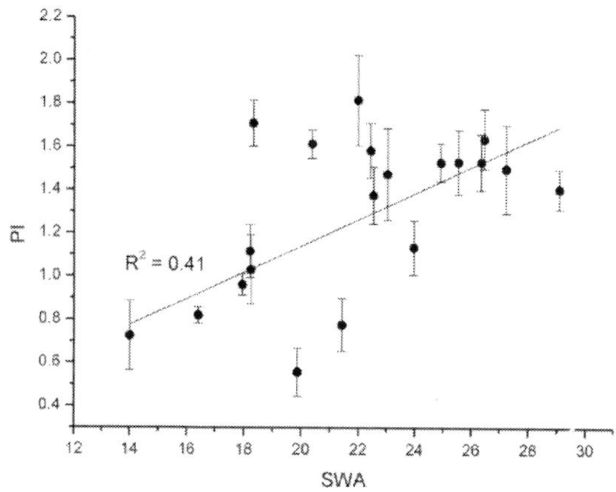

PI
$R^2 = 0.41$
2.2
2.0
1.8
1.6
1.4
1.2
1.0
0.8
0.6
0.4
12 14 16 18 20 22 24 26 28 30
SWA

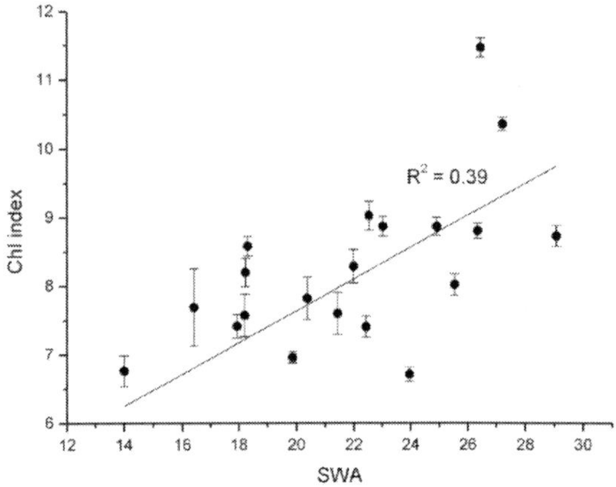


**Figure 2.** Fluorescence traits ($F_0$), the minimal Chl fluorescence; ($F_v/F_m$), the maximal efficiency of PSII photochemistry; (PI), the performance index; (Chl index), chlorophyll index, following the soil water availability during the experiment (experiment 1).

Assimilation rate decreased as predawn leaf potential became more negative with a steep drop at leaf water potential close to −2 MPa (Figure 3).

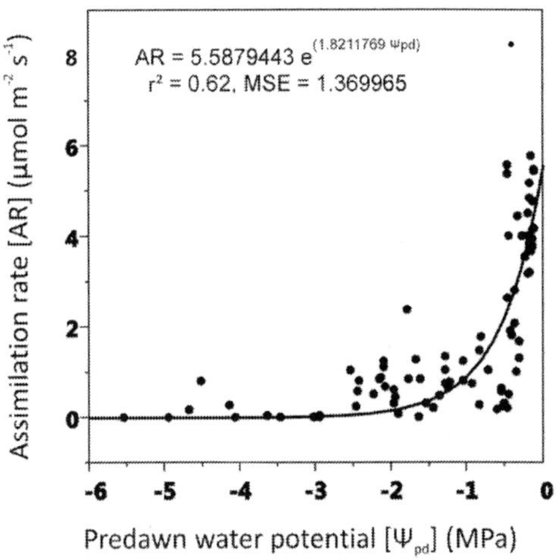


**Figure 3.** Relationships between assimilation rate (AR) and shoot predawn water potential ($\Psi_{pd}$). Non-linear regression model (exponential function) was used for curve-fitting analysis (experiment 1).

## Experiment 2

Leaf traits were significantly different among populations in control conditions (Table 3). However, leaf traits did not show a common pattern among populations, suggesting complex phenotypic effects driven by prevailing environmental conditions at the geographical origin (Table 3). The lowest values of stomatal density and spongy mesophyll thickness were found in PV4 and PV5, of mesophyll thickness in PV4, of stomatal length (polar size) in PV3, and of stomatal width (equatorial size) in PV1.

**Table 3.** Structure of fully expanded foliage (harvested at the beginning of the experiment) of seedlings for each population grown in control conditions

| | PV1 | PV3 | PV4 | PV5 | *F*-value | *p*-level |
|---|---|---|---|---|---|---|
| Stomatal density (n $mm^{-2}$) | 7.00 (0.29) | 6.60 (0.20) | 5.80 (0.21) | 4.60 (0.26) | 20.63 | 0.000 |
| Mesophyll (μm) | 68.84 (2.27) | 72.20 (0.92) | 58.15 (1.53) | 64.96 (1.15) | 16.14 | 0.000 |
| Palisade (μm) | 33.84 (0.85) | 33.15 (0.82) | 26.68 (0.61) | 34.04 (0.49) | 26.54 | 0.000 |
| Spongy (μm) | 23.19 (2.02) | 27.52 (0.87) | 21.21 (1.20) | 19.99 (1.42) | 5.659 | 0.002 |
| Polar size (μm) | 8.97 (0.50) | 8.23 (0.24) | 8.44 (0.49) | 10.29 (0.39) | 5.316 | 0.003 |
| Equatorial size (μm) | 3.51 (0.20) | 3.71 (0.20) | 3.97 (0.18) | 4.46 (0.36) | 2.982 | 0.039 |
| Distance between guard cells (μm) | 11.37 (0.20) | 12.15 (0.33) | 11.51 (0.20) | 11.58 (0.20) | 2.23 | 0.096 |

*Values are means ± standard error for population (n = 5 seedlings). Significance (F-value, P-level) of population effects is shown (experiment 2).*

Photosynthetic response curves to internal $CO_2$ and irradiance (analyzed in control plants) did not show differences among populations (Table 4).

**Table 4.** Populations effects on photosynthesis (A) response curves to internal $[CO_2]$ ($C_i$) ($A/C_i$) and to irradiance (Q) (A/Q)

| | PV1 | PV3 | PV4 | PV5 | ANOVA one way | |
|---|---|---|---|---|---|---|
| | | | | | *F*-value | *P*-level |
| **A/Ci CURVE PARAMETERS** | | | | | | |
| $R_{day}$ | 2.67 (1.71) | 1.46 (0.07) | −0.03 (0.08) | 0.83 (0.41) | 2.00 | 0.232 |
| $V_{cmax}$ | 26.65 (14.32) | 21.61 (1.89) | 20.58 (4.37) | 27.25 (0.33) | 0.17 | 0.912 |
| $J_{max}$ | 31.16 (14.74) | 21.10 (2.93) | 20.39 (3.45) | 26.42 (0.66) | 0.39 | 0.762 |
| TPU | 2.86 (0.58) | 2.51 (0.16) | 2.67 (0.36) | 2.77 (0.20) | 0.17 | 0.911 |
| **A/Q CURVE PARAMETERS** | | | | | | |
| $A_{max}$ | 5.49 (0.43) | 5.21 (0.91) | 6.07 (1.37) | 5.27 (0.83) | 0.26 | 0.854 |
| $\varphi$ | 0.09 (0.01) | 0.09 (0.00) | 0.09 (0.02) | 0.10 (0.01) | 0.63 | 0.614 |
| $R_d$ | −0.19 (0.06) | −0.31 (0.09) | −0.25 (0.20) | −0.21 (0.13) | 0.05 | 0.983 |
| $\Gamma_i$ | 2.67 (1.63) | 2.67 (1.63) | 1.31 (1.60) | 2.65 (1.62) | 0.26 | 0.852 |

*Measurements were made on fully expanded leaves of control plants. Values are the means ± standard error for population (n = 5 seedlings). Significance (F-value; p-level) of population effects for A/Ci and A/Q curve parameters are presented. $R_{day}$, daytime respiration (μmol $CO_2$ $m^{-2}$ $s^{-1}$); $V_{cmax}$, maximum rate of carboxylation (μmol $CO_2$ $m^{-2}$ $s^{-1}$); $J_{max}$, light-saturated rate of electron transport (μmol electrons $m^{-2}$ $s^{-1}$); TPU, rate of triose phosphate utilization (μmol $CO_2$ $m^{-2}$ $s^{-1}$); $A_{max}$, maximum photosynthetic rate (μmol $CO_2$ $m^{-2}$ $s^{-1}$); φ, apparent quantum efficiency (μmol $CO_2$ $μmol^{-1}$ photons); $R_d$, apparent dark respiration (μmol $CO_2$ $m^{-2}$ $s^{-1}$); and $\Gamma_i$, light compensation point (μmol photons $m^{-2}$ $s^{-1}$) (experiment 2).*

Photosynthetic rate, $C_i/C_a$ and the ratio photosynthesis and stomatal conductance ($A/g_s$) significantly differed among populations ($P < 0.001$) (Table 5). Decreasing SWA impacted all considered ecophysiological traits ($P < 0.001$), as well as the interaction population x treatment ($P < 0.001$).

**Table 5.** Daytime gas exchange under saturating light conditions of seedlings for each population grown in different drought thresholds (treatment)

| Gas exchange parameter | SWA (%) | PV1 | PV3 | PV4 | PV5 | ANOVA | *F*-value | *P*-level |
|---|---|---|---|---|---|---|---|---|
| **PHOTOSYNTHESIS RATE (A) (μmol $m^{-2}$ $s^{-1}$)** | | | | | | | | |
| | 30 | 1.10 (0.55) | 0.36 (0.24) | 1.36 (0.65) | 0.52 (0.50) | Population | 8.087 | 0.000 |
| | 40 | 4.04 (0.45) | 3.35 (0.59) | 2.51 (1.06) | 3.56 (0.52) | Treatment | 59.940 | 0.000 |
| | 50 | 5.51 (0.29) | 4.63 (0.44) | 3.81 (0.60) | 4.19 (0.42) | Population × Treatment | 34.806 | 0.000 |
| | Control | 5.26 (0.12) | 4.03 (0.20) | 4.35 (0.32) | 4.30 (0.30) | | | |
| **STOMATAL CONDUCTANCE (gs) (mol $m^{-2}$ $s^{-1}$)** | | | | | | | | |
| | 30 | 0.01 (0.00) | 0.01 (0.00) | 0.02 (0.00) | 0.01 (0.00) | Population | 1.114 | 0.344 |
| | 40 | 0.06 (0.01) | 0.05 (0.01) | 0.04 (0.02) | 0.05 (0.01) | Treatment | 34.965 | 0.000 |
| | 50 | 0.07 (0.00) | 0.07 (0.01) | 0.06 (0.01) | 0.06 (0.01) | Population × Treatment | 18.313 | 0.000 |
| | Control | 0.06 (0.00) | 0.05 (0.00) | 0.06 (0.01) | 0.05 (0.00) | | | |
| **$C_i/C_a$ (μmol $mol^{-1}$)** | | | | | | | | |
| | 30 | 0.69 (0.08) | 0.89 (0.06) | 0.71 (0.08) | 0.87 (0.07) | Population | 6.372 | 0.000 |
| | 40 | 0.61 (0.03) | 0.70 (0.03) | 0.71 (0.04) | 0.65 (0.04) | Treatment | 21.689 | 0.000 |
| | 50 | 0.61 (0.02) | 0.66 (0.03) | 0.70 (0.02) | 0.63 (0.03) | Population × Treatment | 14.201 | 0.000 |
| | Control | 0.57 (0.02) | 0.63 (0.02) | 0.64 (0.02) | 0.59 (0.02) | | | |
| **A/gs (μmol $mol^{-1}$)** | | | | | | | | |
| | 30 | 66.88 (21.33) | 15.27 (15.40) | 61.60 (19.55) | 19.55 (17.95) | Population | 6.148 | 0.000 |
| | 40 | 85.23 (7.52) | 50.90 (16.97) | 60.35 (10.84) | 73.66 (10.68) | Treatment | 21.008 | 0.000 |
| | 50 | 92.79 (3.95) | 73.78 (8.23) | 63.97 (5.59) | 78.00 (7.14) | Population × Treatment | 13.670 | 0.000 |
| | Control | 95.61 (4.06) | 82.75 (5.29) | 78.22 (4.33) | 92.97 (4.93) | | | |

*Photosynthetic rate (A), stomatal conductance ($g_s$), ratio of intercellular ($C_i$) to ambient ($C_a$) [$CO_2$] ($C_i/C_a$), and leaf intrinsic WUE ($A/g_s$) were measured on the first fully expanded leaf (LPI 5) (measurements made at 1000–1200 h). Values are means (± standard error) for population (n = 5 seedlings). Significance of treatment effect (drought treatment expressed at SWA level, ANOVA) for each variable and their interaction is indicated by F-value and P-level (experiment 2).*

Gas exchange was related to SWA through a sigmoidal pattern, describing a threshold-type transition from one phase to the next (Figure 4). The inflection point of the function was observed between 30 and 26% SWA (29.45, 29.42, and 26.15% SWA, for photosynthetic rate, stomatal conductance and Ci/Ca, respectively).

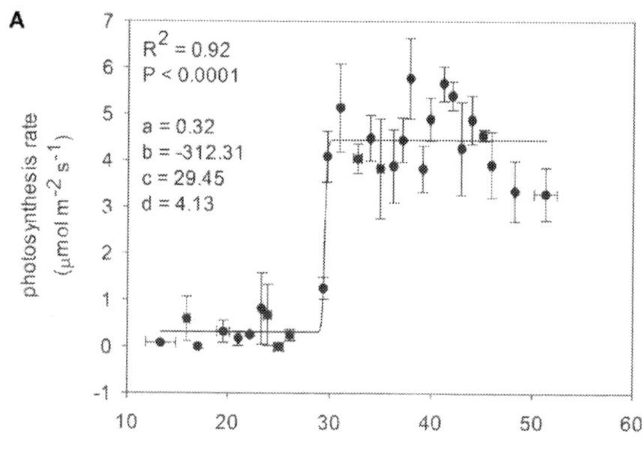

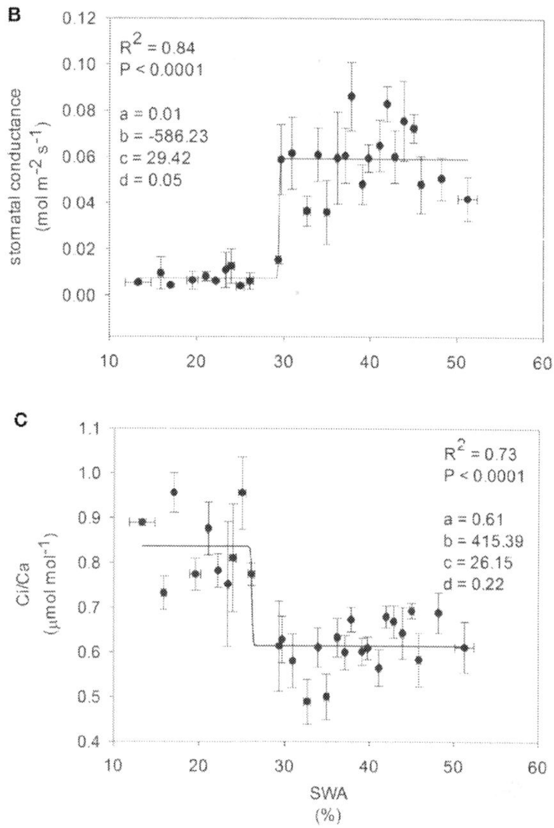


**Figure 4.** Relationship between ecophysiological traits and soil water availability (each point corresponds mean of five plants). Non-linear regression model was used for curve-fitting analysis, such as a four parameter logistic equation, where: (A) is the minimum asymptote; (B)is the hill slope; (C) is the inflection point; (D) is the maximum asymptote; global goodness of fit is defined by R-squared, and significance of the regression by p-level (experiment 2).

# DISCUSSION

## Differences among Populations

Beech seedlings from seeds collected across a climatic gradient were monitored in two common garden experiments conducted in Germany and Italy to quantify the geographic structure of variation in ecophysiological traits and the adaptation potential of populations to drought stress. We hypothesized that differentiation of tolerance/avoidance traits in beech progenies was constitutively linked to the seed source, and tested whether population-specific functional and structural responses to drought stress were revealed when seedlings were exposed to decreasing soil water availability.

The study highlighted intraspecific differentiation in plant performance when seedlings were not exposed to constraints (Tables 2–4), which was related to phenotypic variation among populations (e.g., Thiel et al., 2014). Beech has a wide area of distribution and the gradual change in ecophysiological traits and environmental gradients exhibited by different populations can shape local adaptation. Netzer et al. (2016) found that Greek populations were more tolerant to drought stress than German ones, while a Bulgarian population was the most sensitive. These results confirm once again the relevance of seed source for assessing the response of beech to drought stress. The French population (PV3) showed a distinct behavior in morphological traits; whereas, Danish (PV1) and Spanish (PV7) populations showed the lowest plant height, collar diameter, and leaf number (Table 2). Well-watered conditions resulted in higher plant height in French, Romanian, and Bosnian populations, and higher leaf number in German, French, Romanian, and Bosnian populations, suggesting higher plant growth potential in French, Romanian, and Bosnian populations, but failing to identify a clear population-related trait. Contradictory results on growth differentiation between beech populations with different native precipitation regimes are reported in literature (e.g., Tognetti et al., 1995; Knutzen et al., 2015). Indeed, genetic differentiation between beech stands was found to be located mainly within stands and not between stands within the distribution range of this species (Buiteveld et al., 2007; Meier and Leuschner, 2008). Even though, clear genetic differentiation also among stands based on adaptive genetic markers for populations coming from different climatic conditions was found by Sander et al. (2000) and Dounavi et al. (2016). Results of photosynthetic response curve to internal $CO_2$ concentration and irradiance did not show differences among populations, highlighting common assimilation potential and photosynthetic capacity across populations (Table 4). These results pointed to the lack of intraspecific differences in these traits among populations related to drier or moister native conditions, as observed by Knutzen et al. (2015). The impact of the last glacial period on the areal distribution of beech and the preferred selection of seed sources by forest managers may have concurred in reducing differences among populations in certain functional traits.

Structural leaf traits did not show a common pattern across populations (Table 3). Variations in tissue density, mesophyll thickness, and stomatal features have been related to the influence exerted by environmental conditions on leaf economics via modulation of hydraulic conductance and carbon budget (Brodribb et al., 2013; Sack et al., 2013; Blonder et al., 2015). Thicker leaves (and mesophyll) have long been recognized as structural adaptive trait in dry environments (Vogel, 2009), generally resulting in reductions in light interception and carbon gain expressed on a tissue volume basis. However, thicker palisade layer is a common trait for efficient light interception in mesomorphic sun leaves (Marchi et al., 2008). As efficient light harvesting is expensive in terms of biomass investment in support tissues, the efficiency with which foliage can be supported is a major factor determining species ability to grow under constrained environments (e.g., Niinemets et al., 1998). Intrinsic limitations associated with too high costs for foliar support may constrain more those populations in environments with high irradiance (southern Europe), where carbon for foliage construction is relatively cheaper.

As a major effect of experimental conditions on plant traits was expected to determine the phenotypic plasticity of corresponding traits, a significant interaction between beech populations and greenhouse conditions would indicate that responses to drought depended on the origin of seed sources. Hence, it is important to notice that we considered plasticity of a given trait at the population level, as an average value of the trait across individuals from each population, rather than at the genotype level.

## Drought Stress Impacts

This experiment and the companion study (Bolte et al., 2016) demonstrated the high correlation between physiological responses and water availability in these beech populations. This is consistent with findings in other common garden experiments conducted on beech, which showed considerable differences among populations in adaptation to environmental conditions (Kreyling et al., 2012, 2014; Thiel et al.,

2014). The effect of drought treatment did not elicit marked differences among populations as expected (see Peuke et al., 2002).

When beech seedlings were exposed to water constraints, they adopted stress-compensating mechanisms (e.g., reduction of leaf water potential, regulation of gas exchange, control of plant growth), which differed among populations (Knutzen et al., 2015). Water constraints significantly decreased photosynthetic rate and stomatal conductance to prevent plant dehydration plant (Table 2) (Peuke et al., 2002; Aranda et al., 2015). However, according to Tognetti et al. (1995), there was no clear relationship between plant response and population range, suggesting a variable sensitivity to drought. The limitation of gas diffusion through stomata, before any changes occur in plant water status, could suggest a control of leaf water loss toward isohydric behavior (Attia et al., 2015), thus preserving leaf hydration. It is also possible that plants with highly sensitive stomata (isohydric), closing at relatively high water potentials, but with even more drought-sensitive hydraulic system could show anisohydric behavior. In the present study, $C_i/C_a$-values increased with the reduction of water availability, indicating regulation capacity of water balance achieved by a slow stomatal reaction in response to a decrease in leaf water potential (Lawlor and Tezara, 2009). Yet, a reduction in leaf intrinsic WUE ($A/g_s$) was observed. Drought sensitivity increased in the sequence France Montagne Noir (PV4) < Denmark (PV1) < Romania (PV5) < France Crecy (PV3), with 21, 30, 79, and 81.5% of reduction, respectively, in comparison with control plants. Photosynthesis was substantially reduced and eventually stopped under severe drought, which may affect metabolic patterns and the allocation of carbon to structural traits to varying degrees, ultimately translating into plant death (McDowell et al., 2011). In this framework, the performance of beech seedlings was probably determined by plant size, defining the attitude of a plant to photosynthesize, store nutrients, and mobilize resources, and to growth (e.g., Villar-Salvador et al., 2012). Inhibition of metabolism, with maintenance of light respiration, probably increased $C_i$, defining a greater sensitivity of photosynthetic rate than light respiration to water deficit (Cornic and Fresneau, 2002). Plants that display isohydric characteristics have tight and continuous water potential homeostasis through stomatal control. They constantly regulate their water loss within a certain range to avoid damaging water deficits due to hydraulic failure (Buckley,

2005), though this may inhibit $CO_2$ diffusion into leaves and increase the risk of carbon starvation. The identification of population-specific performance under decreasing SWA, in this as well as in other recent experiments (e.g., Aranda et al., 2015; Knutzen et al., 2015; Dounavi et al., 2016), helps defining the risk of or resilience to local mortality across the native range of beech. However, drought led to heterogeneous and variable response patterns in these populations.

Beech populations variably combined functional traits in response to drought, although there was not a well-defined gradient in drought tolerance among populations. The phenotypic behavior showed a straight reduction of functional processes in response to decreasing SWA. The decrease of leaf water potential with the consequent reduction of assimilation rate under water stress showed a conservative control of water loss in these beech populations (Figure 3), highlighting the capacity of the plant hydraulic system to regulate the supply of water to leaves (Attia et al., 2015). The shape parameter of this relationship described accurately the response dynamics of seedlings to changing SWA (Figure 4). Furthermore, the inflection point of the function allowed the detection of a SWA threshold, where the SWA-value that induced a sharp drop of gas exchange was 28.34% (calculated as the mean of SWA-values, Figure 4). The reduction of SWA induced also the reduction of PI and Chl index and the increase of $F_0$; whereas, the $F_v/F_m$ was not substantially affected by water stress. Overall, these parameters showed relatively high values, illustrating the resilience of the photochemical apparatus with decreasing SWA in beech (but see García-Plazaola and Becerril, 2000). However, when seedlings reached leaf predawn water potential of −2 MPa, gas exchange approached zero, indicating high sensitivity of stomatal behavior in response to extreme drought (in agreement with $\Psi_{pd}$-values as low as −1.76 MPa—Köcher et al., 2009). A parallel reduction in the activity of PSII reaction center, efficiency of light capture, and rate of electron transfer was observed in all populations, which may avoid permanent damage from photoinhibition if stress conditions were quickly released. PV2 maintained the highest values of ETR and NPQ, utilizing more absorbed light energy for photochemistry and allocating more light energy to NPQ pathways than the other populations. The reversible down-regulation of PSII photochemistry would contribute to an enhanced photo-protection in severely stressed beech seedlings (Gallé and Feller, 2007).

Genetic variability and phenotypic plasticity, at morphological, physiological, and phenological level, are the key factors for the adaptation to environmental constraints, and are supposed to be wide in species, such as beech, with ample distribution range (Bussotti et al., 2015). The issue of the genetic background on the enhanced tolerance to drought in beech has been discussed mostly because of the probable evolutionary adaptation of local population by selection processes, which allow only drought-tolerant individuals to inherit genetic and phenotypic traits to next generations (Hampe and Petit, 2005). This is reputed to exhibit the highest resistance to the environmental stress (Thiel et al., 2014; Aranda et al., 2015; Knutzen et al., 2015). It must be pointed out that differences in moisture availability at the native sites may not be high enough to elicit distinct seedling performance among populations, and that factors other than differences in precipitation amount at the origin may be equally important, thus hindering the detection of clear population differences in common garden experiments.

## CONCLUDING REMARKS

A common trend in responses of beech originated from different sites was established in relation to a decreasing of soil water availability. Phenotypic changes of plant traits in relation to SWA thresholds highlighted that decreasing SWA thresholds may favor ecophysiological traits that optimize carbon assimilation to counterbalance the reduced growing period. Functional traits define the contribution of genetic variability and phenotypic plasticity for understanding the adaptive potential to drought stress of beech across its large geographical range of distribution. In this sense, our study did not include populations spanning the entire environmental gradient of beech, which warrants further studies to better elaborate the relationships between ecophysiological acclimation to drought and seed source climate. A deeper understanding of the mechanism of drought tolerance in beech is needed to support strategies of forest management toward assisted selection and seed transfer. Planting of more resistant genotypes into drought-impacted forest stands may help implementing adaptive forest management options to ameliorate the impact to anticipated climate change, but parallel garden experiments are needed and should include more seed sources and ecophysiological traits.

## AUTHOR CONTRIBUTIONS

Substantial contributions to the conception and design of the work: CC, TC, AB, RT. Acquisition, analysis, or interpretation of data for the work: CC, MM, EP, LD, SM, LM, AA, TC, AB, RT. Drafting the work or revising it critically for important intellectual content: CC, MM, EP, LD, SM, LM, AA, TC, AB, RT. Final approval of the version to be published: CC, MM, EP, LD, SM, LM, AA, TC, AB, RT. Agreement to be accountable for all aspects of the work in ensuring that questions related to the accuracy or integrity of any part of the work are appropriately investigated and resolved: CC, MM, EP, LD, SM, LM, AA, TC, AB, RT.

## CONFLICT OF INTEREST STATEMENT

The authors declare that the research was conducted in the absence of any commercial or financial relationships that could be construed as a potential conflict of interest.

## ACKNOWLEDGMENTS

We are grateful to Dinca Lucian (Marin Dracea National Forest Research-Development Institute, Bucharest, Romania), Anders Ræbild (University of Copenhagen, Dept. of Geosciences and Natural Resource Management, Frederiksberg C, Denmark), Martin de Luis [Grupo de Clima, Agua, Cambio Global y Sistemas Naturales, Departamento de Geografía y Ordenación del Territorio, Facultad de Filosofía y Letras, Instituto de Investigación en Ciencias Ambientales (IUCA), Universidad de Zaragoza, Zaragoza, España], Branislav Cvjetkovic (University of Banja Luka, Faculty of Forestry, Banja Luka, Bosnia and Herzegovina) that provided the seeds of some of the studied populations. We are grateful to Dr. Mirko Liesebach and Rainer Ebbinghaus from the Thünen Institute of Forest Genetics (Geß-Hansdorf, Germany) for the cultivation of the beech seedlings used in the Experiment 1 (Germany). This article is based upon work from COST Action FP1106 STReESS, supported by COST (European Cooperation in Science and Technology).

## REFERENCES

1. Aitken, S. N., Yeaman, S., Holliday, J. A., Wang, T. L., and Curtis-McLane, S. (2008). Adaptation, migration or extirpation: climate change out-comes for tree populations. *Evol. Appl.* 1, 95–111. doi: 10.1111/j.1752-4571.2007.00013.x
2. Allen, C. D., Macalady, A. K., Chenchouni, H., Bachelet, D., McDowell, N., Vennetier, M., et al. (2010). A global overview of drought and heat-induced tree mortality reveals emerging climate change risks for forests. *For. Ecol. Manage.* 259, 660–684. doi: 10.1016/j.foreco.2009.09.001
3. Améglio, T., Archer, P., Cohen, M., Valancogne, C., Daudet, F.-A., Dayau, S., et al. (1999). Significance and limits in the use of predawn leaf water potential for tree irrigation. *Plant Soil*207, 155–167. doi: 10.1023/A:1026415302759
4. Aranda, I., Cano, F. J., Gascó, A., Cochard, H., Nardini, A., Mancha, J. A., et al. (2015). Variation in photosynthetic performance and hydraulic architecture across European beech (*Fagus sylvatica* L.) populations supports the case for local adaptation to water stress. *Tree Physiol.* 35, 34–46. doi: 10.1093/treephys/tpu101
5. Attia, Z., Domec, J. C., Oren, R., Way, D. A., and Moshelion, M. (2015). Growth and physiological responses of isohydric and anisohydric poplars to drought. *J. Exp. Bot.* 66, 4373–4381. doi: 10.1093/jxb/erv195
6. Backes, K., and Leuschner, C. (2000). Leaf water relations of competitive *Fagus sylvatica* and *Quercus petraea* trees during 4 years differing in soil drought. *Can. J. For. Res.* 30, 335–346. doi: 10.1139/cjfr-30-3-335
7. Barigah, T. S., Charrier, O., Douris, M., Bonhomme, M., Herbette, S., Améglio, T., et al. (2013). Water stress-induced xylem hydraulic failure is a causal factor of tree mortality in beech and poplar. *Ann. Bot.* 112, 1431–1437. doi: 10.1093/aob/mct204
8. Beck, W., and Heußner, K. U. (2012). "Increased sensitivity in ring width series of common beech after 1990 -climatic impact or normal patterns due to ageing?," in *TRACE - Tree Rings in Archaeology, Climatology and Ecology*, Vol. 10, eds H. Gärtner, P. Rozenberg,

P. Montés, O. Bertel, G. Helle, and I. Heinrich (Potsdam: GFZ Potsdam), 13–19. Scientific Technical Report STR 12/03.

9. Blonder, B., Vasseur, F., Violle, C., Shipley, B., Enquist, B. J., and Vile, D. (2015). Testing models for the leaf economics spectrum with leaf and whole-plant traits in *Arabidopsis thaliana*. *AoB PLANTS* 7:plv049. doi: 10.1093/aobpla/plv049
10. Blume, H.-P., Brümmer, G. W., Fleige, H., Horn, R., Kandeler, E., Kšgel-Knabner, I., et al. (2016). *Soil Science*. Berlin; Heidelberg: Springer.
11. Bolte, A., Czajkowski, T., Cocozza, C., Tognetti, R., De Miguel, M., Pšidová, E., et al. (2016). Desiccation and mortality dynamics in seedlings of different European beech *(Fagus sylvatica*L.) populations under extreme drought conditions. *Front. Plant Sci.* 7:751. doi: 10.3389/fpls.2016.00751
12. Bolte, A., Czajkowski, T., and Kompa, T. (2007). The north-eastern distribution area of European beech – a review. *Forestry* 80, 413–429. doi: 10.1093/forestry/cpm028
13. Borghetti, M., Leonardi, S., Raschi, A., Snyderman, D., and Tognetti, R. (1993). Ecotypic variation of xylem embolism, phenological traits, growth parameters and allozyme characteristics in *Fagus sylvatica*. *Funct. Ecol.* 7, 713–720. doi: 10.2307/2390193
14. Bresson, C. C., Vitasse, Y., Kremer, A., and Delzon, S. (2011). To what extent is altitudinal variation of functional traits driven by genetic adaptation in European oak and beech? *Tree Physiol.* 31, 1164–1174. doi: 10.1093/treephys/tpr084
15. Brodribb, T. J., Jordan, G. J., and Carpenter, R. J. (2013). Unified changes in cell size permit coordinated leaf evolution. *New Phytol.* 199, 559–570. doi: 10.1111/nph.12300
16. Buckley, T. N. (2005). The control of stomata by water balance. *New Phytol.* 168, 275–292. doi: 10.1111/j.1469-8137.2005.01543.x
17. Buiteveld, J., Vendramin, G. G., Leonardi, S., Kamer, K., and Geburek, T. (2007). Genetic diversity and differentiation in European beech (*Fagus sylvatica* L.) stands varying in management history. *For. Ecol. Manage.* 247, 98–106. doi: 10.1016/j.foreco.2007.04.018
18. Bussotti, F., Pollastrini, M., Holland, V., and Brüggemann, W. (2015). Functional traits and adaptive capacity of European forests

to climate change. *Environ. Exp. Bot.* 111, 91–113. doi: 10.1016/j.envexpbot.2014.11.006

19. Caemmerer, S. G. D., and Farquhar, D. D. (1981). Some relationships between the biochemistry of photosynthesis and the gas exchange of leaves. *Planta* 153, 376–387. doi: 10.1007/BF00384257
20. Cassol, D., Silva, F. S. P., Falqueto, A. R., and Bacarin, M. A. (2008). An evaluation of non-destructive methods to estimate total chlorophyll content. *Photosynthetica* 46, 634–636. doi: 10.1007/s11099-008-0109-6
21. Ciais, P., Reichstein, M., Viovy, N., Granier, A., Ogée, J., Allard, V., et al. (2005). Europe-wide reduction in primary productivity caused by the heat and drought in 2003. *Nature* 437, 529–533. doi: 10.1038/nature03972
22. Cornic, G., and Fresneau, C. (2002). Photosynthetic carbon reduction and carbon oxidation cycles are the main electron sinks for Photosystem II activity during a mild drought. *Ann. Bot.*89, 887–894. doi: 10.1093/aob/mcf064
23. Czajkowski, T., Kuhling, M., and Bolte, A. (2005). Impact of the 2003 summer drought on growth of beech sapling natural regeneration (*Fagus sylvatica* L.) in north-eastern Central Europe. *Allg. Forst-u. Jagdzt.* 176, 133–143.
24. Dounavi, A., Netzer, F., Celepirovic, N., Ivanković, M., Burger, J., Figueroa, A. G., et al. (2016). Genetic and physiological differences of European beech provenances (*F. sylvatica L.*) exposed to drought stress. For. Ecol. Manage. 361, 226–236. doi: 10.1016/j.foreco.2015.11.014
25. Farquhar, G. D., von Caemmerer, S., and Berry, J. A. (1980). A biochemical model of photosynthetic $CO_2$ assimilation in leaves of C3 species. *Planta* 149, 78–90. doi: 10.1007/BF00386231
26. Fischer, A., Marshall, P., and Camp, A. (2013). Disturbances in deciduous temperate forest ecosystems of the northern hemisphere: their effects on both recent and future forest development. *Biodiv. Conserv.* 22, 1863–1893. doi: 10.1007/s10531-013-0525-1
27. Gallé, A., and Feller, U. (2007). Changes of photosynthetic traits in beech saplings (*Fagus sylvatica*) under severe drought stress and

during recovery. *Physiol. Plant.* 131, 412–421. doi: 10.1111/j.1399-3054.2007.00972.x

28. García-Plazaola, J. I., and Becerril, J. M. (2000). Effects of drought on photoprotective mechanisms in European beech (*Fagus sylvatica* L.) seedlings from different provenances. *Trees*14, 485–490. doi: 10.1007/s004680000068
29. Geßler, A., Keitel, C., Kreuzwieser, J., Matyssek, R., Sailer, J., and Rennenberg, H. (2007). Potential risks for European beech *(Fagus sylvatica* L.) in a changing climate. *Trees* 21, 1–11. doi: 10.1007/s00468-006-0107-x
30. Granier, A., Reichstein, M., Brßda, N., Janssens, I. A., Falge, E., Ciais, P., et al. (2007). Evidence for soil water control on carbon and water dynamics in European forests during the extremely dry year: 2003. *Agr. For. Meteorol.* 143, 123–145. doi: 10.1016/j.agrformet.2006.12.004
31. Hampe, A., and Petit, R. J. (2005). Conserving biodiversity under climate change: the rear edge matters. *Ecol. Lett.* 8, 461–467. doi: 10.1111/j.1461-0248.2005.00739.x
32. Hovenden, M. J., and Vander Schoor, J. K. (2003). Nature vs nurture in the leaf morphology of Southern beech, *Nothofagus cunninghamii* (*Nothofagaceae*). *New Phytol.* 161, 585–594. doi: 10.1046/j.1469-8137.2003.00931.x
33. Hultine, K. R., and Marshall, J. D. (2001). A comparison of three methods for determining the stomatal density of pine needles. *J. Exp. Bot.* 52, 369–373. doi: 10.1093/jexbot/52.355.369
34. Jump, A. S., Hunt, J. M., and Peñuelas, J. (2006). Rapid climate change-related growth decline at the southern range edge of *Fagus sylvatica. Glob. Change Biol.* 12, 2163–2174. doi: 10.1111/j.1365-2486.2006.01250.x
35. Knutzen, F., Meier, I. C., and Leuschner, C. (2015). Does reduced precipitation trigger physiological and morphological drought adaptations in European beech *(Fagus sylvatica* L.)? Comparing provenances across a precipitation gradient. *Tree Physiol.* doi: 10.1093/treephys/tpv057
36. Köcher, P., Gebauer, T., Horna, V., and Leuschner, C. (2009). Leaf water status and stem xylem flux in relation to soil drought in five

temperate broad-leaved tree species with contrasting water use strategies. *Ann. For. Sci.* 66, 101. doi: 10.1051/forest/2008076

37. Kreyling, J., Buhk, C., Backhaus, S., Hallinger, M., Huber, G., Huber, L., et al. (2014). Local adaptations to frost in marginal and central populations of the dominant forest tree *Fagus sylvatica* L.as affected by temperature and extreme drought in common garden experiments. *Ecol. Evol.* 4, 594–605. doi: 10.1002/ece3.971
38. Kreyling, J., Thiel, D., Nagy, L., Jentsch, A., Huber, G., Konnert, M., et al. (2012). Late frost sensitivity of juvenile *Fagus sylvatica* L. differs between southern Germany and Bulgaria and depends on preceding air temperature. *Eur. J. For. Res.* 131, 717–725. doi: 10.1007/s10342-011-0544-y
39. Lawlor, D. W., and Tezara, W. (2009). Causes of decreased photosynthetic rate and metabolic capacity in water-deficient leaf cells: a critical evaluation of mechanisms and integration of processes. *Ann. Bot.* 103, 561–579. doi: 10.1093/aob/mcn244
40. Leuzinger, S., Zotz, G., Asshoff, R., and Körner, C. (2005). Responses of deciduous forest trees to severe drought in Central Europe. *Tree Physiol.* 25, 641–650. doi: 10.1093/treephys/25.6.641
41. Marchi, S., Tognetti, R., Minnocci, A., Borghi, M., and Sebastiani, L. (2008). Variation in mesophyll anatomy and photosynthetic capacity during leaf development in a deciduous mesophyte fruit tree (*Prunus persica*) and an evergreen sclerophyllous Mediterranean shrub (*Olea europaea*). *Trees* 22, 559–571. doi: 10.1007/s00468-008-0216-9
42. McDowell, N. G. (2011). Mechanisms linking drought, hydraulics, carbon metabolism, and mortality. *Plant Physiol.* 155, 1051–1059. doi: 10.1104/pp.110.170704
43. McDowell, N. G., Beerling, D. J., Breshears, D. D., Fisher, R. A., Raffa, K. F., Stitt, M., et al. (2011). The interdependence of mechanisms underlying climate-driven vegetation mortality. *Trends Ecol. Evol.* 10, 523–532. doi: 10.1016/j.tree.2011.06.003
44. Meier, I. C., and Leuschner, C. (2008). Genotypic variation and phenotypic plasticity in the drought response of fine roots of European beech. *Tree Physiol.* 28, 297–309. doi: 10.1093/treephys/28.2.297
45. Netzer, F., Thöm, C., Celepirovic, N., Ivankovic, M., Dounavi, A.,

Simon, J., et al. (2016). Drought effects on C, N and P nutrition and the antioxidative system of beech (*Fagus sylvatica* L.) seedlings depend on geographic origin. *J. Plant Nutr. Soil Sci.* 179, 136–150. doi: 10.1002/jpln.201500461

46. Niinemets, Ü. (2010). Responses of forest trees to single and multiple environmental stresses from seedlings to mature plants: past stress history, stress interactions, tolerance and acclimation. *For. Ecol. Manage.* 260, 1623–1639. doi: 10.1016/j.foreco.2010.07.054
47. Niinemets, Ü., Kull, O., and Tenhunen, J. D. (1998). An analysis of light effects on foliar morphology, physiology, and light interception in temperate deciduous woody species of contrasting shade tolerance. *Tree Physiol.* 18, 681–696. doi: 10.1093/treephys/18.10.681
48. Ohlemuller, R., Gritti, E. S., Sykes, M. T., and Thomas, C. D. (2006). Quantifying components of risk for European woody species under climate change. *Glob. Change Biol.* 12, 1788–1799. doi: 10.1111/j.1365-2486.2006.01231.x
49. Peuke, A. D., Schraml, C., Hartung, W., and Rennenberg, H. (2002). Identification of drought sensitive beech ecotypes by physiological parameters. *New Phytol.* 154, 373–387. doi: 10.1046/j.1469-8137.2002.00400.x
50. Piovesan, G., Biondi, F., Di Filippo, A., Alessandrini, A., and Maugeri, M. (2008). Drought-driven growth reduction in old beech (*Fagus sylvatica*) forests of the central Apennines, Italy. *Glob. Change Biol.* 14, 1–17. doi: 10.1111/j.1365-2486.2008.01570.x
51. Poorter, H., Niinemets, U., Walter, A., Fiorani, F., and Schurr, U. (2010). A method to construct dose–response curves for a wide range of environmental factors and plant traits by means of a meta-analysis of phenotypic data. *J. Exp. Bot.* 61, 2043–2055. doi: 10.1093/jxb/erp358
52. Prioul, J. L., and Chartier, P. (1977). Partitioning of transfer and carboxylation components of intracellular resistance to photosynthetic $CO_2$ fixation: a critical analysis of the methods used. *Ann. Bot.* 41, 789–800.
53. Pšidová, E., Ditmarová, L., Jamnická, G., Kurjak, D., Majerová, J., Cazajkowski, T., et al. (2015). Photosynthetic response of beech

seedlings of different origin to water deficit. *Photosynthetica* 53, 187–194. doi: 10.1007/s11099-015-0101-x

54. Reid, C. P. P., Crowley, D. E., Kim, H. J., Powell, P. E., and Szaniszlo, P. J. (1984). Utilization of iron by oat when supplied as ferrated synthetic chelate or as ferrated hydroxamate siderophore. *J. Plant Nutr.* 7, 437–447. doi: 10.1080/01904168409363210
55. Rennenberg, H., Loreto, F., Polle, A., Brilli, F., Fares, S., Beniwal, R. S., et al. (2006). Physiological responses of forest trees to heat and drought. *Plant Biol.* 8, 556–571. doi: 10.1055/s-2006-924084
56. Rose, L., Leuschner, C., Kšckemann, B., and Buschmann, H. (2009). Are marginal beech (*Fagus sylvatica* L.) provenances a source for drought tolerant ecotypes? *Eur. J. For. Res.* 128, 335–343. doi: 10.1007/s10342-009-0268-4
57. Sack, L., Scoffoni, C., John, G. P., Poorter, H., Mason, C. M., Mendez-Alonzo, R., et al. (2013). How do leaf veins influence the worldwide leaf economic spectrum? *Review and synthesis. J. Exp. Bot.* 64, 4053–4080. doi: 10.1093/jxb/ert316
58. Sala, A., Piper, F., and Hoch, G. (2010). Physiological mechanisms of drought-induced tree mortality are far from being resolved. *New Phytol.* 186, 274–281. doi: 10.1111/j.1469-8137.2009.03167.x
59. Sander, T., Kšnig, S., Rothe, G. M., Jahnsen, A., and Weisgerber, H. (2000). Genetic variation of European beech (*Fagus sylvatica* L.) along an altitudinal transect at mount Vogelsberg in Hesse, Germany. *Molec. Ecol.* 9, 1349–1361. doi: 10.1046/j.1365-294x.2000.01014.x
60. Scherrer, D., Bader, M. K., and Koerner, C. (2011). Drought-sensitivity ranking of deciduous tree species based on thermal imaging of forest canopies. *Agric. For. Meteorol.* 151, 1632–1640. doi: 10.1016/j.agrformet.2011.06.019
61. Sharkey, T. D. (1985). $O_2$-insensitive photosynthesis in C3 plants. *Plant Physiol.* 78, 71–75. doi: 10.1104/pp.78.1.71
62. Sork, V. L., Aitken, S. N., Dyer, R. J., Eckert, A. J., Legendre, P., and Neale, D. B. (2013). Putting the landscape into the genomics of trees: approaches for understanding local adaptation and population responses to changing climate. *Tree Genet. Genomes* 9, 901–911. doi: 10.1007/s11295-013-0596-x
63. Stojnić, S., Orlović, S., Miljković, D., Galić, Z., Kebert, M., and

von Wuehlisch, G. (2015). Provenance plasticity of European beech leaf traits under differing environmental conditions at two Serbian common garden sites. *Eur. J. For. Res.* 134, 1109–1125. doi: 10.1007/s10342-015-0914-y

64. Tegel, W., Seim, A., Hakelberg, D., Hoffmann, S., Panev, M., Westphal, T., et al. (2013). A recent growth increase of European beech (*Fagus sylvatica* L.) at its Mediterranean distribution limit contradicts drought stress. *Eur. J. For. Res.* 133, 61–71. doi: 10.1007/s10342-013-0737-7
65. Thiel, D., Kreyling, J., Backhaus, S., Beierkuhnlein, C., Buhk, C., Egen, K., et al. (2014). Different reactions of central and marginal provenances of *Fagus sylvatica* to experimental drought. *Eur. J. For. Res.* 133, 247–260. doi: 10.1007/s10342-013-0750-x
66. Thomas, S. C. (2011). Genetic vs. *phenotypic responses of trees to altitude. Tree Physiol.* 31, 1161–1163. doi: 10.1093/treephys/tpr105
67. Tognetti, R., Johnson, J. D., and Michelozzi, M. (1995). The response of European beech (*Fagus sylvatica* L.) seedlings from two Italian populations to drought and recovery. *Trees* 9, 348–354. doi: 10.1007/BF00202499
68. Tognetti, R., Johnson, J. D., and Michelozzi, M. (1997). Ecophysiological responses of *Fagus sylvatica* seedlings to changing light conditions. I. interactions between photosynthetic acclimation and photoinhibition during simulated canopy gap formation. *Physiol. Plant.* 101, 115–123. doi: 10.1111/j.1399-3054.1997.tb01827.x
69. Tognetti, R., Lombardi, F., Lasserre, B., Cherubini, P., and Marchetti, M. (2014). Tree-ring stable isotopes reveal twentieth-century increases in water-use efficiency of Fagus sylvatica and Nothofagus spp. in Italian and Chilean mountains. *PLoS ONE* 9:e113136. doi: 10.1371/journal.pone.0113136
70. Tognetti, R., Sebastiani, L., and Minnocci, A. (2004). Gas exchange and foliage characteristics of two poplar clones grown in soil amended with industrial waste. *Tree Physiol.* 24, 75–82. doi: 10.1093/treephys/24.1.75
71. Van der Werf, G. W., Sass-Klaasen, U., and Mohren, G. M. J. (2007). The impact of the 2003 summer drought on the intra-annual growth pattern of beech (*Fagus sylvatica* L.) and oak (*Quercus robur* L.) on

a dry site in the Netherlands. *Dendrochronologia* 25, 103–112. doi: 10.1016/j.dendro.2007.03.004

72. Veihmeyer, F. J., and Hendrickson, A. H. (1927). The relation of soil moisture to cultivation and plant growth. *Proc. 1st Intern. Congr. Soil Sci.* 3, 498–513.
73. Villar-Salvador, P., Puértolas, J., Cuesta, B., Peñuelas, J. L., Uscola, M., Heredia-Guerrero, N., et al. (2012). Increase in size and nitrogen concentration enhances seedling survival in Mediterranean plantations. Insights from an ecophysiological conceptual model of plant survival. *New For.* 43, 755–770. doi: 10.1007/s11056-012-9328-6
74. Vitasse, Y., Lenz, A., Kollas, C., Randin, C. F., Hoch, G., and Körner, C. (2014). Genetic vs. *non-genetic responses of leaf morphology and growth to elevation in temperate tree species. Funct. Ecol.* 28, 243–252. doi: 10.1111/1365-2435.12161
75. Vogel, S. (2009). Leaves in the lowest and highest winds: temperature, force and shape. *New Phytol.* 183, 13–26. doi: 10.1111/j.1469-8137.2009.02854.x
76. Wullschleger, S. D. (1993). Biochemical limitations to carbon assimilation in C3 plants–a retrospective analysis of the A/Ci curves from 109 species. *J. Exp. Bot.* 44, 907–920. doi: 10.1093/jxb/44.5.907

CHAPTER

7

# TRANSCRIPTIONAL REGULATION OF DROUGHT RESPONSE: A TORTUOUS NETWORK OF TRANSCRIPTIONAL FACTORS

Dhriti Singh and Ashverya Laxmi

National Institute of Plant Genome Research, New Delhi, India

Drought is one of the leading factors responsible for the reduction in crop yield worldwide. Due to climate change, in future, more areas are going to be affected by drought and for prolonged periods. Therefore, understanding the mechanisms underlying the drought response is one of the major scientific concerns for improving crop yield. Plants deploy diverse strategies and mechanisms to respond and tolerate drought stress. Expression of numerous genes is modulated in different plants under drought stress that help them to optimize their growth and development. Plant hormone abscisic acid (ABA) plays a major role in plant response and tolerance by regulating the expression of many genes under drought stress. Transcription factors being the major regulator of gene expression play a crucial role in stress response. ABA regulates the expression of most of many genes leading to some important physiological as well as biochemical changes that help plant to survive under stress (Umezawa et al., 2010). Molecular and genomic analyses have revealed the existence of ABA-independent signal transduction pathway in conjunction to the ABA-dependent signal transduction pathway during drought stress (Yamaguchi-Shinozaki and Shinozaki, 2006).

## TRANSCRIPTIONAL REGULATORY NETWORK

Plants respond to various environmental stresses including drought through changes ranging from physiological to molecular level. These changes help plants to optimize their growth and stress resistance. Drought stress changes the expression of many genes that are thought to play an important role in stress response and tolerance. Many of these genes have been identified and characterized (Yamaguchi-Shinozaki and Shinozaki, 2006; Todaka et al., 2015). Microarray analyses by various groups have revealed thousands of genes that are upregulated and downregulated in response to drought stress. A significant number of drought-inducible genes are also induced by high salinity, suggesting a cross-talk between drought and salt stress. Comparatively lesser number of drought-inducible genes are induced by cold stress (Yamaguchi-Shinozaki and Shinozaki, 2006). There is a very small overlap of only 27 genes that were found to be commonly induced in microarray studies (Bray, 2004). This lack of commonality may be attributed to the fact that different sets of probes were used during these microarray experiments and variations in conditions of plant growth and stress. Recently, 17 microarray experiments of *Arabidopsis*, rice, wheat, and barley were compared using a novel **C**ross-**S**pecies meta-**A**nalysis of progressive Drought stress at the reproductive stage (CSA:Drought); and 225 differentially expressed genes were identified that were shared across studies and taxa (Shaar-Moshe et al., 2015).

Stress inducible genes in *Arabidopsis* can be classified into two categories: functional and regulatory genes (Yamaguchi-Shinozaki and Shinozaki, 2006). Genes encoding proteins required for cellular stress tolerance fall into the former category, for example, LEA (late embryogenesis abundant) proteins, molecular chaperones, reactive oxygen species detoxifying enzymes, and sugars or proline biosynthetic enzymes. Whereas, genes encoding proteins that are involved in signal transduction and gene expression come under the latter category, such as protein kinases, components of ABA signaling, enzymes for lipid signaling, and various transcription factors (Yamaguchi-Shinozaki and Shinozaki, 2006).

As stated above, plant hormone ABA plays an important role in response to water deficit including regulation of transcriptional network (Yamaguchi-Shinozaki and Shinozaki, 2006). A large number of genes that are induced by water deficit are also highly induced by exogenous application of ABA. Conversely, there are several genes that are induced by water deficit but are not affected by exogenous ABA. These findings suggested that transcriptional response to water deficit is regulated by both ABA-dependent and ABA-independent signal transduction pathways (Yamaguchi-Shinozaki and Shinozaki, 2006).

Transcription factors are regulatory proteins that can modulate expression of a specific set of genes through binding to their promoter. They play an important role in converting the stress-induced signals to cellular responses. A single transcription factor can modulate the expression of a number of genes. A collection of genes under regulation of the same regulatory protein is called regulons. Plants activate many regulons under drought and other stresses to optimize plant growth, some of them have been well determined in *Arabidopsis* (Nakashima et al., 2009). Both ABA-dependent and ABA-independent pathways regulate the transcriptional response by affecting one or more regulons active under drought stress (Nakashima et al., 2009). In the following section, we will discuss in brief about water deficit induced regulons and pathways affecting them.

## AREB/ABF Regulon

Abscisic acid-responsive element binding protein (AREB) /ABF (ABRE binding factor) regulon function in ABA-dependent regulation of gene expression under drought stress (Nakashima et al., 2009; Yoshida et al., 2015), (Figure 1). Many genes that are affected by water deficit also respond to the exogenous application of ABA (Nakashima et al., 2009). Promoter analysis revealed that most of these ABA-responsive genes are regulated by ABRE (ABA-responsive element) in their promoter region. ABRE is a conserved, 8 bp long *cis* element (PyACGTGG/TC) with a core ACGT sequence (Nakashima et al., 2009; Fujita et al., 2011). A single copy of an ABRE is not sufficient to induce ABA-responsive gene expression. To function as an active *cis*-acting element, ABRE requires in proximity other copies of ABRE or another specific *cis*-acting element, which is termed as the coupling element. Certain sequences have been shown to function as coupling elements such as, CE1 (coupling element1) and CE3 (coupling element3); and DRE (dehydration-responsive element)/CRT (C-repeat) *cis* element (major *cis* element in ABA-independent pathway; Shen et al., 1996; Narusaka et al., 2003). CEs are usually GC-rich sequence and they are similar to ABRE (Yamaguchi-Shinozaki and Shinozaki, 2006).

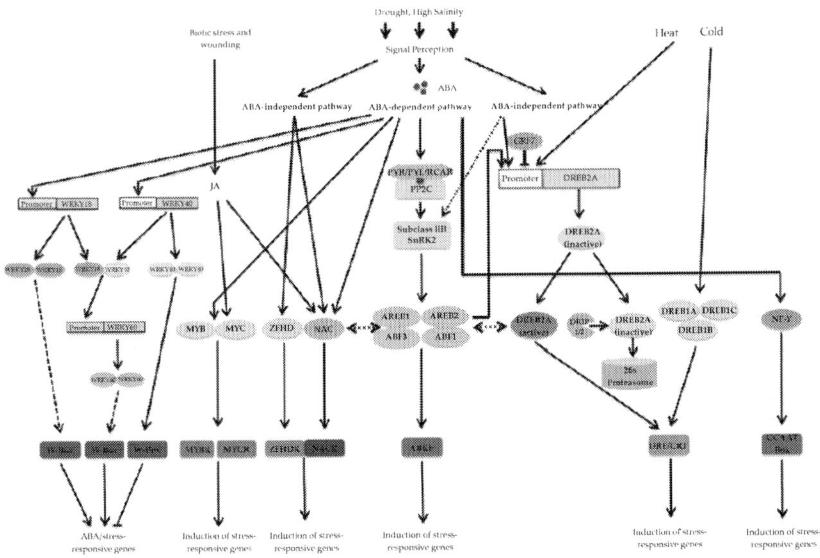


**Figure 1.** Major transcriptional regulatory networks of transcription factors involved in drought stress. Drought signal perception leads to activation of both abscisic acid (ABA)-dependent and ABA-independent pathways. In ABA-dependent pathway, accumulation of ABA leads to activation of sub class III SnRK2s through PYR/PYL/RCAR-PP2C receptor complex. Four transcription factors ABA-responsive element (ABRE) binding protein 1 (AREB1), AREB2, ABRE binding factor 3 (ABF3), and ABF1 are mainly phosphorylated through sub class III SnRK2s under drought stress and regulate most of the downstream genes by binding to the ABRE *cis*-element present in their promoter region. In addition to this, ABA also modulates the activity of MYB/MYCs, NACs, WRKYs, and NF-Y transcription factors. MYC2 and NAC proteins are also involved in JA signaling. ABA signal perception leads to induction of WRKY18

and WRKY40 and their product could bind to W-box present in WRKY60 and thereby induce it. DREB2A plays a pivotal role in ABA-independent gene expression regulation under drought stress. *DREB2A* expression is regulated by GRF7 under unstressed condition. Additionally, DRIPs regulate levels of DREB2A protein under unstressed condition. DREB2A also participate in gene expression regulation under heat stress. Three DREB1 proteins, DREB1A, DREB1B, and DREB1C are the key factors regulating gene expression under cold stress. Some NAC transcription factors also regulate gene expression in ABA-independent manner. Recently, it has been reported that AREB/ABFs can induce DREB2A and AREB/ABFs interact with DREB2A, based on these observations, ABA-dependent and ABA-independent pathways are thought to crosstalk under drought stress. Apart from this, some recent reports suggest a possible interaction between NACs and AREB/ABFs. Transcription factors and DNA-binding proteins are shown in colored ellipses. Dashed lines indicate possible although unconfirmed routes.

Abscisic acid-responsive element binding protein or ABF transcription factors were found to bind to ABRE element in yeast one-hybrid screening. These are major transcription factors that bind to ABRE and regulate ABA-responsive gene expression (Choi et al., 2000; Uno et al., 2000), (Figure 1). AREB/ABF is a subfamily of the basic leucine zipper (bZIP) family that consists of 9 members in *Arabidopsis.* All AREB/ABF transcription factors contain four conserved domains in addition to bZIP domain (Fujita et al., 2011, 2013). In *Arabidopsis, AREB1/ABF2, AREB2/ABF4, ABF1*, and *ABF3* are expressed mainly in the vegetative tissues in response to ABA and osmotic stress (Fujita et al., 2011). In contrast, some other members are expressed during seed maturation such as *Arabidopsis ABI5, AREB3, DPBF2*, and *EEL* (Finkelstein and Lynch, 2000; Lopez-Molina and Chua, 2000; Bensmihen et al., 2002). Transgenic plants overexpressing *AREB1/ABF2, AREB2/ABF4*, or *ABF3*exhibit enhanced drought tolerance and increased ABA sensitivity (Kang et al., 2002; Fujita et al., 2005). Triple AREB/ABF mutant *areb1 areb2 abf3* exhibits reduced tolerance to drought and decreased sensitivity to exogenous ABA compared to that of wild-type, single mutants, or double mutants (Yoshida et al., 2010). Transcriptome analysis of the triple mutant under osmotic stress conditions showed reduced levels of many osmotic stress-inducible genes (Yoshida et al., 2010). Recently, ABF1 has also been reported to play an important role in ABA-mediated gene expression under drought stress. Although, it is expressed in lesser quantity as compared to those three AREB/ABFs, the *areb1 areb2 abf3 abf1*quadruple mutant plants show an increase in drought sensitivity and decreased ABA sensitivity in comparison to *areb1 areb2 abf3* mutant. In quadruple mutant many dehydration-inducible genes including LEA protein genes and transcription factors show reduced expression (Yoshida et al., 2015). Both triple mutants, as well as quadruple mutants, exhibited decreased inflorescence heights as compared to wild-type; as well as delayed bolting. Except for this, all of them showed phenotypes similar to wild-type (Yoshida et al., 2010, 2015). Thus, these four AREB/ABFs have been shown to be central transcription factors that cooperatively function in ABA-dependent transcriptional activation through their ABREs under these abiotic stress conditions (Figure 1).

Abscisic acid-responsive element binding protein/ABF trans cription factors are fully activated only after phosphory lation of their conserved regions (Fujita et al., 2011). The phosphorylation is catalyzed by serine/threonine kinase SnRK2s (SNF1-related protein kinase) that are induced by ABA. The importance of phosphorylation was suggested by the observation that overexpression of *AREB1/ABF2* activated downstream gene expression only when phosphorylated active form of this gene was used (Furihata et al., 2006). Thus, under water deficit, cellular ABA concentration is increased that is recognized by the ABA receptors PYR/PYL/RCARs (pyrabactin resistance/pyrabactin resistance1- like/regulatory component of ABA receptors) leading to the inhibition of phosphatase activity of PP2C (protein phosphatase 2C). PP2C is a negative regulator of ABA signaling which dephosphorylates and thereby inactivates subclass III SnRK2s. Subsequently, released subclass III SnRK2s get accumulated in the cell and phosphorylate AREB/ABF thereby inducing expression of AREB/ABF regulon genes (reviewed in Cutler et al., 2010; Umezawa et al., 2010), (Figure 1). There are 10 members of SnRK2 family in *Arabidopsis* that are divided into three groups including three members in subclass III SnRK2s, SRK2D/SnRK2.2, SRK2E/OST1/SnRK2.6, and SRK2I/SnRK2.3 (Umezawa et al., 2010). Out of these 10 SnRK2s, nine are activated by osmotic stress. However, subclass III SnRK2s are also strongly induced by ABA and mediate most of the ABA responses (Yoshida et al., 2014). Subclass III SnRK2s have been shown to phosphorylate AREB/ABFs *in vitro* as well as co-localize and interact with them in plant cell nuclei (Furihata et al., 2006; Fujii et al., 2007; Fujita et al., 2009; Yoshida et al., 2010). Studies involving triple mutant *srk2d/e/i* showed that expression of most of the AREB1/ABF2, AREB2/ABF4, and ABF3 regulated genes are highly reduced and ABA-dependent phosphorylation of AREB/ABFs is completely abolished (Fujii and Zhu, 2009; Fujita et al., 2009). All these results indicate that subclass III SnRK2s regulate ABA-responsive gene expression under drought stress by phosphorylating AREB/ABFs.

## DREB1/CBF and DREB2 Regulons

Dehydration-responsive element binding protein 1 (DREB1)/CBF (C-repeat binding factor) and DREB2 regulons function in ABA-independent regulation of gene expression under drought stress (Nakashima et al., 2009), (Figure 1). In *Arabidopsis*, the *RD29A/COR78/LTI78* gene is induced by drought and cold. This gene has been found to be ABA-inducible but at the same time it can be induced by drought and cold stress in mutants defective in ABA biosynthesis and signaling (Yamaguchi-Shinozaki and Shinozaki, 2006). The analysis of its promoter, together with expression studies suggested that the dehydration inducibility of this gene is regulated by both ABA-independent and ABA-dependent pathways through different *cis*-acting elements (Yamaguchi-Shinozaki and Shinozaki, 2006).

The promoter of *RD29A* gene contains DRE/CRT *cis* element in addition to ABRE (Yamaguchi-Shinozaki and Shinozaki, 1994, 2005). DRE element is responsible for ABA-independent induction of many genes in response to osmotic and cold stress in many plants including *Arabidopsis* (Yamaguchi-Shinozaki and Shinozaki, 2005). DRE is a conserved 9 bp long (TACCGACAT) *cis* element. Unlike ABRE, a single copy of DRE is sufficient to induce genes under osmotic and cold stress (Yamaguchi-Shinozaki and Shinozaki, 1994). Similar *cis*-acting elements, named C-repeat (CRT) and low-temperature responsive element (LTRE), were identified in low-temperature-inducible genes. These sequences share a common core sequence, A/GCCGAC, which is referred to as the DRE/CRT core sequence (Baker et al., 1994; Jiang et al., 1996; Thomashow, 1999). DRE/CRTs are found in promoters of many stress-inducible genes.

Dehydration-responsive element binding protein 1/CBF and DREB2 transcription factors recognize DRE/CRT and activate downstream genes

(Figure 1). Both DREB1/CBF and DREB2 belong to the plant-specific AP2 (APETALA2)/ ERF (ethylene-responsive element-binding factor) family having AP2/ERF DNA-binding motif. There are 145 members in AP2/ERF transcription factor family in *Arabidopsis* (Sakuma et al., 2002). DREB transcription factors constitute a subfamily of AP2/ERF family. *Arabidopsis* has six and eight *DREB1/CBF*-type and *DREB2*-type genes, respectively. Among them, three DREB1/CBF-type transcription factors, DREB1A/CBF3, DREB1B/CBF1, and DREB1C/CBF2 are rapidly induced by low temperature and act as major transcription factors that activate gene transcription through DRE/CRT in response to cold stress (Nakashima et al., 2009), (Figure 1). Transgenic plants overexpressing *DREB1* show enhanced tolerance to cold and accumulate osmoprotectants such as proline and various sugars (Gilmour et al., 2000). These transgenic plants also exhibit "dwarf" phenotype and pronounced prostrate growth habits. They have shorter petiole in comparison to wild-type plants with leaves having bluish-green tint. Additionally, plants overexpressing *DREB1* show delayed bolting and flowering as well as lower yield in comparison to wild-type plants (Gilmour et al., 2000).

In contrast to DREB1, two DREB2-type transcription factors, DREB2A and DREB2B, are highly induced in response to osmotic stress conditions and are considered to be involved in DRE-mediated gene transcription in response to water deficit (Liu et al., 1998), (Figure 1). Later, it was found that DREB2A plays a pivotal role in gene expression regulation under salt stress whereas DREB2B regulates gene expression in response to drought stress (Nakashima et al., 2009). However, weak induction of several *DREB1* genes such as *DREB1D/CBF4, DREB1E/DDF2*, and *DREB1F/DDF1* under dehydration stress suggests that DREB1/CBF and DREB2 regulons interact with each other (Haake et al., 2002; Sakuma et al., 2002; Magome et al., 2004).

Although DREB2A and DREB1A were isolated together (Liu et al., 1998), later, it was discovered that both of them have slight difference in their downstream genes (Maruyama et al., 2009). Microarray analysis has suggested that products of most of the genes downstream to DREAB1A and DREB2A have similar putative functions, but carbohydrate metabolism genes have different expression pattern in DREB1A and DREB2A transgenic plants (Maruyama et al., 2009). Plants overexpressing *DREB1A* exhibit changes in expression of genes responsible for starch degradation, sucrose metabolism and sugar alcohol synthesis similar to that observed in dehydration and cold stress. These changes lead to accumulation of many kinds of sugar and sugar alcohols in plants that might be responsible for enhanced dehydration and cold stress tolerance. In contrast plants overexpressing *DREB2A-CA* (constitutively active form of *DREB2A*) do not exhibit the increase in these metabolites level (Maruyama et al., 2009). The reason for this might be the difference in their DNA-binding specificity (Nakashima et al., 2009). DREB1A has a high affinity to A/GCCGACNT sequences, whereas DREB2A preferentially binds ACCGAC motifs (Sakuma et al., 2006a) resulting in slight variation in target genes.

Although DREB2A regulates the expression of many genes involved in stress response and tolerance, it causes growth retardation and reduced reproduction rate in plants, therefore, its expression is tightly regulated (Yoshida et al., 2014). *DREB2A* expression is negatively regulated by GRF7 (growth-regulating factor7). GRF proteins are a family of putative transcription factors that consist of nine members in *Arabidopsis* (Kim et al., 2003). Among these nine members, GRF7 inhibits expression of *DREB2A* under normal conditions by binding to its short promoter region (Figure 1). Both knockdown and knockout mutants of *GRF7* exhibit enhanced DREB2A expression under non-stressed condition (Kim et al., 2012a). These plants also exhibit enhanced salinity tolerance and retarded growth. Microarray analysis revealed that a large number of osmotic stress-responsive genes were upregulated in *grf7* knockout mutants under non-stressed condition. These shreds of evidence suggest that GRF7 regulates a large number of osmotic stress responsive genes by regulating *DREB2A* expression (Kim et al., 2012a).

In addition to transcriptional regulation, DREB2A is also regulated at post-transcriptional level. Transgenic plant overexpressing DREB1/CBF under *Arabidopsis* stress-responsive RD29A promoter showed strong tolerance to stresses mainly against cold (Kasuga et al., 1999). However, transgenic plant overexpressing *DREB2A* did not exhibit any significant phenotypic change (Liu et al., 1998). Domain analysis revealed that central region of DREB2A protein has a negative regulatory domain (NRD), and removal of the NRD from DREB2A converts the protein into its constitutively active form (DREB2A-CA). The DREB2A-CA proteins are more stable in the nucleus than the wild-type protein. In contrast to wild-type *DREB2A*, overexpression of *DREB2A-CA* shows enhanced drought tolerance as well as a slight improvement in cold stress tolerance and upregulation of many stress-inducible genes (Sakuma et al., 2006a). Transgenic plants overexpressing *DREB2A-CA* exhibited growth retardation in comparison to wild-type as well as rounded, slightly darker leaves with short petiole. Additionally, extent of retardation and these mentioned phenotypes was in correlation to the expression of transgene (Sakuma et al., 2006a). All these reports suggest that stability control is a posttranslational regulatory mechanism of DREB2A.

Furthermore, DRIP1 (DREB2A-interacting protein 1), a ubiquitin E3 ligase is thought to degrade the leaky expression under normal conditions (Qin et al., 2008). DRIP1 harbors a C3HC4-type RING domain and has been found to interact with DREB2A in yeast two-hybrid screening. Transgenic *Arabidopsis* plants overexpressing *DRIP1* exhibit delayed expression of DREB2A-regulated drought responsive genes, however, double knockout mutants of *DRIP1* and its homolog *DRIP2* exhibit increased expression of these genes. DRIP1 and DRIP2 function as E3 ubiquitin ligase and target DREB2A to 26S proteasome proteolysis and thereby negatively regulate the expression of drought-responsive genes (Qin et al., 2008), (Figure 1). Recently, it has been shown that stress signal play important role in stabilization of DREB2A (Morimoto et al., 2013). In *drip1drip2* mutant, DREB2A protein levels were induced rapidly under heat stress suggesting their involvement in DREB2A degradation but the protein levels reduced in the mutant by prolonged heat stress (Morimoto et al., 2013). It has also been shown that stabilization of DREB2A is important but not sufficient for induction of downstream genes (Morimoto et al., 2013). All these results suggest that apart from DRIPs some other factors might be involved in degradation and activation of DREB2A.

Recently, transcription factor ERF53 (ethylene response factor53) and two homologous C3HC4-type RING E3 ligases, RGLG2 (RING domain ligase2) and RGLG1 has been identified that also function similar to DREB2A and DRIPs and regulate drought stress-responsive genes (Cheng et al., 2012). ERF53 is an AP2/ERF transcription factor that belongs to the non-DREB2 subfamily. *AtERF53* expression increase significantly under drought and high salinity but it exhibit mild induction to exogenous ABA (Cheng et al., 2012; Hsieh et al., 2013). Overexpression of *AtERF53*induces unstable drought tolerance. AtERF53 has been found to interact with RGLG2 and RGLG1, both of which act as E3 ubiquitin ligase and target it to proteasomal degradation. AtERF53-GFP fusion protein accumulates more stably in double mutant *rglg1 rglg2* leading to enhanced drought stress tolerance. All these reports suggest that AtERF53 and RGLGs function in combination to regulate osmotic stress-responsive genes (Cheng et al., 2012; Hsieh et al., 2013).

Dehydration-responsive element binding protein 2A regulon also operates under heat shock stress (Figure 1). Overexpression studies of *DREB2A-CA* showed improved thermotolerance by inducing expression of heat shock related genes (Sakuma et al., 2006b; Mizoi et al., 2012). Similarly, transgenic plants overexpressing *DREB2C* showed increased expression of heat shock stress-inducible genes, leading to thermotolerance (Lim et al., 2007). Therefore, it is clear that DRE/CRT is involved in gene expression not only during dehydration but also under conditions of low and high temperature.

## NAC Regulon

NAM, ATAF, and CUC (NAC) are plant-specific transcription factors that constitute one of the largest families of plant transcription factors. There are more than a hundred members in *Arabidopsis* and rice that have been classified in 10 groups based on their phylogenetic relationship (Jensen et al., 2010). NAC transcription factors have a highly conserved N-terminal DNA-binding domain and variable C-terminal region this C-terminal region is thought to play a crucial role in determination of their target genes (Nuruzzaman et al., 2013). NAC transcription factors are involved in various developmental processes from shoot meristem development to auxin signaling (Olsen et al., 2005). There are reports demonstrating their involvement in abiotic and biotic stress signaling (reviewed in Olsen et al., 2005; Nakashima et al., 2012; Nuruzzaman et al., 2013). NAC transcription factors involved in stress response and tolerance have been classified in stress-responsive NAC (SNAC) group.

The *ERD1* (*early response to dehydration1*) gene encodes a ClpA homolog of *Arabidopsis* (Nakashima et al., 1997). ERD1 was found to be induced by osmotic stress such as dehydration and salt stress as well as during senescence; however, exogenous ABA application could not strongly induce it (Nakashima et al., 1997). Two different *cis*-acting elements, a MYC-like sequence (CATGTG) and a 14-bp *rps1* site 1-like sequence are necessary for dehydration induced expression of *ERD1* (Simpson et al., 2003). Three NAC transcription factors; ANAC019, ANAC055, and ANAC072/RD26 were reported to bind MYC-like sequence present in the promoter of ERD1 (Tran et al., 2004). These three proteins are included in the SNAC group of NAC transcription factors (Nakashima et al., 2012). Detailed DNA binding assay of these NAC transcription factors determined NACRS (NAC recognition sequence) ANNNNNTCNNNNNNNACACGCATGT, containing CATGT and harboring CACG as the core DNA-binding site (Tran et al., 2004), (Figure 1). These NAC genes were found to be expressed within 1–2 h of ABA treatment suggesting that they are induced through ABA-independent pathway under drought stress (Tran et al., 2004). Transgenic plants overexpressing *ANAC019* and *ANAC072* showed phenotype and the time course of growth similar to that of vector control (Tran et al., 2004). In contrast, plants overexpressing *ANAC055* exhibited growth rate similar to that of vector control until they reached rosette stage; after this point, plants in which the expression of transgene was at the medium level, showed a little delay in bolting as compared to vector control whereas; plants in which the transgene was overexpressed at high level remained at rosette stage for an additional few days before first bolting (Tran et al., 2004). Transgenic plant overexpressing *ANAC019, ANAC055*, or *ANAC072/RD26* induced expression of many stress-inducible genes but failed to up-regulate *ERD1*(Fujita et al., 2004; Tran et al., 2004).

Subsequently, a zinc-finger homeodomain (ZFHD) transcription factor, ZFHD1, was identified as a transcriptional activator that recognizes the 14-bp rps1 site1-like sequence (CACTAAATTGTCAC) and this sequence was termed as ZFHDR (zinc finger homeodomain recognition sequence; Tran et al., 2007), (Figure 1). Expression of the *ERD1* gene was induced only when both the NAC and ZFHD proteins were overproduced simultaneously in a transgenic plant. Thus, these transcription factors cooperatively activate the transcription of the *ERD1* gene (Tran et al., 2007). Studies suggest that in addition to the role of ZFHD1 in cooperation of NAC transcription factor it can also function as transcriptional activator alone (Tran et al., 2007).

Overexpression of several stress-responsive NAC factors in *Arabidopsis* and rice has imparted drought tolerance in transgenic plants. For example, transgenic plants overexpressing *ANAC072/RD26* exhibited enhanced drought tolerance as well as increased sensitivity to ABA (Fujita et al., 2004; Tran et al., 2004). Microarray analysis of *ANAC072/RD26* overexpressing plants showed upregulation of stress-inducible genes and ABA-responsive genes suggesting that RD26 is involved in regulation of drought-responsive genes in ABA-dependent manner (Fujita et al., 2004). Similarly, overexpression of another NAC factor *ATAF1* resulted in improved drought tolerance (Wu et al., 2009). Recently, NAC genes such as *ANAC096* and *ANAC016* are associated with drought response and tolerance (Xu et al., 2013; Sakuraba et al., 2015). In case of rice, overexpression of *NAC* genes such as *SNAC1, OsNAC6/SNAC2, OsNAC5*, and *OsNAC10* improved drought tolerance (reviewed in Nakashima et al., 2014). NTLs (NAC with transmembrane motif 1-like) such as NTL4 and NTL6 are also involved in drought stress. Transgenic plants overexpressing *NTL6* and *ntl4* null mutants exhibit drought tolerance, suggesting that these two work antagonistic to each other during drought stress (Lee and Park, 2012; Kim et al., 2012b).

Several stress-inducible NAC genes are also induced by jasmonates and/or during senescence in *Arabidopsis* and rice (Figure 1). Thus, these stress-responsive NAC transcription factors not only function in the transcriptional response to abiotic stress conditions, including water stress, but are likely involved in the cross talk between abiotic and biotic stress responses (Nakashima et al., 2012).

## Other Transcriptional Pathways Involved in Water Stress Responses

In addition to the above-mentioned pathways of *cis*-acting elements and transcription factors, many other transcriptional pathways function in water stress responses. The *Arabidopsis RD22* gene is inducible by drought stress in ABA-dependent manner (Yamaguchi-Shinozaki and Shinozaki, 1993). Although it is induced by ABA, it does not have ABRE *cis*-element in its promoter region; in spite of that its expression is regulated by two *cis*-acting elements, MYC and MYB recognition elements (Abe et al., 1997). A MYC like transcription factor, MYC2, and a MYB transcription factor, MYB2 bind to these *cis*-acting elements and cooperatively activate the transcription of this gene (Abe et al., 1997, 2003), (Figure 1). Transgenic plants overexpressing *AtMYC2* and

*AtMYB2* exhibit higher ABA sensitivity as well as osmotic tolerance (Abe et al., 2003). Transgenic plants overexpressing *AtMYC2* show morphology similar to wild-type. However, overexpression of *AtMYB2* causes growth retardation in transgenic plants growing on soil. Similarly, overexpression of both transcription factors causes severe growth retardation in plants growing on soil (Abe et al., 2003). Although, *AtMYC2* overexpressing plants have the morphology similar to wild-type, they have characteristically irregular shaped leaf epidermal cells. In contrast, plants overexpressing *AtMYB2* and both transcription factors have leaf epidermal as well as parenchymal cells similar to wild-type but smaller in shape (Abe et al., 2003). Microarray analysis has suggested that their target genes include many ABA-inducible genes. Conversely, a mutation in MYC2 decreased the expression of target genes, including *RD22* (Abe et al., 2003). All these results suggest that in addition to ABRE mediated gene regulation, MYB and MYC transcription factors regulate gene expression in response to ABA under drought stress (Figure 1).

MYB transcription factors are one of the largest transcription factor families in plants that have characteristic MYB domain in their DNA binding region (Lindemose et al., 2013). Analysis of transcriptome data present in GENEVESTIGATOR database showed that 51% of *Arabidopsis MYB*genes are upregulated and 41% are downregulated under drought stress (Baldoni et al., 2015). Furthermore, many MYB genes have been found to be involved in drought stress responses (reviewed in Baldoni et al., 2015).

MYC transcription factors belong to bHLH (basic-helix-loop-helix) transcription factor family of plants that have a characteristic bHLH domain (Kazan and Manners, 2013). Guard cell transcriptome analysis showed that ABA-responsive genes having MYC-binding motifs in the promoter region are present in large number in these cells (Wang et al., 2011). MYC2 protein has emerged as a master player in jasmonic acid signaling as well as cross-talk between jasmonic acid and ABA signaling (Kazan and Manners, 2013).

The involvement of WRKY transcription factors in drought stress has been reported recently. WRKY transcription factors constitute a family that has one or two WRKY domains which is involved in DNA binding. WRKY transcription factors bind to a conserved sequence named as W box and regulate gene expression (Rushton et al., 2010), (Figure 1).

WRKY transcription factors are involved in various plant processes including biotic stress responses (Ulker and Somssich, 2004; Rushton et al., 2010). Recently, they have been reported to be involved in abiotic stress responses (Rushton et al., 2010; Banerjee and Roychoudhury, 2015). Various WRKY transcription factors have been found to be involved in ABA signaling. WRKY18 and WRKY60 act as positive regulators of ABA signaling during seed germination, and stress response while WRKY40 has the opposite effect on ABA signaling. WRKY18 and WRKY60 act as weak transcriptional activator whereas WRKY40 binds to the promoters of multiple stress-inducible transcription factor genes, including *DREB1A/CBF3, DREB2A*, and *MYB2*, and represses their expression (Chen et al., 2010; Shang et al., 2010). ABA signal perception leads to induction of WRKY18 and WRKY40 and their product could bind to W-box present in WRKY60 promoter and thereby induce it (Chen et al., 2010), (Figure 1). In another report, WRKY gene, *WRKY63/ ABO3* (*ABA Overly Sensitive3*) has been demonstrated to be involved in drought responses. *abo3* mutant exhibits hypersensitive response for ABA in the seedling stage as well as reduced drought tolerance. WRKY63 has been shown to bind to the promoter of *AREB1/ABF2* and thereby positively regulating its expression (Ren et al., 2010). Apart from these WRKY genes, many other have been reported to be involved in drought and salt stress responses (Bakshi and Oelmüller, 2014; Banerjee and Roychoudhury, 2015)

In addition to these, NF-Y (nuclear factor-Y) transcription factors also take part in drought stress response and tolerance mechanisms. NF-Y transcription factors are heterotrimeric proteins with three distinct subunits NF-YA, NF-YB, and NF-YC and bind to CCAAT box in the promoter region of target genes (Figure 1). NF-Y transcription factors are crucial factors in nodulation in nitrogen-fixing plants and nitrogen assimilation but there are several reports suggesting their role in stress responses especially in drought stress response and tolerance (Li et al., 2008; Petroni et al., 2012; Laloum et al., 2013; Xu et al., 2014; Quach et al., 2015). Under drought stress, *AtNF-YA5* has been shown to be upregulated in ABA-dependent manner in leaf and roots of *Arabidopsis* plants (Li et al., 2008), (Figure 1). Transgenic plants overexpressing *AtNF-YA5* exhibited improved drought resistance and reduced water loss; whereas, *Atnf-ya5* mutant plants were found to be hypersensitive to drought (Li et al., 2008).

## INTERACTION BETWEEN DIFFERENT TRANSCRIPTION FACTORS INVOLVED IN DROUGHT STRESS RESPONSE

Evidence suggests that drought-responsive transcription factors work cooperatively to regulate gene expression. The subclass III SnRK2s play a very important role in this context. These subclass III SnRK2s are induced by both ABA and drought stress. In-gel kinase assay using ABA-insensitive and –deficient mutants has suggested that drought stress activate SnRK2s independent of ABA (Yoshida et al., 2006; Boudsocq et al., 2007), (Figure 1). Transcriptome analysis of *srk2d/e/i* triple mutants suggests that they modulate the expression of genes involved in ABA-dependent as well as ABA-independent pathways (Fujita et al., 2009). Recently, phosphoproteome study has shown that 5 min of ABA treatment results in phosphorylation of phosphopeptide corresponding subclass III SnRK2s, but osmotic stress fails to do the same. However, short osmotic stress led to phosphorylation of a phosphopeptide corresponding to subclass I SnRK2s, which are not activated by ABA, and phosphopeptides corresponding to MAP3K and MAP4K (E Stecker et al., 2014). Novel proteins are thought to be involved in osmotic stress-dependent activation of SnRK2s.

As mentioned earlier, the target site of DREB/CBF, i.e., DRE/CRT motif also function as a coupling element for ABRE and is present in many ABA-inducible drought responsive genes (Narusaka et al., 2003). AREB/ABF proteins have been shown to physically interact with DREB/CBFs including DREB1A, DREB2A, and DREB2C (Lee et al., 2010), (Figure 1). Recent reports have shown that ABRE sequence in the promoter region is required for induction of *DREB2A* under osmotic stress. Furthermore, transient-expression analyses coupled with ChIP (Chromatin Immunoprecipitation) assays has shown AREB/ABFs such as AREB1, AREB2, and ABF3 can bind to the promoter of DREB2A and thereby induce them in an ABRE-dependent manner (Kim et al., 2011), (Figure 1). In *grf7* mutant expression of ABA-inducible genes and osmotic stress-responsive genes is upregulated (Kim et al., 2012a). All these reports suggest a complex interaction between the AREB and DREB regulons that need to be further studied in order to create a more comprehensive picture of transcriptional regulation under drought stress.

There are several reports indicating the interaction between AREB/ABFs and NACs. SNAC transcription factor *ATAF1* has been reported to bind to the promoter of NCED3 and thereby regulating ABA hormone levels, giving rise to a probability those SNACs may be involved in regulation of ABA-dependent gene expression of AREB/ABF regulons (Jensen et al., 2013). Conversely, ABRE sequences have been reported in the promoter region of SNAC genes (Nakashima et al., 2012). ANAC096 directly interacts with ABF2 and ABF4 but not with ABF3. Evidence suggests that ANAC096 acts cooperatively with ABFs in activation of ABA-inducible drought responsive genes (Xu et al., 2013), (Figure 1). Sakuraba et al. (2015) found that ANAC016 negatively regulates drought stress tolerance. They also found that ANAC016 directly binds to the promoter of *AREB1* and represses its expression. In addition to these two, five other drought-responsive NAC transcription factors (NAC019, NAP, NAC053/NTL4, NAC055, and NAC072) have been found to be associated with ABA signaling (Tran et al., 2004; Lee and Park, 2012; Zhang and Gan, 2012); suggesting they might act as additional tier in regulation of ABA-dependent drought responsive genes.

## CONCLUSION

Plant response to drought is a complex process comprising many changes from morphological to molecular level. Under drought stress, many transcription factors operate both exclusively and cooperatively forming a web of interactions. In this review, we have summarized major transcription factors that play a pivotal role in drought stress response and tolerance.

Drought activates many pathways in plants that have been broadly classified in two categories, i.e., ABA-dependent pathways and ABA-independent pathways. AREB/ABFs, DREBs and NACs are the vital transcription factors regulating a large fraction of drought inducible genes (Figure 1). Along with these, some other transcription factors such as MYB/MYC factors, WRKY and NF-Y have also been demonstrated to be involved in one or more drought responsive mechanisms (Abe et al., 1997, 2003; Yamaguchi-Shinozaki and Shinozaki, 2006; Baldoni et al., 2015). Four AREB/ABFs, AREB1/ABF2, AREB2/ABF4, ABF3, and ABF1 are the central players of ABA-mediated regulation of gene expression. These AREB/ABFs carry out ABA-regulated responses through binding to the ABRE *cis*-elements present in the promoter region of target genes (Yoshida et al., 2015). On the other hand, DREB1/CBF and DREB2 transcription factors regulate gene expression in ABA-independent manner. Three DREB1/CBFs, DREB1A/CBF3, DREB1B/CBF1, and DREB1C/CBF2 regulate cold-responsive gene expression whereas, DREB2A and DREB2B are mainly involved in regulation of osmotic stress-responsive gene expression (Nakashima et al., 2009). Additionally, heat stress-inducible genes are also regulated by DREB2A (Mizoi et al., 2012). However, under unstressed conditions, DREB2A levels are tightly regulated at the transcriptional and post-transcriptional levels through GRF7 and DRIPs, respectively (Yoshida et al., 2014). These DREBs bind to DRE/CRT element present in the promoters of genes acting downstream to them (Nakashima et al., 2009). Different NAC factors regulate drought-inducible gene expression by binding to NACRS *cis* elements (Nakashima et al., 2012, 2014).

Drought-responsive transcription factors interact with each other as well as components of other stress pathways resulting in overlap of target genes of these pathways. In this context, subclass III SnRK2s might act as a nodal point as these kinases can regulate the expression of AREB/ABFs as well as DREB2A (Yoshida et al., 2014). Furthermore, there are several pieces of evidence suggesting that AREB/ABF transcription factors interact with DREB and NAC transcription factors (Nakashima et al., 2014).

Transcription factors are potent candidates for engineering stress tolerant plants as a single transcription factor can modulate a large set of genes. Many of drought-responsive transcription factors have been used to improve drought tolerance in different crops such as rice, wheat, soybean, and maize (Nakashima et al., 2014; Krannich et al., 2015).

All these transcription factors along with other interacting partners constitute a complex network that has been extensively studied but not yet completely understood. In future, further studies about transcriptional regulatory network should provide a more comprehensive picture of the pathways as well as cross-talks. Identification of crucial factors of these pathways along with evolution of new technologies to produce genetically engineered plants should lead to development of plants with improved drought tolerance under field condition with minimal negative effects on crop yield and development.

## CONFLICT OF INTEREST STATEMENT

The authors declare that the research was conducted in the absence of any commercial or financial relationships that could be construed as a potential conflict of interest.

## ACKNOWLEDGMENT

The authors are thankful to University Grants Commission for research fellowship to DS.

## REFERENCES

1. Abe, H., Urao, T., Ito, T., Seki, M., Shinozaki, K., and Yamaguchi-Shinozaki, K. (2003). *Arabidopsis* AtMYC2 (bHLH) and AtMYB2 (MYB) function as transcriptional activators in abscisic acid signaling. *Plant Cell* 15, 63–78. doi: 10.1105/tpc.006130
2. Abe, H., Yamaguchi-Shinozaki, K., Urao, T., Iwasaki, T., Hosokawa, D., and Shinozaki, K. (1997). Role of *arabidopsis* MYC and MYB homologs in drought- and abscisic acid-regulated gene expression. *Plant Cell* 9, 1859–1868.
3. Baker, S. S., Wilhelm, K. S., and Thomashow, M. F. (1994). The 5′-region of *Arabidopsis thaliana* cor15a has cis-acting elements that confer cold-, drought- and ABA-regulated gene expression. *Plant Mol. Biol.* 24, 701–13. doi: 10.1007/BF00029852
4. Bakshi, M., and Oelmüller, R. (2014). WRKY transcription factors: jack of many trades in plants. *Plant Signal. Behav.* 9:e27700 doi: 10.4161/psb.27700
5. Baldoni, E., Genga, A., and Cominelli, E. (2015). Plant MYB transcription factors: their role in drought response mechanisms. *Int. J. Mol. Sci.* 16, 15811–15851. doi: 10.3390/ijms160715811
6. Banerjee, A., and Roychoudhury, A. (2015). WRKY proteins: signaling and regulation of expression during abiotic stress responses. *Sci. World J.* 2015:807560. doi: 10.1155/2015/807560
7. Bensmihen, S., Rippa, S., Lambert, G., Jublot, D., Pautot, V., Granier, F., et al. (2002). The homologous ABI5 and EEL transcription factors function antagonistically to fine-tune gene expression during late embryogenesis. *Plant Cell* 14, 1391–1403. doi: 10.1105/tpc.000869

8. Boudsocq, M., Droillard, M. J., Barbier-Brygoo, H., and Laurière, C. (2007). Different phosphorylation mechanisms are involved in the activation of sucrose non-fermenting 1 related protein kinases 2 by osmotic stresses and abscisic acid. *Plant Mol. Biol.* 63, 491–503. doi: 10.1007/s11103-006-9103-1
9. Bray, E. A. (2004). Genes commonly regulated by water-deficit stress in *Arabidopsis thaliana*. *J. Exp. Bot.* 55, 2331–2341. doi: 10.1093/jxb/erh270
10. Bray, E. A., Bailey-Serres, J., and Weretilnyk, E. (2000). "Responses to abiotic stresses," in *Biochemistry and Molecular Biology of Plants*, eds B. B. Buchanan, W. Gruissem, and R. L. Jones (Rockville, MD: American Society of Plant Physiologists), 1158–1203.
11. Chen, H., Lai, Z., Shi, J., Xiao, Y., Chen, Z., and Xu, X. (2010). Roles of *Arabidopsis* WRKY18, WRKY40 and WRKY60 transcription factors in plant responses to abscisic acid and abiotic stress. *BMC Plant Biol.* 10:281. doi: 10.1186/1471-2229-10-281
12. Cheng, M. C., Hsieh, E. J., Chen, J. H., Chen, H. Y., and Lin, T. P. (2012). *Arabidopsis* RGLG2, functioning as a RING E3 ligase, interacts with AtERF53 and negatively regulates the plant drought stress response. *Plant Physiol.* 158, 363–375. doi: 10.1104/pp.111.189738
13. Choi, H., Hong, J., Ha, J., Kang, J., and Kim, S. Y. (2000). ABFs, a family of ABA-responsive element binding factors. *J. Biol. Chem.* 275, 1723–1730. doi: 10.1074/jbc.275.3.1723
14. Cutler, S. R., Rodriguez, P. L., Finkelstein, R. R., and Abrams, S. R. (2010). Abscisic acid: emergence of a core signaling network. *Annu. Rev. Plant Biol.* 61, 651–679. doi: 10.1146/annurev-arplant-042809-112122
15. E Stecker, K., Minkoff, B. B., and Sussman, M. R. (2014). Phosphoproteomic analyses reveal early signaling events in the osmotic stress response. *Plant Physiol.* 165, 1171–1187. doi: 10.1104/pp.114.238816
16. Finkelstein, R. R., and Lynch, T. J. (2000). The *Arabidopsis abscisic* acid response gene ABI5 encodes a basic leucine zipper transcription factor. *Plant Cell* 12, 599–609. doi: 10.1105/tpc.12.4.599
17. Fujii, H., Verslues, P. E., and Zhu, J. K. (2007). Identification of

two protein kinases required for abscisic acid regulation of seed germination, root growth, and gene expression in *Arabidopsis*. *Plant Cell* 19, 485–494. doi: 10.1105/tpc.106.048538

18. Fujii, H., and Zhu, J. K. (2009). *Arabidopsis* mutant deficient in 3 abscisic acid-activated protein kinases reveals critical roles in growth, reproduction, and stress. *Proc. Natl. Acad. Sci. U.S.A.* 106. 8380–8385. doi: 10.1073/pnas.0903144106
19. Fujita, M., Fujita, Y., Maruyama, K., Seki, M., Hiratsu, K., Ohme-Takagi, M., et al. (2004). A dehydration-induced NAC protein, RD26, is involved in a novel ABA-dependent stress-signaling pathway. *Plant J.* 39, 863–876. doi: 10.1111/j.1365-313X.2004.02171.x
20. Fujita, Y., Fujita, M., Satoh, R., Maruyama, K., Parvez, M. M., Seki, M., et al. (2005). AREB1 is a transcription activator of novel ABRE-dependent ABA signaling that enhances drought stress tolerance in *Arabidopsis*. *Plant Cell* 17, 3470–3488. doi: 10.1105/tpc.105.035659
21. Fujita, Y., Fujita, M., Shinozaki, K., and Yamaguchi-Shinozaki, K. (2011). ABA-mediated transcriptional regulation in response to osmotic stress in plants. *J. Plant Res.* 124, 509–525. doi: 10.1007/s10265-011-0412-3
22. Fujita, Y., Nakashima, K., Yoshida, T., Katagiri, T., Kidokoro, S., Kanamori, N., et al. (2009). Three SnRK2 protein kinases are the main positive regulators of abscisic acid signaling in response to water stress in *Arabidopsis*. *Plant Cell Physiol.* 50, 2123–2132. doi: 10.1093/pcp/pcp147
23. Fujita, Y., Yoshida, T., and Yamaguchi-Shinozaki, K. (2013). Pivotal role of the AREB/ABF-SnRK2 pathway in ABRE-mediated transcription in response to osmotic stress in plants. *Physiol. Plant.* 147, 15–27. doi: 10.1111/j.1399-3054.2012.01635.x
24. Furihata, T., Maruyama, K., Fujita, Y., Umezawa, T., Yoshida, R., Shinozaki, K., et al. (2006). Abscisic acid-dependent multisite phosphorylation regulates the activity of a transcription activator AREB1. *Proc. Natl. Acad. Sci. U.S.A.* 103, 1988–1993. doi: 10.1073/pnas.0505667103
25. Gilmour, S. J., Sebolt, A. M., Salazar, M. P., Everard, J. D., and Thomashow, M. F. (2000). Overexpression of the *Arabidopsis* CBF3

transcriptional activator mimics multiple biochemical changes associated with cold acclimation. *Plant Physiol.* 124, 1854–1865. doi: 10.1104/pp.124.4.1854

26. Haake, V., Cook, D., Riechmann, J. L., Pineda, O., Thomashow, M. F., and Zhang, J. Z. (2002). Transcription factor CBF4 is a regulator of drought adaptation in *Arabidopsis*. *Plant Physiol.* 130, 639–648. doi: 10.1104/pp.006478
27. Hirayama, T., and Shinozaki, K. (2010). Research on plant abiotic stress responses in the post-genome era: past, present and future. *Plant J.* 61, 1041–1052. doi: 10.1111/j.1365-313X.2010.04124.x
28. Hsieh, E. J., Cheng, M. C., and Lin, T. P. (2013). Functional characterization of an abiotic stress-inducible transcription factor AtERF53 in *Arabidopsis thaliana*. *Plant Mol. Biol.* 82, 223–237. doi: 10.1007/s11103-013-0054-z
29. Ingram, J., and Bartels, D. (1996). The molecular basis of dehydration tolerance in plants. *Annu. Rev. Plant Physiol. Plant Mol. Biol.* 47, 377--403. doi: 10.1146/annurev.arplant.47.1.377
30. Iuchi, S., Kobayashi, M., Taji, T., Naramoto, M., Seki, M., Kato, T., et al. (2001). Regulation of drought tolerance by gene manipulation of 9-cis-epoxycarotenoid dioxygenase, a key enzyme in abscisic acid biosynthesis in *Arabidopsis*. *Plant J.* 27, 325–333. doi: 10.1046/j.1365-313x.2001.01096.x
31. Jensen, M. K., Kjaersgaard, T., Nielsen, M. M., Galberg, P., Petersen, K., O'Shea, C., et al. (2010). The *Arabidopsis thaliana* NAC transcription factor family: structure-function relationships and determinants of ANAC019 stress signalling. *Biochem. J.* 426, 183–196. doi: 10.1042/BJ20091234
32. Jensen, M. K., Lindemose, S., de Masi, F., Reimer, J. J., Nielsen, M., Perera, V., et al. (2013). ATAF1 transcription factor directly regulates abscisic acid biosynthetic gene NCED3 in *Arabidopsis thaliana*. *FEBS Open. Biol.* 3, 21–7. doi: 10.1016/j.fob.2013.07.006
33. Jiang, C., Iu, B., and Singh, J. (1996). Requirement of a CCGAC cis-acting element for cold induction of the BN115 gene from winter *Brassica napus*. *Plant Mol. Biol.* 30, 679–684. doi: 10.1007/BF00049344

34. Kang, J. Y., Choi, H. I., Im, M. Y., and Kim, S. Y. (2002). *Arabidopsis* basic leucine zipper proteins that mediate stress-responsive abscisic acid signaling. *Plant Cell* 14, 343–357. doi: 10.1105/tpc.010362

35. Kasuga, M., Liu, Q., Miura, S., Yamaguchi-Shinozaki, K., and Shinozaki, K. (1999). Improving plant drought, salt, and freezing tolerance by gene transfer of a single stress-inducible transcription factor. *Nat. Biotechnol.* 17, 287–291. doi: 10.1038/7036

36. Kazan, K., and Manners, J. M. (2013). MYC2: the master in action. *Mol. Plant.* 6, 686–703. doi: 10.1093/mp/sss128

37. Kim, J. H., Choi, D., and Kende, H. (2003). The AtGRF family of putative transcription factors is involved in leaf and cotyledon growth in *Arabidopsis*. *Plant J.* 36, 94–104. doi: 10.1046/j.1365-313X.2003.01862.x

38. Kim, J. S., Mizoi, J., Kidokoro, S., Maruyama, K., Nakajima, J., Nakashima, K., et al. (2012a). *Arabidopsis* growth-regulating factor7 functions as a transcriptional repressor of abscisic acid- and osmotic stress-responsive genes, including DREB2A. *Plant Cell* 24, 3393–3405. doi: 10.1105/tpc.112.100933

39. Kim, M. J., Park, M. J., Seo, P. J., Song, J. S., Kim, H. J., and Park, C. M. (2012b). Controlled nuclear import of the transcription factor NTL6 reveals a cytoplasmic role of SnRK2.8 in the drought-stress response. *Biochem. J.* 448, 353–363. doi: 10.1042/BJ20120244

40. Kim, J. S., Mizoi, J., Yoshida, T., Fujita, Y., Nakajima, J., Ohori, T., et al. (2011). An ABRE promoter sequence is involved in osmotic stress-responsive expression of the DREB2A gene, which encodes a transcription factor regulating drought-inducible genes in *Arabidopsis*. *Plant Cell. Physiol.* 52, 2136–2146. doi: 10.1093/pcp/pcr143

41. Krannich, C. T., Maletzki, L., Kurowsky, C., and Horn, R. (2015). Network candidate genes in breeding for drought tolerant crops. *Int. J. Mol. Sci.* 16, 16378–16400. doi: 10.3390/ijms160716378

42. Krasensky, J., and Jonak, C. (2012). Drought, salt, and temperature stress-induced metabolic rearrangements and regulatory networks. *J. Exp. Bot.* 63, 1593–1608. doi: 10.1093/jxb/err460

43. Laloum, T., De Mita, S., Gamas, P., Baudin, M., and Niebel, A. (2013). CCAAT-box binding transcription factors in plants: Y so many?

*Trends Plant Sci.* 18, 157–166. doi: 10.1016/j.tplants.2012.07.004

44. Lee, S., and Park, C. M. (2012). Regulation of reactive oxygen species generation under drought conditions in *Arabidopsis*. *Plant Signal. Behav.* 7, 599–601. doi: 10.4161/psb.19940
45. Lee, S. J., Kang, J. Y., Park, H. J., Kim, M. D., Bae, M. S., Choi, H. I., et al. (2010). DREB2C interacts with ABF2, a bZIP protein regulating abscisic acid-responsive gene expression, and its overexpression affects abscisic acid sensitivity. *Plant Physiol.* 153, 716–727. doi: 10.1104/pp.110.154617
46. Li, W. X., Oono, Y., Zhu, J., He, X. J., Wu, J. M., Iida, K., et al. (2008). The *Arabidopsis* NFYA5 transcription factor is regulated transcriptionally and posttranscriptionally to promote drought resistance. *Plant Cell* 20, 2238–2251. doi: 10.1105/tpc.108.059444
47. Lim, C. J., Hwang, J. E., Chen, H., Hong, J. K., Yang, K. A., Choi, M. S., et al. (2007). Over-expression of the *Arabidopsis* DRE/CRT-binding transcription factor DREB2C enhances thermotolerance. *Biochem. Biophys. Res. Commun.* 362, 431–436. doi: 10.1016/j.bbrc.2007.08.007
48. Lindemose, S., O'Shea, C., Jensen, M. K., and Skriver, K. (2013). Structure, function and networks of transcription factors involved in abiotic stress responses. *Int. J. Mol. Sci.* 14, 5842–5878. doi: 10.3390/ijms14035842
49. Liu, Q., Kasuga, M., Sakuma, Y., Abe, H., Miura, S., Yamaguchi-Shinozaki, K., et al. (1998). Two transcription factors, DREB1 and DREB2, with an EREBP/AP2 DNA binding domain separate two cellular signal transduction pathways in drought- and low-temperature-responsive gene expression, respectively, in *Arabidopsis*. *Plant Cell* 10, 1391–1406. doi: 10.1105/tpc.10.8.1391
50. Lopez-Molina, L., and Chua, N. H. (2000). A null mutation in a bZIP factor confers ABA-insensitivity in *Arabidopsis thaliana*. *Plant Cell Physiol.* 41, 541–547. doi: 10.1093/pcp/41.5.541
51. Magome, H., Yamaguchi, S., Hanada, A., Kamiya, Y., and Oda, K. (2004). Dwarf and delayed-flowering 1, a novel *Arabidopsis mutant* deficient in gibberellin biosynthesis because of overexpression of a putative AP2 transcription factor. *Plant J.* 37, 720–729. doi: 10.1111/j.1365-313X.2003.01998.x

52. Maruyama, K., Takeda, M., Kidokoro, S., Yamada, K., Sakuma, Y., Urano, K., et al. (2009). Metabolic pathways involved in cold acclimation identified by integrated analysis of metabolites and transcripts regulated by DREB1A and DREB2A. *Plant Physiol.* 150, 1972–1980. doi: 10.1104/pp.109.135327

53. Mizoi, J., Shinozaki, K., and Yamaguchi-Shinozaki, K. (2012). AP2/ERF family transcription factors in plant abiotic stress responses. *Biochim. Biophys. Acta* 1819, 86–96. doi: 10.1016/j.bbagrm.2011.08.004

54. Morimoto, K., Mizoi, J., Qin, F., Kim, J. S., Sato, H., Osakabe, Y., et al. (2013). Stabilization of *Arabidopsis* DREB2A is required but not sufficient for the induction of target genes under conditions of stress. *PLoS ONE* 8:e80457. doi: 10.1371/journal.pone.0080457

55. Nakashima, K., Ito, Y., and Yamaguchi-Shinozaki, K. (2009). Transcriptional regulatory networks in response to abiotic stresses in *Arabidopsis* and grasses. *Plant Physiol.* 149, 88–95. doi: 10.1104/pp.108.129791

56. Nakashima, K., Kiyosue, T., Yamaguchi-Shinozaki, K., and Shinozaki, K. (1997). A nuclear gene, erd1, encoding a chloroplast-targeted Clp protease regulatory subunit homolog is not only induced by water stress but also developmentally up-regulated during senescence in *Arabidopsis thaliana. Plant J.* 12, 851–861. doi: 10.1046/j.1365-313X.1997.12040851.x

57. Nakashima, K., Takasaki, H., Mizoi, J., Shinozaki, K., and Yamaguchi-Shinozaki, K. (2012). NAC transcription factors in plant abiotic stress responses. *Biochim. Biophys. Acta* 1819, 97–103. doi: 10.1016/j.bbagrm.2011.10.005

58. Nakashima, K., Yamaguchi-Shinozaki, K., and Shinozaki, K. (2014). The transcriptional regulatory network in the drought response and its crosstalk in abiotic stress responses including drought, cold, and heat. *Front. Plant Sci.* 5:170. doi: 10.3389/fpls.2014.00170

59. Narusaka, Y., Nakashima, K., Shinwari, Z. K., Sakuma, Y., Furihata, T., Abe, H., et al. (2003). Interaction between two cis-acting elements, ABRE and DRE, in ABA-dependent expression of *Arabidopsis* rd29A gene in response to dehydration and high-salinity stresses. *Plant J.* 34, 137–148. doi: 10.1046/j.1365-313X.2003.01708.x

60. Nuruzzaman, M., Sharoni, A. M., and Kikuchi, S. (2013). Roles of NAC transcription factors in the regulation of biotic and abiotic stress responses in plants. *Front. Microbiol.* 4:248. doi: 10.3389/fmicb.2013.00248
61. Olsen, A. N., Ernst, H. A., Leggio, L. L., and Skriver, K. (2005). NAC transcription factors: structurally distinct, functionally diverse. *Trends Plant Sci.* 10, 79–87. doi: 10.1016/j.tplants.2004.12.010
62. Petroni, K., Kumimoto, R. W., Gnesutta, N., Calvenzani, V., Fornari, M., Tonelli, C., et al. (2012). The promiscuous life of plant NUCLEAR FACTOR Y transcription factors. *Plant Cell* 24, 4777–4792. doi: 10.1105/tpc.112.105734
63. Qin, F., Sakuma, Y., Tran, L. S., Maruyama, K., Kidokoro, S., Fujita, Y., et al. (2008). *Arabidopsis* DREB2A-interacting proteins function as RING E3 ligases and negatively regulate plant drought stress-responsive gene expression. *Plant Cell* 20, 1693–1707. doi: 10.1105/tpc.107.057380
64. Quach, T. N., Nguyen, H. T., Valliyodan, B., Joshi, T., Xu, D., and Nguyen, H. T. (2015). Genome-wide expression analysis of soybean NF-Y genes reveals potential function in development and drought response. *Mol. Genet. Genomics* 290, 1095–1115. doi: 10.1007/s00438-014-0978-2
65. Ren, X., Chen, Z., Liu, Y., Zhang, H., Zhang, M., Liu, Q., et al. (2010). ABO3, a WRKY transcription factor, mediates plant responses to abscisic acid and drought tolerance in *Arabidopsis*. *Plant J.* 63, 417–429. doi: 10.1111/j.1365-313X.2010.04248.x
66. Rushton, P. J., Somssich, I. E., Ringler, P., and Shen, Q. J. (2010). WRKY transcription factors. *Trends Plant Sci.* 15, 247–258. doi: 10.1016/j.tplants.2010.02.006
67. Sakuma, Y., Liu, Q., Dubouzet, J. G., Abe, H., Shinozaki, K., and Yamaguchi-Shinozaki, K. (2002). DNA-binding specificity of the ERF/AP2 domain of *Arabidopsis* DREBs, transcription factors involved in dehydration- and cold-inducible gene expression. *Biochem. Biophys. Res. Commun.* 290, 998–1009. doi: 10.1006/bbrc.2001.6299
68. Sakuma, Y., Maruyama, K., Osakabe, Y., Qin, F., Seki, M.,

Shinozaki, K., et al. (2006a). Functional analysis of an *Arabidopsis* transcription factor, DREB2A, involved in drought-responsive gene expression. *Plant Cell* 18, 1292–1309. doi: 10.1105/tpc.105.035881

69. Sakuma, Y., Maruyama, K., Qin, F., Osakabe, Y., Shinozaki, K., and Yamaguchi-Shinozaki, K. (2006b). Dual function of an *Arabidopsis* transcription factor DREB2A in water-stress-responsive and heat-stress-responsive gene expression. *Proc. Natl. Acad. Sci. U.S.A.* 103, 18822–18827. doi: 10.1073/pnas.0605639103
70. Sakuraba, Y., Kim, Y. S., Han, S. H., Lee, B. D., and Paek, N. C. (2015). The *Arabidopsis* transcription factor NAC016 promotes drought stress responses by repressing AREB1 transcription through a trifurcate feed-forward regulatory loop involving NAP. *Plant Cell* 27, 1771–1787. doi: 10.1105/tpc.15.00222
71. Shaar-Moshe, L., Hübner, S., and Peleg, Z. (2015). Identification of conserved drought-adaptive genes using a cross-species meta-analysis approach. *BMC Plant Biol.* 15:111. doi: 10.1186/s12870-015-0493-6
72. Shang, Y., Yan, L., Liu, Z. Q., Cao, Z., Mei, C., Xin, Q., et al. (2010). The Mg-chelatase H subunit of *Arabidopsis* antagonizes a group of WRKY transcription repressors to relieve ABA-responsive genes of inhibition. *Plant Cell* 22, 1909–1935. doi: 10.1105/tpc.110.073874
73. Shen, Q., Zhang, P., and Ho, T. H. (1996). Modular nature of abscisic acid (ABA) response complexes: composite promoter units that are necessary and sufficient for ABA induction of gene expression in barley. *Plant Cell* 8, 1107–1119. doi: 10.1105/tpc.8.7.1107
74. Shinozaki, K., and Yamaguchi-Shinozaki, K. (2000). Molecular responses to dehydration and low temperature: differences and cross-talk between two stress signaling pathways. *Curr. Opin. Plant Biol.* 3, 217–223. doi: 10.1016/S1369-5266(00)00067-4
75. Simpson, S. D., Nakashima, K., Narusaka, Y., Seki, M., Shinozaki, K., and Yamaguchi-Shinozaki, K. (2003). Two different novel cis-acting elements of erd1, a clpA homologous *Arabidopsis*gene function in induction by dehydration stress and dark-induced senescence. *Plant J.* 33, 259–270. doi: 10.1046/j.1365-313X.2003.01624.x
76. Thomashow, M. F. (1999). PLANT COLD ACCLIMATION: freezing tolerance genes and regulatory mechanisms. *Annu. Rev.*

*Plant Physiol. Plant Mol. Biol.* 50, 571–599. doi: 10.1146/annurev.arplant.50.1.571

77. Todaka, D., Shinozaki, K., and Yamaguchi-Shinozaki, K. (2015). Recent advances in the dissection of drought-stress regulatory networks and strategies for development of drought-tolerant transgenic rice plants. *Front. Plant Sci.* 18:684. doi: 10.3389/fpls.2015.00084
78. Tran, L. S., Nakashima, K., Sakuma, Y., Osakabe, Y., Qin, F., Simpson, S. D., et al. (2007). Co-expression of the stress-inducible zinc finger homeodomain ZFHD1 and NAC transcription factors enhances expression of the ERD1 gene in *Arabidopsis*. *Plant J.* 49, 46–63. doi: 10.1111/j.1365-313X.2006.02932.x
79. Tran, L. S., Nakashima, K., Sakuma, Y., Simpson, S. D., Fujita, Y., Maruyama, K., et al. (2004). Isolation and functional analysis of *Arabidopsis* stress-inducible NAC transcription factors that bind to a drought-responsive cis-element in the early responsive to dehydration stress 1 promoter. *Plant Cell* 16, 2481–2498. doi: 10.1105/tpc.104.022699
80. Ulker, B., and Somssich, I. E. (2004). WRKY transcription factors: from DNA binding towards biological function. *Curr. Opin. Plant Biol.* 7, 491–498. doi: 10.1016/j.pbi.2004.07.012
81. Umezawa, T., Nakashima, K., Miyakawa, T., Kuromori, T., Tanokura, M., Shinozaki, K., et al. (2010). Molecular basis of the core regulatory network in ABA responses: sensing, signaling and transport. *Plant Cell Physiol.* 51, 1821–1839. doi: 10.1093/pcp/pcq156
82. Uno, Y., Furihata, T., Abe, H., Yoshida, R., Shinozaki, K., and Yamaguchi-Shinozaki, K. (2000). *Arabidopsis* basic leucine zipper transcription factors involved in an abscisicacid-dependent signal transduction pathway under drought and high-salinity conditions. *Proc. Natl. Acad. Sci. U.S.A.* 97, 11632–11637. doi: 10.1073/pnas.190309197
83. Wang, R. S., Pandey, S., Li, S., Gookin, T. E., Zhao, Z., Albert, R., et al. (2011). Common and unique elements of the ABA-regulated transcriptome of *Arabidopsis* guard cells. *BMC Genomics* 12:216. doi: 10.1186/1471-2164-12-216

84. Wu, Y., Deng, Z., Lai, J., Zhang, Y., Yang, C., Yin, B., et al. (2009). Dual function of *Arabidopsis* ATAF1 in abiotic and biotic stress responses. *Cell Res.* 19, 1279–1290. doi: 10.1038/cr.2009.108
85. Xu, L., Lin, Z., Tao, Q., Liang, M., Zhao, G., Yin, X., et al. (2014). Multiple NUCLEAR FACTOR Y transcription factors respond to abiotic stress in *Brassica napus* L. *PLoS ONE* 30:e111354. doi: 10.1371/journal.pone.0111354
86. Xu, Z. Y., Kim, S. Y., Hyeon do, Y., Kim, D. H., Dong, T., Park, Y., et al. (2013). The *Arabidopsis* NAC transcription factor ANAC096 cooperates with bZIP-type transcription factors in dehydration and osmotic stress responses. *Plant Cell* 25, 4708–4724. doi: 10.1105/tpc.113.119099
87. Yamaguchi-Shinozaki, K., and Shinozaki, K. (1993). The plant hormone abscisic acid mediates the drought-induced expression but not the seed-specific expression of rd22, a gene responsive to dehydration stress in *Arabidopsis thaliana. Mol. Gen. Genet.* 238, 17–25.
88. Yamaguchi-Shinozaki, K., and Shinozaki, K. (1994). A novel cis-acting element in an *Arabidopsis* gene is involved in responsiveness to drought, low-temperature, or high-salt stress. *Plant Cell* 6, 251–264. doi: 10.1105/tpc.6.2.251
89. Yamaguchi-Shinozaki, K., and Shinozaki, K. (2005). Organization of cis-acting regulatory elements in osmotic- and cold-stress-responsive promoters. *Trends Plant Sci.* 10, 88–94. doi: 10.1016/j.tplants.2004.12.012
90. Yamaguchi-Shinozaki, K., and Shinozaki, K. (2006). Transcriptional regulatory networks in cellular responses and tolerance to dehydration and cold stresses. *Annu. Rev. Plant Biol.* 57, 781–803. doi: 10.1146/annurev.arplant.57.032905.105444
91. Yoshida, R., Umezawa, T., Mizoguchi, T., Takahashi, S., Takahashi, F., and Shinozaki, K. (2006). The regulatory domain of SRK2E/OST1/SnRK2.6 interacts with *ABI*1 and integrates abscisic acid (ABA) and osmotic stress signals controlling stomatal closure in *Arabidopsis. J. Biol. Chem.* 281, 5310–5318. doi: 10.1074/jbc.M509820200
92. Yoshida, T., Fujita, Y., Maruyama, K., Mogami, J., Todaka,

D., Shinozaki, K., et al. (2015). Four *Arabidopsis* AREB/ABF transcription factors function predominantly in gene expression downstream of SnRK2 kinases in abscisic acid signalling in response to osmotic stress. *Plant Cell Environ.* 38, 35–49. doi: 10.1111/pce.12351

93. Yoshida, T., Fujita, Y., Sayama, H., Kidokoro, S., Maruyama, K., Mizoi, J., et al. (2010). AREB1, AREB2, and ABF3 are master transcription factors that cooperatively regulate ABRE-dependent ABA signaling involved in drought stress tolerance and require ABA for full activation. *Plant J.* 61, 672–685. doi: 10.1111/j.1365-313X.2009.04092.xYoshida, T., Mogami, J., and Yamaguchi-Shinozaki, K. (2014). ABA-dependent and ABA-independent signaling in response to osmotic stress in plants. *Curr. Opin. Plant Biol.* 21, 133–139. doi: 10.1016/j.pbi.2014.07.009

94. Zhang, K., and Gan, S. S. (2012). An abscisic acid-AtNAP transcription factor-SAG113 protein phosphatase 2C regulatory chain for controlling dehydration in senescing *Arabidopsis leaves. Plant Physiol.* 158, 961–969. doi: 10.1104/pp.111.190876

CHAPTER

# 8

# NETWORK CANDIDATE GENES IN BREEDING FOR DROUGHT TOLERANT CROPS

Christoph Tim Krannich, Lisa Maletzki, Christina Kurowsky and Renate Horn

Institute of Biological Sciences, Department of Plant Genetics, University of Rostock, Albert-Einstein-Str. 3, 18059 Rostock, Germany

## ABSTRACT

Climate change leading to increased periods of low water availability as well as increasing demands for food in the coming years makes breeding for drought tolerant crops a high priority. Plants have developed diverse strategies and mechanisms to survive drought stress. However, most of these represent drought escape or avoidance strategies like early flowering or low stomatal conductance that are not applicable in breeding for crops with high yields under drought conditions. Even though a great deal of research is ongoing, especially in cereals, in this regard, not all mechanisms involved in drought tolerance are yet understood. The identification of candidate genes for drought tolerance that have a high potential to be used for breeding drought tolerant crops represents a challenge. Breeding for drought tolerant crops has to focus on acceptable yields under water-limited conditions and not on survival. However, as more and more knowledge about the complex networks and the cross talk during drought is available, more options are revealed. In addition, it has to be considered that conditioning a crop for drought tolerance might require the production of metabolites and might cost the plants energy and resources that cannot be used in terms of yield. Recent research indicates that yield penalty exists and efficient breeding for drought tolerant crops with acceptable yields under well-watered and drought conditions might require uncoupling yield penalty from drought tolerance.

**Keywords:** Candidate genes, breeding, yield penalty, drought tolerance, genetic variants

## INTRODUCTION

Producing enough food for the growing world population will be a challenge in the next 50 years [1]. Regarding abiotic stress factors, drought is probably the environmental factor that has worldwide the strongest negative impact on crop productivity [2]. Several factors will worsen the situation in the coming years [3]: (1) climate change will lead to differences in rain patterns all over the world with reduced rainfalls and more extensive drought periods, especially in subtropical areas; (2) continuous loss of arable land by soil erosion, urbanization and salinization of irrigated fields will reduce land available for agriculture; and (3) crop production, especially in the developing countries, is more and more relocated into more drought-prone areas with poor soil qualities. Even relatively mild droughts, such as experienced in the UK, can reduce yields obtained on rain-fed agriculture compared to maximal yields (under irrigated conditions or in years with more rain) by as much as 30%–60% [4].

Agricultural drought addresses the situation in which the water available to the plants through rainfall and/or irrigation is insufficient to meet the transpiration needs of the crop, which leads to yield reductions [5]. The crop yield gap (Yg) is defined as the difference between the yield potential Yp (irrigated systems) or water-limited yield potential Yw (rainfed) and the average actual farm yield (Ya) [6]. The agricultural drought will differ between crops because of two major factors (crop water demand and crop water supply) [7].

Drought tolerance is a very complex trait as it depends not only on the severity of the drought (mild or severe), but also on the developmental stage of the plant as well as the duration [8,9]. From the physiological point of view, survival (and/or recovery) is the major aim in plant stress tolerance, whereas from the agricultural point of view crop yield is the trait that determines a drought tolerant crop [10]. Different types of drought tolerance can be present within a crop like in wheat (*Triticum aestivum*). The genotype RAC875 and Excalibur demonstrated different strategies for drought tolerance under the same drought scenario [11]. RAC875 showed a more conservative strategy by producing fewer tillers, higher number of grains per tiller and by revealing a moderate osmotic adjustment. RAC875 plants stored more water-soluble carbohydrates and the leaves were more waxed and thicker, both under well-water and drought conditions. On the other hand, Excalibur showed a more responsive strategy by higher osmotic adjustment, low abscisic acid (ABA), high stomatal conductance and rapid recovery after stress. Leaves rolled under dry conditions, higher number of tillers were produced and aborted under water-limitations, but the loss was compensated by the production of more grains per tiller after recovery. Both strategies for drought tolerance would be interesting from the breeding perspective, especially as these two types may perform differently under other drought scenarios.

Plant hormones are involved in every aspect of plant development as well as in the reaction of plants to abiotic and biotic stress [12,13]. The overlap in hormone-regulated pathways and interactions indicate a complex network of extensive cross talk between the different hormone signaling pathways during adaptation to drought [12,14]. Downstream of the signaling pathways more than 50 families of transcription factors (TFs), e.g., ABA response element binding factors (AREBs), basic region/leucine zipper (bZIP) TFs, drought responsive element binding proteins (DREBs), ethylene response factors (ERFs), MYC/myeloblastosis (MYB) TFs, and no apical meristem/ATAF/CUP-shaped cotyledons (NACs) function as integrators of the complex network of drought regulated genes [15,16,17].

For the proof-of-concept as well as functional analysis, transgenic approaches have been successfully used for a better understanding of the networks underlying drought tolerance and for identifying candidate genes for breeding drought tolerant crops. However, in most cases only the survival (and/or recovery) under drought has been addressed, but not the yield aspect. Our knowledge about drought tolerance now needs to be applied to produce drought tolerant crops with improved yields (and yield stability).

In this review, we will highlight some candidate genes out of the complex network engaged in drought tolerance, which involves biosynthesis of plant hormones and their signaling pathways, osmoprotective strategies as well as the detoxification of reactive oxygen species (ROS), that seem to be most promising to be analyzed with regard to the genetic variation in drought tolerant and drought sensitive plant material.

## ROLE OF PLANT HORMONES IN DROUGHT RESPONSE

### Abscisic Acid (ABA) Biosynthesis, Catabolism and Signaling Pathways Are Major Players under Drought Conditions

Abscisic acid (ABA) plays an important role in adaptive responses to environmental stresses such as drought, but also in developmental processes such as seed maturation, dormancy and senescence [18,19]. Genes involved in ABA biosynthesis, catabolism and signaling network represent interesting candidate genes for breeding crops improved in drought tolerance.

ABA synthesis starts in the plastids with zeaxanthin, which is a product of the 2-*C*-methyl-D-erythritol-4-phosphate (MEP) pathway [20]. The enzyme zeaxanthin epoxidase (ZEP) converts the $C_{40}$ carotenoid zeaxanthin via the intermediate antheraxanthin to all-trans-violaxanthin [18]. The reverse reaction is catalyzed by the violaxanthin de-epoxidase (VDE). This so-called xanthophyll cycle is supposed to be relevant for the protection of photosystems from photooxidative damage under intense light conditions [21,22]. In the following, all-trans-violaxanthin can be transformed to either 9-*cis*-violaxanthin or 9′-violaxanthin by a yet unknown enzyme. The rate-limiting step of ABA biosynthesis is the formation of xanthoxin via oxidative cleavage from either of these precursors by the enzyme 9-*cis*-epoxycarotenoid dioxygenase (NCED). Overexpression of NCED leads to higher levels of ABA, a reduction of the transpiration rate in leaves and thus enhances drought tolerance [23,24]. But also overexpression of *ZEP* in *Arabidopsis thaliana* conferred greater tolerance indicating that this enzyme might be limiting with regard to some stress response [25]. Xanthoxin is transported into the cytosol were it is converted to abscisic aldehyde by one of the short chain dehydrogenase/reductase (SDR) family, e.g., abscisic acid 2 (*ABA2*) in *Arabidopsis* [26]. The last step in the synthesis of ABA is catalyzed by abscisic aldehyde oxidase (AAO). This AAO requires a sulphurylated form of a molybdenum cofactor (MoCo) for its activity [27,28]. The MoCo sulfurase is encoded by low expression of osmotically responsive genes (*LOS5*)/abscisic acid 3 (*ABA3*) gene [27,29]. Overexpression of the molybdenum cofactor sulfurase gene enhanced drought tolerance in soybean [30] by increasing ABA accumulation, reducing water loss through reduced opening of the stomata and increased activities of superoxide dismutase (SOD), peroxidase (POD) and catalase (CAT). This overexpression resulted in considerably higher yields in form of more shoot and root biomass as well as more pods under water limited conditions. Yields were comparable to the wild type under well-watered conditions. Also for rice (*Oryza sativa*) [31] and for maize (*Zea mays*) [32], enhanced drought tolerance combined with improved yield under drought stress was observed in transgenic plants overexpressing *LOS5/ABA3*. Two catabolic pathways, hydroxylation predominantly by ABA 8′-hydroxylase and conjugation by ADP glucosyltransferase, are known for ABA [33,34]. ABA levels are maintained through the balance between *de novo*biosynthesis and catabolism, rather than only biosynthesis [35]. Drought leads to upregulation of *CYP707A* genes encoding for ABA 8′-hydroxylases, which is further increased upon rehydration. The catabolic products like ABA glucosyl ester and phaseic acid are stored in the vacuole or the apoplast pool. Under drought conditions, the enzyme β-glucosidase can cleave the conjugated ABA and release ABA, which can then be transported to the guard cell to induce stomatal closure [36].

The ABA signal transduction pathway has been well analyzed using *A. thaliana* as model plant [20,37,38]. The signaling pathway comprises three major components: a group of ABA receptors in the cytosol that binds to protein phosphatase 2C (PP2C) in the presence of ABA and (sucrose non-fermenting) SNF1-related protein kinase 2 (SnRK2). The ABA receptors represent a complex of pyrabactin resistance (PYR), PYR-like (PYL) and regulatory component of ABA receptor (RCAR). All components of the complex are encoded by multigene families, which results in redundancy in function [19,39]. Variation in one of the genes might not exhibit a great effect and thereby reduces the chance to use variation in these genes to breed for drought tolerance. Quadruple mutants had to be applied to demonstrate the role of the receptors. In the presence of ABA, PP2C is bound by the ABA receptor PYR/PYL/RCAR, which frees SnRK2 to be phosphorylated. PP2C like ABA insensitive 1 (ABI1), ABI2 or other similar ones belong to group A of the protein phosphatase gene family [40]. SnRKs are serine/threonine protein kinases, which are divided in three groups: SnRK1, SnRK2 and SnRK3 [41]. The SnRK2 family represents the key regulator of plant response to abiotic stress. Despite the multigene families involved, overexpression of a single *PP2C* gene, e.g., *OsPP108* from rice, leads to low sensitivity to ABA and high drought tolerance in *Arabidopsis* [42]. Also constitutive overexpression of PYL5 in rice enhanced drought tolerance [43]. However, this was accompanied with deleterious effects on total seed yield. Overexpression of PYL1/RCAR, PYL5/RCAR8 or PYL8/RCAR3 gave more tolerance to drought in *Arabidopsis* [20]. Table 1 summarizes the potential candidate genes from the ABA biosynthesis, catabolism and signaling pathway that might be evaluated in breeding for drought tolerant plants.

ABA plays the key role in plant adaptation to water-limited conditions. The fastest reaction of the plants under drought conditions is an increase in ABA level, which triggers the expression of ABA-responsive genes and induces stomatal closure [12,44]. As a consequence water loss via transpiration is reduced. Overexpression of enzymes of the biosynthesis pathway, e.g., ZEP or NCED leads to an increased ABA level [23,25], which improves drought tolerance by reducing transpiration. Growth regulation under water-limited conditions is also primarily mediated by ABA and ethylene [14,45]. In addition, XERICO (Greek for "drought tolerant"), a RING (really interesting new gene)-H2 zinc transcription factor, functions as a node between abiotic stress response and development by linking gibberellic acid (GA) and the ABA signaling pathway [46,47]. The effect of different ABA levels on yields under well-watered and drought conditions have still to be studied [14]. In addition, a better understanding of the cross talk with other plant hormone signaling pathways might offer the chance to improve breeding for drought tolerant crops.

**Table 1.** Candidate genes of the abscisic acid (ABA) biosynthesis, catabolism and signaling

| Gene | Gene Function | Abiotic Stress Tolerance | Reference |
|---|---|---|---|
| Zeaxanthin epoxidase (*ZEP*) | ABA biosynthesis | Improved tolerance to drought stress and salt | [22,25] |
| 9- *cis*-Epoxycarotenoid dioxygenase (*NCED*) | ABA biosynthesis | Improved drought tolerance by overexpression in tomato | [23,48] |
| Molybdenum cofactor sulfurase (*LOS5*/*ABA3*) | ABA biosynthesis | Increased drought tolerance in maize and soybean combined with higher yields | [30,32] |
| ABA 8′-hydroxylase | ABA catabolism | Upregulated during drought stress and rehydration: reduction in ABA level | [35] |
| Abscisic acid receptor (*PYL*) | ABA signaling network | Enhanced drought and salinity tolerance | [19,43] |
| Protein phosphatase 2C (*PP2C*) | ABA signaling network | Induced by drought, repressed by oxidative stress and heat shock | [49,50] |
| Serine/threonine protein kinase (*SnRK2*) | ABA signaling network | Enhanced tolerance to drought, salt, and freezing stress | [51,52] |
| SCARECROW (*SCR*) | Transcription factor, structural root differentiation | Binding to regulatory regions of stress-responsive genes; Regulating abscisic acid responses in *Arabidopsis* | [51] |

## Role of Ethylene Biosynthesis, Catabolism and Signaling Pathway for Drought Tolerance

The gaseous plant hormone ethylene has been reported to be involved in a large variety of developmental processes, including seed germination and ripening, as well as abiotic or biotic stress responses, e.g., stomatal closure. Both ethylene biosynthesis and signaling are important components of plant responses to different abiotic stresses such as drought and salt. As there are numerous genes involved in ethylene synthesis and signaling, it is important to identify candidate genes, that are significantly correlated to stress responses and therefore useful for breeding drought tolerant crops (Table 2).

**Table 2.** Candidate genes for ethylene synthesis and signaling

| Gene | Gene Function | Abiotic Stress Tolerance | Reference |
|---|---|---|---|
| 1-Aminocyclopropane-1-carboxylic acid oxidase (*ACO*) | Ethylene biosynthesis | Induced expression during wilting in tea plants promotes decreased metabolism and energy consumption by shedding of older leaves | [53] |
| 1-Aminocyclopropane-1-carboxylic acid synthase (*ACS*) | Ethylene biosynthesis | Delayed senescence in absence of *ACS* in maize | [54] |
| 1-Aminocyclopropane-1-carboxylic acid synthase 7 (*ACS7*) | Ethylene biosynthesis | *ACS7*-knock-out promotes tolerance to salt, osmotic and heat stresses | [55] |
| Ethylene-overproducer 1 (*ETO1*) | Limitation of ACS in ethylene biosynthesis | *ETO1*-deficient mutants in *Arabidopsis* exhibited delayed stomatal closure in response to drought | [56] |
| Ethylene overproducer 1-like (*ETOL1*) | Ethylene biosynthesis and energy metabolism | *ETOL1*-overexpressing rice more tolerant to drought and submergence | [57] |
| Ethylene response 1 (*ETR1*) | Ethylene signaling | Cross-link between ethylene and $H_2O_2$ signal transduction in*Arabidopsis*, stomatal closure; inhibition of seed germination under salt stress | [58,59] |
| Ethylene response 2 (*ETR2*) | Ethylene signaling | *ETR2* promotes seed germination under NaCl stress in *Arabidopsis* | [59] |

| | | | |
|---|---|---|---|
| Ethylene insensitive 2 (*EIN2*) | Ethylene signaling | *EIN2* downregulated in *Arabidopsis* during salt and osmotic stress, *EIN2*-deficient mutants sensitive to salt and osmotic stress | [60] |
| Ethylene insensitive 3 (*EIN3*) | Downstream element of ethylene signaling pathway | *EIN3*-deficient mutants in *Arabidopsis* exhibited oxidative stress after exposure to salt stress | [61] |
| Ethylene response factor 1 (*ERF1*) | Jasmonate and ethylene signaling | *ERF1*-overexpressing *Arabidopsis* more tolerant to drought and salt stress | [62] |
| Ethylene responsive element binding protein 1 (*EREBP*) | Connection between different stress signal transduction pathways | Expression induced in oil palm fruits in response to abiotic stress (drought, cold and salinity) | [63] |
| Ethylene response factor 5 (*ERF5*) | Vegetative growth and plant development | *ERF5*-overexpressing tomatoes more tolerant to drought and salt stress | [64] |
| Ethylene response factor 6 (*ERF6*) | Cell proliferation and leaf growth | *ERF5*/*ERF6*-deficient *Arabidopsis* mutants less affected by stress | [65] |
| Jasmonate and ethylene response factor 1 (*JERF1*) | Activation of stress responsive genes, proline synthesis and ABA biosynthesis key enzymes | *JERF1*-overexpressing rice more tolerant to drought stress | [66] |

Two main groups of enzymes, comprising 1-aminocyclopropane 1-carboxylic acid (ACC) synthases (ACSs) and ACC oxidases (ACOs) are indispensable for ethylene biosynthesis. While ACSs are responsible for converting *S*-adenosyl methionine (SAM) to ACC, the family of ACOs catalyzes the conversion of ACC to ethylene [67]. The conversion of SAM to ACC represents the rate-limiting step that is catalyzed by different isoforms of ACS. In addition, 1-malonyl aminocyclopropane-1-carboxylic acid (MACC) transferase (MACCT) is considered to be a sink for ethylene, as its activity converts ACC into the non-volatile MACC [68]. In tea (*Camellia sinensis*) plants, a gene of the ACO-group is connected with drought tolerance [53]. At the wilting stage the ACO expression increased, possibly in order to decrease energy consumption and metabolism through shedding of older leaves. In addition, the frequently observed involvement of ethylene in leaf senescence can also be assigned to another biosynthetic enzyme, in this case a member of the ACS family. In *Arabidopsis*, ACS7-deficient mutants exhibited increased tolerance to salt stress [55]. On the molecular level, ACS7 loss-of-function mutants had a higher expression of several salt responsive genes including sodium/hydrogen exchanger 1 (*NHX1*), which encodes a tonoplast sodium/proton antiporter [69]. In ACS6-deficient maize mutants leaf senescence under normal conditions was delayed and drought-induced senescence was inhibited [54]. Furthermore it was demonstrated, that the leaves maintained physiological and biochemical function under drought stress. In fact ACS6-deficient plants even exhibited a significantly higher $CO_2$ assimilation rate, which appears to be a major advantage in terms of photosynthetic activity under stress conditions as a substantial factor for crop yield.

In ethylene biosynthesis ACS is the rate-limiting enzyme and therefore is strictly regulated in order to maintain a distinct level of ethylene production. Ethylene overproducer 1 (ETO1) serves as a main component in ACS regulation through promoting the degradation of ACS by a proteasome dependent pathway [56]. ETO1-deficient mutants in *Arabidopsis*however exhibited distinct responses upon drought stress. Ethylene interferes with ABA mediated closure of stomata. Thereby mutants showed a delayed stomatal closure, resulting in a higher degree of water loss by transpiration during drought [70]. Furthermore, an ethylene overproducer 1-like (ETOL1) protein, with a homologous function to ETO1, has also been reported to benefit drought tolerance during distinct developmental stages in rice (*Oryza sativa*) [57].

The ethylene signaling pathway is more complex, due to the large number of genes that are involved and the multiple processes and responses that are regulated [39,71]. The signaling network includes the receptor complexes I (ETR1/ERS1) and II (ETR2/ERS2/EIN4), that are membrane associated and connected to a negative constitutive triple response 1 (CTR1) protein kinase. Complex I consists of ethylene response 1 (ETR1) and ethylene response sensor 1 (ERS1), while complex II has three components consisting of ETR2, ERS2 and ethylene insensitive 4 (EIN4). In absence of ethylene CTR1 negatively regulates the ethylene insensitive 2 (EIN2) protein. After binding of ethylene to one of the receptor subfamilies, a signal peptide (EIN2C) is cleaved from EIN2 and transported into the nucleus, where the ethylene insensitive 3 (EIN3)/ethylene insensitive like 1 (EIL1) dependent gene transcription is activated [71]. Different elements of this proposed ethylene signaling pathway were reported to be associated with stress responses after drought and salt stress.

ETR1 represents an important link between ethylene and $H_2O_2$ signaling, as *Arabidopsis* mutants deficient of ETR1 were insensitive to $H_2O_2$ and thereby lacking ethylene mediated stomatal closure [58]. Under salt stress conditions ETR1-deficient mutants in *Arabidopsis* germinated faster and with a shorter delay in the onset of germination [59]. In contrast, ETR2 inhibits ETR1 and stimulates seed germination under salt stress underlining the different receptor functions.

Moving downstream, the signaling pathway EIN2 facilitates the signal transduction from the cytosol into the nucleus. During salt and osmotic stress EIN2 has been reported to be downregulated in *Arabidopsis*, while EIN2-deficient mutants were highly sensitive to salt and osmotic stress and showed increased ABA levels [60]. This effect can possibly be considered as crosstalk between ethylene and ABA signaling that also integrates stress signals. After binding of ethylene to a receptor subfamily and EIN2-mediated transport of the signal into the nucleus, the downstream transcription factors such as EIN3 and other ethylene insensitive-like proteins (EILs) are activated [72].

Specifically, EIN3 is essential for ethylene signaling, as EIN3-deficient mutants exhibited severe cell growth inhibition as well as accelerated senescence. The expression of EIN3 during salt stress is induced, suggesting that EIN3 could be a key factor in salt stress signaling [61]. In addition, EIN3-deficient mutants in *Arabidopsis* exposed to salt stress also showed enhanced oxidative stress, underlining the important role in response to salt stress. Another class of ethylene dependent genes, that have received major attention, are ethylene responsive element binding factors (EREBFs), often also referred to as ethylene response factors (ERFs), which like EIN3 represent downstream components of the signaling pathway. The unique feature of this ERF family is a highly conserved DNA binding domain. Several ERFs were reported to participate in drought response in different plants [73,74].

Rice mutants overexpressing the response factor jasmonate and ethylene response factor 1 (JERF1), which encodes a tomato ERF-protein, exhibited enhanced drought response [66]. This effect is mediated by the expression of stress responsive genes, including key enzymes of proline-biosynthesis. Furthermore, also ABA-biosynthesis was activated by JERF1, which illustrates the high impact of a single ERF on diverse signaling and biosynthesis pathways. Another example of crosstalk mediated by a single ERF was observed in *Arabidopsis*. ERF1 represents a component of both jasmonate (JA) and ethylene signaling and is highly induced under drought and salt stress conditions [62]. The ERF1-overexpressing mutants showed improved tolerance to drought, salt and heat. As the stress induction was inhibited by ABA, ERF1 integrates signals of different plant hormones, resulting in multiple stress responses.

Concerning leaf growth inhibition under osmotic stress, the redundant transcription factors ERF5 and ERF6 were reported to be central components [65]. ERF5/ERF6-deficient mutants in *Arabidopsis* exhibited better growth under short and long term osmotic stress. ERF6 regulates gibberellin-biosynthesis (GA) through transcriptional induction of GA 2-oxidase 6 (GA2-OX6). Thus GA-breakdown is triggered, leading to stabilization of DELLA proteins, known to negatively influence growth. ERF5-overexpressing tomatoes (*Solanum lycopersicum*) were more tolerant to abiotic stresses including drought [64]. As the expression of ERF5 was also induced by ABA, this transcription factor could also depict another cross-link between ethylene and ABA-signaling. In oil palm (*Elaeis guineensis*) a new member of the ERF family, which appears to be closely related to several stress responses including drought, has been identified [63]. This ethylene responsive element binding protein (EREBP) was induced after the onset of different stresses including drought, salt and cold. Furthermore, increased expression has also been observed after treatment with different plant hormones including ethylene and ABA. These results suggest a possible cross-link between different stress response signaling pathways, mediated by a single transcription factor.

In summary, ethylene metabolism and signaling influence a large variety of stress responses in plants. Particularly the ethylene signaling pathway appears to be a major cross-link between ethylene and other plant hormones, e.g., ABA and GA. The reported interactions between ethylene and other plant hormones also benefit immediate stress responses such as stomatal closure as well as long term adaptations, e.g., improved growth under drought conditions. Especially candidate genes that are related to maintenance of growth under low water conditions during different developmental stages are agronomically relevant as they provide an efficient resource for crop improvement and counteract yield penalties.

# PROTECTION OF THE CELLS AGAINST OSMOLYTIC AND OXIDATIVE DAMAGES

## Osmolytes: Glycine Betaine, Proline and Trehalose

In order to protect the membrane-integrity and thereby an aqueous environment for organelles, plants possess various mechanisms to control water efflux and influx. As a part of controlling water movement, the biosynthesis of different substances that function as osmotically active solutes (osmolytes) is induced [75]. Especially during stress responses and acclimation osmolytes are essential for maintaining cellular functions as well as protein and enzyme activity. Different osmolytes in plants have been reported to be associated with drought and salt stress, including proline, glycine betaine and trehalose. Enzymes involved in the metabolism of these osmolytes represent potential candidates for breeding drought tolerant plants (Table 3).

**Table 3.** Candidate genes for osmolyte biosynthesis and cell protection

| Gene | Gene Function | Abiotic Stress Tolerance | Reference |
|---|---|---|---|
| Betaine aldehyde dehydrogenase (*BADH*) | Osmolyte biosynthesis | BADH from spinach in potato improved drought tolerance | [76] |
| $\Delta^1$-pyrroline-5-carboxylate synthetase (*P5CS*) | Osmolyte biosynthesis | Proline: most frequent osmolyte in water-stressed plants | [77] |

| | | | |
|---|---|---|---|
| Proline oxidase/dehydrogenase 1 (*PDH1*) | Osmolyte biosynthesis | Proline: most frequent osmolyte in water-stressed plants | [77] |
| $\Delta^1$-pyrroline-5-carboxylate reductase (*P5CR*) | Osmolyte biosynthesis | Proline: most frequent osmolyte in water-stressed plants; enhanced drought tolerance in soybean | [78] |
| Trehalose-phosphate synthase 1 (*TPS1*) | Osmolyte biosynthesis | Transgenic expression in potato improved drought tolerance | [79] |
| DNA-binding protein 4 | Transcription factor | Induced by drought | [51] |
| Dehydration-responsive element binding factor 1 (*DREB1B*) | Transcription factor | *Arabidopsis* gene in transgenic potato improved drought tolerance | [80] |

A transgenic potato (*Solanum tuberosum*) cultivar, containing a betaine aldehyde dehydrogenase (*BADH*) gene from spinach (*Spinacia oleracea*) under the control of a stress induced *Arabidopsis* promoter, has been reported to exhibit improved growth after induction of BADH by NaCl and drought stress [76].

Another highly relevant osmolyte in plants is the aforementioned proline, which accumulates under different stress conditions [81]. The role of the amino acid proline in plant cells is considered to be diverse and includes contributing to osmotic adjustment and stabilization of sub-cellular structures [82]. The enzyme $\Delta^1$-pyrroline-5-carboxylate synthetase (P5CS1) is a major component in proline biosynthesis. A recent study with P5CS1-deficient *Arabidopsis* mutants indicated that proline synthesis is required in order to maintain growth at low water availability [77]. In addition, the study also confirmed the necessity of proline-catabolism, as proline dehydrogenase 1 (PDH1)-deficient *Arabidopsis* mutants, with blocked proline catabolism, exhibited decreased root growth, fresh weight and dry weight. Furthermore additional components of the proline biosynthetic pathway have also been reported to be associated with stress responses. In transgenic soybean (*Glycine max*) with a $\Delta^1$-pyrroline-5-carboxylate reductase (*P5CR*) gene and the antisense construct from *Arabidopsis*, it was found that proline might enhance survival during drought stress [78]. The *P5CR* gene and antisense construct were manipulated using an inducible heat shock promoter (IHSP). In contrast to wild-type (WT) plants, transformed plants failed to survive drought stress at 37 °C, suggesting proline as an essential component of tolerance to drought.

Two transgenic potato lines, which expressed a trehalose-6-phosphate synthase (*TPS1*) gene from yeast (*Saccharomyces cerevisiae*) were found to be more effective in keeping water and acceptable levels of photosynthesis during drought compared to WT-plants [79]. However under optimal growth conditions the transgenic lines were less efficient in terms of $CO_2$ fixation than control plants, as the stomatal densities were reduced.

## Aquaporins

A subfamily of major intrinsic proteins (MIPs), referred to as aquaporins (AQPs), function as intercellular water channels and thereby enable rapid water and solute transport between cells (Table 4). Since the first AQP was discovered more than 20 years ago [83], a large variety of AQPs in different plants and tissues were reported. The family of AQPs can further be sub-classified into plasma membrane intrinsic proteins (PIPs), tonoplast intrinsic proteins (TIPs), nodulin-26 like intrinsic proteins (NIPs), small basic intrinsic proteins (SIPs) and x intrinsic proteins (XIPs) [84]. During water limiting conditions, with low water availability and increased transpiration, an effective water transport across membranes is indispensable for plants in order to maintain growth and favourable metabolic levels in an unfavourable environment.

In rice (*Oryza sativa*) plants grown in the presence of $NH_4$, expression levels of AQPs were elevated in roots, resulting in an increased root protoplast water permeability and root hydraulic conductivity [85]. Based on these findings root water uptake and drought resistance could be improved by efficient nitrogen supply.

**Table 4.** Candidate genes for aquaporins

| Gene | Gene Function | Abiotic Stress Tolerance | Reference |
|---|---|---|---|
| *PIP2;1* | Water transport | *PIP2;1* transcription down-regulated in *Arabidopsis*; *PIP2;1* degradation benefits drought tolerance | [86,87] |
| *GoPIP1* | Water transport | *GoPIP1* overexpressing *Arabidopsis* mutants more sensitive to drought | [88] |
| *NtPIP1;1*, *NtPIP12;1* | Water transport | Downregulation of *NtPIP1;1* and *NtPIP2;1* in tobacco after drought stress | [89] |
| *NtAQP1* | Water transport | *NtAQP1* transcription upregulated in tobacco after drought stress | [89] |
| *MaPIP1;1* | Water transport | *MaPIP1;1* transcription in banana upregulated during drought stress; *MaPIP1;1* overexpressing *Arabidopsis* mutants more tolerant to drought | [90] |

Furthermore, *Arabidopsis* mutants overexpressing a RING membrane-anchor E3 ubiquitin ligase homolog gene (*RmA1H1*), that causes degradation of the AQP *PIP2;1* and inhibits its transport, exhibited significantly increased drought tolerance compared to wild-type (WT) plants [91]. Consistent with these findings the transcription level of *PIP2;1* also decreased upon dehydration stress [86,87].

In addition, *Arabidopsis* mutants overexpressing the *GoPIP1* gene from the forage crop *Galega orientalis* were more sensitive to drought and exhibited lower leaf water content as well as faster water loss through leaves and reduced rosette weight [88]. These results suggest that, especially under abiotic stress conditions, AQPs are associated with drought sensitivity. In tobacco (*Nicotiana tabacum*) the transcript levels of two AQP genes (*NtPIP1;1*, *NtPIP2;1*) were significantly decreased after drought stress [89]. At the same time the transcription level of a third AQP gene (*NtAQP1*) was significantly increased after drought stress, which might indicate different functions of AQPs during exposure to drought. Another example of a high impact AQP gene was reported in banana plants (*Musa acuminata*), where transcript levels of *MaPIP1;1* were elevated during drought stress [90]. Furthermore *MaPIP1;1* overexpressing *Arabidopsis* mutants exhibited better growth, higher survival rates and reduced water loss rates than WT plants upon drought stress [90].

In conclusion, the functions of AQPs during responses to drought stress can be characterized as diverse. AQPs are able to facilitate both sensitivity as well as tolerance to drought, probably pointing to heterogeneous functions in plants. These different functions elucidate the importance of AQPs during abiotic stress responses and illustrate their potential value for breeding drought tolerant crops [92].

## Protection against Reactive Oxygen Species

Reactive oxygen species (ROS) in form of oxygen radicals or $H_2O_2$ are formed under drought conditions and lead to cell damage via oxidation or membrane damage. Candidate genes, which can lead to an enhanced protection against ROS during drought periods, are shown in an overview in Table 5. Detection of genetic variation in these genes within crops would allow an improved breeding for stress tolerance improving the conditions and survival of plants under suboptimal water conditions.

**Table 5.** Candidate genes for detoxification

| Gene | Gene Function | Abiotic Stress Tolerance | Reference |
|---|---|---|---|
| Aldehyde dehydrogenase family 7 member (*ALDH7*) | Oxygen radical detoxification | Extenuated oxidative stress | [51,93] |
| Ascorbate peroxidase (*APX*) | $H_2O_2$ metabolism | Preventing oxidative stress | [94] |
| Glutathione reductase (*GR*) | $H_2O_2$ metabolism | Induced by oxidative stress | [95] |
| Superoxide dismutase (*SOD*) | Oxygen radical detoxification | Improved drought tolerance; Induced by drought | [96–98] |

The expression of aldehyde dehydrogenases (ALDHs) is upregulated under stress situations like dehydration, salinity and oxidative stress. ALDHs are able to convert highly reactive aldehydes and hence extenuate oxidative stress [99].

In a transcription profiling study of two drought tolerant Andean native potato clones (*Solanum tuberosum* subsp. *andigena*) it was reported that a member of the aldehyde dehydrogenase family 7 (ALDH7) was induced under drought stress conditions [51]. These findings are in conformity with functional analyses of an ALDH7 gene member (*GmPP55*) from soybean (*Glycine max*). Transgenic *Arabidopsis* and tobacco (*Nicotiana tabacum*) plants exhibited improved tolerance to $H_2O_2$ as well as salt and drought conditions during different developmental stages [93].

A key enzyme in the detoxification of ROS is ascorbate peroxidase (APX), which functions as a $H_2O_2$ reductant [100]. APX utilizes ascorbate, an electron donor, which enables the reduction of $H_2O_2$ to water and monodehydroascorbate under oxidative stress conditions. Monodehydroascorbate can be converted into ascorbate or dehydroascorbate. Dehydroascorbate is reduced to ascorbate by the action of dehydroascorbate reductase, which requires glutathione. This implies that the activity of ascorbate peroxidase prevents oxidative stress by $H_2O_2$ detoxification [101]. Enhanced tolerance to various stresses including salt, water and PEG-induced stress was reported from transgenic tobacco plants overproducing an *Arabidopsis* gene, encoding a cytosolic APX, in the chloroplast [94]. These results suggest that APXs are essential in plants for maintaining optimal photosynthetic rates during abiotic stress and therefore benefit survival and growth under water limiting conditions by the detoxification of ROS.

Studies of different wheat cultivars revealed another component in the detoxification process of ROS that might enable improved stress tolerance. Glutathione reductase (GR) is a part of the ascorbate-glutathione cycle; in a drought tolerant wheat cultivar, GR activity was enhanced in leaves and roots in response to osmotic stress, which might facilitate protection against ROS [95]. Hence the importance of the ascorbate-glutathione cycle as a major component is elucidated, as different components might serve as candidate genes in drought-tolerant crop breeding.

Additionally superoxide dismutases (SODs) are essential enyzmes during detoxification of ROS in plants. Depending on the metal co-factor four isoforms have been classified: copper zinc SODs (Cu/Zn-SODs), iron SODs (Fe-SODs), manganese SODs (Mn-SODs), and nickel SODs (Ni-SODs) [102]. The latter type has so far only been identified in prokaryotes.

Alfalfa (*Medicago sativa*) overexpressing a SOD of the Mn-type from *Nicotiana plumbaginifolia* has been reported to exhibit enhanced vigor after water deficit as well as improved herbage yield [96]. In addition to these findings, it has been discovered that in sunflower (*Helianthus annuus*) two genes encoding Cu/Zn-SODs were significantly induced upon drought stress treatments [97]. As one of the SOD genes has also been reported to be downregulated in a drought sensitive genotype [98], a positive correlation between the expression of SODs and improved stress tolerance appears to be evident.

Conclusively, results of expression profiling as well as transgenic approaches demonstrated the importance of ROS-detoxification in plants during drought stress. Different candidate genes that are involved in the detoxification process have been reported to benefit plant growth as well as survival upon abiotic stress and therefore provide an excellent resource for improving drought tolerance in crop breeding.

## STAY GREEN TRAIT

Extended periods of photosynthesis are advantageous in order to increase crop yields, whereas premature senescence due to water limitations might lead to reduced productivity [103,104,105]. Stay-green mutants that retain their greenness, which can be visually accessed or measured by photometric devices, have been connected with drought tolerance. Five types A–E have been defined for stay-green mutants [24]. The first two types A and B are so called functional stay-green mutants [106]. Type A is characterized by a delayed initiation of senescence compared to the wild type whereas type B shows a prolonged senescence. In both cases, the duration of photosynthesis is elongated. The other three types C–E belong to the non-functional stay-green mutants, also called cosmetic stay-green mutants [106]. Here the chlorophyll breakdown is impaired. In type C only the degradation of the chlorophyll is delayed, while the photosynthetic activity has already been reduced. In the type D the cells die before or in the middle of the senescence period and plants appear green. The type E is characterized by higher chlorophyll level, which needs longer time for degradation. Functional stay-green mutants have been successfully selected by conventional breeding that show higher productivity, enhanced drought tolerance and better performance under low nitrogen supply [103]. Functional stay-greens are genotypes in which the transition point between C-capture and N-remobilization is delayed, or the transition occurs on time but the subsequent yellowing and N-remobilization run slowly [24,105].

Ethylene plays a major role in leaf senescence. The dominant ethylene-insensitive mutant of *ETR1* (*etr-1*) leads to a stay-green phenotype [107], as well as a mutant of *EIN2*, a positive regulator of the ethylene signaling pathway (*ore3*, also called *oresara3*) [108]. In *Arabidopsis*, stay-green mutants of the *ACS* gene family showed reduced ethylene production as the conversion of SAM to ACC represents the rate-limiting step. Apart from ethylene, ABA promotes leaf senescence. *SAG113*encodes a member of the protein phosphatase 2C family, which is part of the ABA signaling pathway. Mutants of *SAG113*(*sag113*) show reduced ABA sensitivity. This stay-green phenotype is probably due to the ABA-mediated leaf senescence that is induced by loss of turgor [109]. All these mutants are examples for functional stay-green mutants. Furthermore, overexpression of master regulators of stress responses, C-repeat binding factor 2 (CBF2)/DREB1C and CBF3/DREB1A, results in a stay-green phenotype with enhanced stress resistance [110].

Rolando *et al.* [111] investigated the chlorophyll content and its degradation in three potato varieties under limited irrigation. Lower chlorophyll degradation rates were observed in the genotypes Sarnav and Unica that yielded higher final tuber biomass. The short-term increase of chlorophyll after drought induction was regarded as the consequence of the halted leaf growth representing the conservative survival strategy, while the riskier genotypes respond with less water-use efficiency strategy [111]. This strategy resulted in reduced yield losses under moderate stress conditions. For cassava (*Manihot esculenta* Crantz), a relatively high genetic correlation was observed between leaf retention and root yield [112]. Cassava clones with leaf retention produced more total fresh biomass and 33% more root dry matter than plants without the trait. Increased root yields under well-watered and water-limited conditions were associated with total biomass production and increased harvest index [112].

In cereals, major differences occur between the crop species with regard to the source-sink relations during the grain-filling period [103]. In maize, grain yield appears to be dependent on maintenance of high activity of the source, *i.e.*, on keeping the photosynthetic activity in the leaves up [113], whereas this has only a slight effect in wheat. In maize, inbred lines and hybrids could be identified that show delayed senescence in combination with higher grain yield [114,115,116]. The subtropical white dent maize lines CML444 and SC-Malawi also displayed higher yields under low water regime [117]. In sorghum, stay-green hybrids produced 47% more biomass under terminal drought [118,119,120]. A new wheat stay-green mutant *tasg1* showed an enhanced antioxidative capacity that leads to a delayed senescence and better drought resistance [121]. The efficient removal of ROS by the ascorbate-glutathione cycle under limited water conditions might be the key factor for the observed enhanced stress tolerance. In conclusion, in cereals the stay-green trait seems to be favorable in maize and sorghum, but not that much in wheat and barley [103].

The best option for breeding using stay-green mutants with regard to high yields under drought as well as well-water conditions would be to look for functional stay-green mutants that demonstrate normal onset of senescence combined with slow kinetics of senescence or delayed onset of senescence, but normal kinetics of senescence. This can be coupled with, e.g. early flowering as stay-green stands for a prolonged time of green leaf area after anthesis. Drought tolerance is an important breeding goal that can gain from screening for stay-green mutants.

## CONCLUSIONS

Research on abiotic stress caused by limited water supply has unraveled complex networks and cross talks between plant hormones and metabolic pathways. The strategies that have evolutionarily evolved for the survival of plants [122] need now to be employed to guarantee high crop yields and yield stability under drought conditions [14]. Some strategies might be combined with negative agricultural traits or might require too much energy and resources during the development of the plants so that drought tolerance is linked with reduced yields. Tolerance might be only favorable under conditions of limited water supply [123], while under well-water conditions drought tolerant crops might show lower yields. Recent studies on drought tolerance in potato have shown that drought tolerance may be linked to yield penalty [124]. Under field conditions water is at least for some time freely available and additional genotype x environment interactions also affect the yield giving a different picture than under controlled conditions [10,125].

Transgenic approaches have been useful in proof-of-concept studies to understand the molecular mechanisms underlying drought tolerance. However, the majority of transgenic approaches have not been successful in regard to obtaining high yielding drought tolerant crops because strong constitutively expressed promoters instead of drought-regulated or tissue-specific ones have been used to overexpress candidate genes for drought tolerance [10,126]. Unwanted pleiotropic effects as well as yield penalties for employing metabolically cost-intensive pathways not required under well-watered conditions were observed [127]. In addition, in most cases survival (and/or recovery) of drought in model plants like *Arabidopsis* was investigated, not the effect on yield and productivity in crops [128]. Different genetic engineering strategies for abiotic stress and target genes are outlined by Cabello *et al.* [127]. Transgenic approaches using drought inducible promoters, tissue-specific expression [126] or synthetic biology approaches [127] might be successful in the future in improving drought tolerance in crops.

Breeding for drought tolerance is challenging, but absolutely essential under the expected climate changes that might lead to more frequent periods of low water supply, especially in subtropical regions. The extensive compilation of plant drought stress genes and their homologs in nine species in the new database (DB), DroughtDB represents a helpful tool to understand and to develop concepts for drought tolerant crops [129]. As a next step, mining the natural genetic variation in candidate genes for drought tolerance in germplasm material or induced genetic variation in ethyl methansulfonate (EMS) mutagenized populations by using next generation sequencing technologies [130] or ecotilling [131] would be desirable to associate changes on the molecular level with drought relevant traits in order to develop markers that facilitate selection for drought tolerance in breeding programs. On the other hand, the whole reprogramming plants undergo under drought and rehydration can be analyzed using omics technologies [9,132]. On the whole genome level genomic selection might offer a possibility to breed efficiently for drought tolerance by selecting favorable individuals on the basis of the genomic estimated breeding values (GEBVs) [133]. Modelling the behavior of a crop under drought conditions will also help to improve crop productivity under water saving irrigation systems [6,7,134].

Even though an enormous amount of knowledge with regard to drought tolerance has been gained in the recent years, we are still far from understanding all underlying mechanisms and the signaling pathways involved. A huge effort is now required to transfer this knowledge to the various crops. Field trials estimating yields under well-watered and drought conditions have to be performed to successfully breed for drought tolerance in crops.

## ACKNOWLEDGMENTS

Work on drought tolerance in potato is funded in our laboratory by the grant No. 22014713 of the Fachagentur Nachwachsende Rohstoffe (Federal Ministry of Food and Agriculture), Germany.

## AUTHOR CONTRIBUTIONS

All authors contributed to varying degrees to the production of the review: Christoph Tim Krannich: ethylene, aquaporins, osmolytes, ROS; Lisa Maletzki: most of the tables; Christina Kurowsky: references, graphical representation; Renate Horn: abstract, introduction, ABA, stay-green, conclusions, supervision.

## REFERENCES

1. Eckardt, N.A.; Cominelli, E.; Galbiati, M.; Tonelli, C. The future of science: Food and water for life. *Plant Cell* 2009, *21*, 368–372.
2. Chaves, M.M.; Pereira, J.S.; Maroco, J.; Rodrigues, M.L.; Ricardo, C.P.P.; Osorio, M.L.; Carvalho, I.; Faria, T.; Pinheiro, C. How plants cope with water stress in the field? Photosynthesis and growth. *Ann. Bot.* 2002, *89*, 907–916.
3. Mba, C.; Guimaraes, E.P.; Gosh, K. Re-orienting crop improvement for the changing climatic conditions of the 21st century. *Agric. Food Secur.* 2012, *1*, 7.
4. Dodd, I.C.; Whalley, W.R.; Ober, E.S.; Parry, M.A.J. Genetic and management approaches to boost UK wheat yields by ameliorating water deficits. *J. Exp. Bot.* 2012, *62*, 5241–5248.

5. Tuberosa, R. Phenotyping for drought tolerance of crops in the genomic era. *Front. Physiol.* 2012, *3*, 347.
6. Grassini, P.; van Bussel, L.G.J.; van Wart, J.; Wolf, J.; Claessens, L.; Yang, H.; Boogaard, H.; de Groot, H.; van Ittersum, M.K.; Gassman, K.G. How good is good enough? Data requirements for reliable crop yield simulations and yield-gap analysis. *Field Crops Res.* 2015, *177*, 49–63.
7. Mishra, A.K.; Ines, A.V.M.; Das, N.N.; Khedun, P.C.; Singh, V.P.; Sivakumar, B.; Hansen, J.W. Anatomy of the local-scale drought: Application of assimilated remote sensing products, crop model, and statistical methods to an agricultural drought study. *J. Hydrol.* 2015, *526*, 15–29.
8. Daryanto, S.; Wang, L.; Jacinthe, P.-A. Global synthesis of drought effects on food legume production. *PLoS ONE*2015, *10*, e0127401.
9. Zhang, J.-Y.; Cruz de Carvalho, M.H.; Torres-Jerez, I.; Kang, Y.; Allen, S.N.; Huhman, D.V.; Tang, Y.; Murray, J.; Summer, L.W.; Udvardi, M.K. Global reprogramming of transcription and metabolism in *Medicago truncatula* during progressive drought and rewatering. *Plant Cell Environ.* 2014, *37*, 2553–2576.
10. Reguera, M.; Peleg, Z.; Blumwald, E. Targeting metabolic pathways for genetic engineering abiotic stress-tolerance in crops. *Biochim. Biophys. Acta* 2012, *1819*, 186–194.
11. Fleury, D.; Jefferies, S.; Kuchel, H.; Langridge, P. Genetic and genomic tools to improve drought tolerance in wheat. *J. Exp. Bot.* 2010, *61*, 3211–3222.
12. Peleg, Z.; Blumwald, E. Hormone balance and abiotic stress tolerance in crop plants. *Curr. Opin. Plant Biol.* 2011, *14*, 290–295.
13. Wang, C.; Liu, Y.; Li, S.-S.; Han, G.-Z. Insights into the origin and evolution of the plant hormone signaling machinery. *Plant Physiol.* 2015, *167*, 872–886.
14. Albacete, A.A.; Martínez-Andújar, C.; Pérez-Alfocea, F. Hormonal and metabolic regulation of source-sink relations under salinity and drought: From plant survival to crop yield stability. *Biotechnol. Adv.* 2014, *32*, 12–30.
15. Cramer, G.R.; Urano, K.; Delrot, S.; Pezzotti, M.; Shinozaki, K.

Effects of abiotic stress on plants: A systems biology perspective. *BMC Plant Biol.* 2011, *11*, 63.

16. Hirayama, T.; Shinozaki, K. Research on plant abiotic stress responses in the post-genome era: Past, present and future. *Plant J.* 2010, *61*, 1041–1052.
17. Nakashima, K.; Yamaguchi-Shinozaki, K.; Shinozaki, K. The transcriptional regulatory network in the drought response and its crosstalk in abiotic stress responses including drought, cold, and heat. *Front. Plant Sci.* 2014, *5*, 170.
18. Finkelstein, R. Abscisic acid synthesis and response. *Arabidopsis Book* 2013, *11*, e0166.
19. Miyakawa, T.; Fujita, Y.; Yamaguchi-Shinozaki, K.; Tanokura, M. Structure and function of abscisic acid receptors. *Trends Plant Sci.* 2013, *18*, 259–266.
20. Mehrotra, R.; Bhalothia, P.; Bansal, P.; Basantani, M.K.; Bharti, V.; Mehrotra, S. Abscisic acid and abiotic stress tolerance—Different tiers of regulation. *J. Plant Physiol.* 2014, *171*, 486–496.
21. Eskling, M.; Arvidsson, P.-O.; Åkerlund, H.E. The xanthophyll cycle, its regulation and components. *Physiol. Plant.* 1997, *100*, 806–816.
22. Thompson, A.J.; Jackson, A.C.; Parker, R.A.; Morpeth, D.R.; Burbidge, A.; Taylor, I.B. Abscisic acid biosynthesis in tomato: Regulation of zeaxanthin epoxidase and 9-*cis*-epoxycarotenoid dioxygenase mRNAs by light/dark cycles, water stress and abscisic acid. *Plant Mol. Biol.* 2000, *42*, 833–845.
23. Iuchi, S.; Kobayashi, M.; Taji, T.; Naramoto, M.; Seki, M.; Kato, T.; Tabata, S.; Kakubari, Y.; Yamaguchi-Shinozaki, K.; Shinozaki, K. Regulation of drought tolerance by gene manipulation of 9-*cis*-epoxycarotenoid dioxygenase, a key enzyme in abscisic acid biosynthesis in *Arabidopsis*. *Plant J.* 2001, *27*, 325–333.
24. Thomas, H.; Howarth, C.J. Five ways to stay green. *J. Exp. Bot.* 2000, *51*, 329–337.
25. Park, H.Y.; Seok, H.Y.; Park, B.K.; Kim, S.H.; Goh, C.H.; Lee, B.-H.; Lee, C.-H.; Moon, Y.-H. Overexpression of *Arabidopsis* ZEP enhances tolerance to osmotic stress. *Biochem. Biophys. Res. Commun.* 2008, *375*, 80–85.

26. Gonzalez-Guzman, M.; Apostolova, N.; Belles, J.M.; Barrero, J.M.; Piqueras, P.; Ponce, M.R.; Micol, J.L.; Serrano, R.; Rodríguez, P.L. The short-chain alcohol dehydrogenase ABA2 catalyzes the conversion of xanthoxin to abscisic aldehyde. *Plant Cell* 2002, *14*, 1833–1846.
27. Bittner, F.; Oreb, M.; Mendel, R.R. ABA3 is a molybdenum cofactor sulfurase required for activation of aldehyde oxidase and xanthine dehydrogenase in *Arabidopsis thaliana*. *J. Biol. Chem.* 2001, *276*, 40381–40384.
28. Seo, M.; Aoki, H.; Koiwai, H.; Kamiya, Y.; Nambara, E.; Koshiba, T. Comparative studies on the *Arabidopsis* aldehyde oxidase (AAO) gene family revealed a major role of AAO3 in ABA biosynthesis in seeds. *Plant Cell Physiol.* 2004, *45*, 1694–1703.
29. Xiong, L.M.; Lee, H.; Ishitani, M.; Zhu, J.K. The *ArabidopsisLOS5/ABA3* locus encodes a molybdenum cofactor sulfurase and modulates cold stress- and osmotic stress-responsive gene expression. *Plant Cell* 2001, *13*, 2063–2083.
30. Li, Y.; Zhang, J.; Zhang, J.; Hao, L.; Hua, J.; Duan, L.; Zhang, M.; Li, Z. Expression of an *Arabidopsis* molybdenum factor sulphurase gene in soybean enhances drought tolerance and increases yield under field conditions. *Plant Biotechnol. J.* 2013, *11*, 747–758.
31. Xiao, B.Z.; Chen, X.; Xiang, C.B.; Tang, N.; Zhang, Q.F.; Xiong, L.Z. Evaluation of seven function-known candidate genes for their effects on improving drought resistance of transgenic rice under field conditions. *Mol. Plant* 2009, *2*, 73–83.
32. Lu, Y.; Li, Y.; Zhang, J.; Xiao, Y.; Yue, Y.; Duan, L.; Zhang, M.; Li, Z. Overexpression of *Arabidopsis* molybdenum cofactor sulfurase gene confers drought tolerance in maize (*Zea mays* L.). *PLoS ONE* 2013, *8*, e52126.
33. Cutler, A.J.; Krochko, J.E. Formation and breakdown of ABA. *Trends Plant Sci.* 1999, *4*, 472–478.
34. Nambara, E.; Marion-Poll, A. Abscisic acid biosynthesis and catabolism. *Annu. Rev. Plant Biol.* 2005, *56*, 165–185.
35. Kushiro, T.; Okamoto, M.; Nakabayashi, K.; Yamagishi, K.; Kitamura, S.; Asami, T.; Hirai, N.; Koshiba, T.; Kamiya, Y.; Nambara, E. The *Arabidopsis* cytochrome P450 *CYP707A* encodes

ABA 8′-hydroxylases: Key enzymes in ABA catabolism. *EMBO J.* 2004, *23*, 1647–1656.

36. Lee, K.H.; Piao, H.L.; Kim, H.Y.; Choi, S.M.; Jiang, F.; Hartung, W.; Hwang, I.; Kwak, J.M.; Lee, I.J.; Hwang, I. Activation of glucosidase via stress-induced polymerization rapidly increases active pools of abscisic acid. *Cell* 2006, *126*, 1109–1110.
37. Cutler, S.R.; Rodriguez, P.L.; Finkelstein, R.R.; Abrams, S.R. Abscisic acid: Emergence of a core signaling network. *Annu. Rev. Plant Biol.* 2010, *61*, 651–660.
38. Umezawa, T.; Nakashima, K.; Miyakawa, T.; Kuromori, T.; Tanokura, M.; Shinozaki, K.; Yamaguchi-Shinozaki, K. Molecular basis of the core regulatory network in ABA responses: Sensing, signaling and transport. *Plant Cell Physiol.*2010, *51*, 1821–30.
39. Park, S.Y.; Fung, P.; Nishimura, N.; Jensen, D.R.; Hiroaki, F.; Zhao, Y.; Lumba, S.; Santiago, J.; Rodrigues, A.; Chow, T.-F.F.; *et al.* Abscisic acid inhibits PP2Cs via the PYR/PYL family of ABA binding START proteins. *Science* 2009, *324*, 1068–1070.
40. Schweighofer, A.; Hirt, H.; Meskiene, I. Plant PP2C phosphatases: Emerging functions in stress signaling. *Trends Plant Sci.* 2004, *9*, 236–240.
41. Kulik, A.; Wawer, I.; Krzywinska, E.; Bucholc, M.; Dobrowolska, G. SnRK2 protein kinases—Key regulators of plant response to abiotic stresses. *OMICS* 2011, *15*, 859–860.
42. Singh, A.; Jha, S.K.; Bagri, J.; Pandey, G.K. ABA inducible rice protein phosphatase 2C confers ABA insensitivity and abiotic stress tolerance in *Arabidopsis*. *PLoS ONE* 2015, *10*, e0125168.
43. Kim, H.; Lee, K.; Hwang, H.; Bhatnagar, N.; Kim, D.-Y.; Yoon, I.S.; Buyn, M.-O.; Kim, S.T.; Jung, K.-H.; Kim, B.G. Overexpression of *PYL5* in rice enhances drought tolerance, inhibits growth, and modulates gene expression. *J. Exp. Bot.* 2014, *65*, 453–464.
44. Shinozaki, K.; Yamaguchi-Shinozaki, K. Gene networks involved in drought stress response and tolerance. *J. Exp. Bot.*2007, *58*, 221–227.
45. Skirycz, A.; Claeys, H.; de Bodt, S.; Oikawa, A.; Shinoda, S.; Andriankaja, M.; Maleux, K.; Eloy, N.B.; Coppens, F.; Yoo, S.-D.; *et al.* Pause-and-Stop: The effect of osmotic stress on cell

proliferation during early leaf development in Arabidopsis and the role for ethylene signaling in cell cycle arrest. *Plant Cell* 2011, *23*, 1876–1888.

46. Ko, J.H.; Yang, S.H.; Han, K.H. Upregulation of an *Arabidopsis* RING-H2 gene, *XERICO*, confers drought tolerance through increased abscisic acid biosynthesis. *Plant J.* 2006, *47*, 343–355.
47. Golldack, D.; Li, C.; Mohan, H.; Probst, N. Tolerance to drought and salt stress in plants: Unraveling the signaling networks. *Front. Plant Sci.* 2014, *5*, 151.
48. Thompson, A.J.; Andrews, J.; Mulholland, B.J.; McKee, J.M.T.; Hilton, H.W.; Horridge, J.S.; Farquhar, G.D.; Smeeton, R.C.; Smillie, I.R.A.; Black, C.R.; *et al.* Overproduction of abscisic acid in tomato increases transpiration efficiency and root hydraulic conductivity and influences leaf expansion. *Plant Physiol.* 2007, *143*, 1905–1917.
49. Rodriguez, P.L. Protein phosphatase 2C (PP2C) function in higher plants. *Plant Mol. Biol.* 1998, *38*, 919–927.
50. Vranová, E.; Langebartels, C.; van Montagu, M.; Inzé, D.; van Camp, W. Oxidative stress, heat shock and drought differentially affect expression of a tobacco protein phosphatase 2C. *J. Exp. Bot.* 2000, *51*, 1763–1764.
51. Schafleitner, R.; Rosales, R.O.G.; Gaudin, A.; Aliaga, C.A.A.; Martinez, G.N.; Marca, L.R.T.; Bolivar, L.A.; Delgado, F.M.; Simon, R.; Bonierbale, M. Capturing candidate drought tolerance traits in two native Andean potato clones by transcription profiling of field grown plants under water stress. *Plant Physiol. Biochem.* 2007, *45*, 673–690.
52. Mao, X.; Zhang, H.; Tian, S.; Chang, X.; Jing, R. TaSnRK2.4, an SNF1-type serine/threonine protein kinase of wheat (*Triticum aestivum* L.), confers enhanced multistress tolerance in *Arabidopsis*. *J. Exp. Bot.* 2010, *61*, 683–696.
53. Gupta, S.; Bharalee, R.; Bhorali, P.; Bandyopadhyay, T.; Gohain, B.; Agarwal, N.; Ahmed, P.; Saikia, H.; Borchetia, S.; Kalita, M.C.; *et al.* Identification of drought tolerant progenies in tea by gene expression analysis. *Funct. Integr. Genomics* 2012, *12*, 543–563.

54. Young, T.E.; Meeley, R.B.; Gallie, D.R. ACC synthase expression regulates leaf performance and drought tolerance in maize. *Plant J.* 2004, *40*, 813–825. [Google Scholar]

55. Dong, H.; Zhen, Z.; Peng, J.; Chang, L.; Gong, Q.; Wang, N.N. Loss of *ACS7* confers abiotic stress tolerance by modulating ABA sensitivity and accumulation in *Arabidopsis*. *J. Exp. Bot.* 2011, *62*, 4875–4887.

56. Wang, K.L.-C.; Yoshida, H.; Lurin, C.; Ecker, J.R. Regulation of ethylene gas biosynthesis by the *Arabidopsis* ETO1 protein. *Nature* 2004, *428*, 945–950.

57. Du, H.; Wu, N.; Cui, F.; You, L.; Li, X.; Xiong, L. A homolog of ETHYLENE OVERPRODUCER, OsETOL1, differentially modulates drought and submergence tolerance in rice. *Plant J.* 2014, *78*, 834–849.

58. Desikan, R.; Last, K.; Harrett-Williams, R.; Tagliavia, C.; Harter, K.; Hooley, R.; Hancock, J.T.; Neill, S.J. Ethylene-induced stomatal closure in *Arabidopsis* occurs via AtrbohF-mediated hydrogen peroxide synthesis. *Plant J.* 2006, *47*, 907–916.

59. Wilson, R.L.; Kim, H.; Bakshi, A.; Binder, B.M. The ethylene receptors ETHYLENE RESPONSE1 and ETHYLENE RESPONSE2 have contrasting roles in seed germination of *Arabidopsis* during salt stress. *Plant Physiol.* 2014, *165*, 1353–1366.

60. Wang, Y.; Liu, C.; Li, K.; Sun, F.; Hu, H.; Li, X.; Zhao, Y.; Han, C.; Zhang, W.; Duan, Y.; *et al. Arabidopsis* EIN2 modulates stress response through abscisic acid response pathway. *Plant Mol. Biol.* 2007, *64*, 633–644.

61. Asensi-Fabado, M.A.; Cela, J.; Müller, M.; Arrom, L.; Chang, C.; Munné-Bosch, S. Enhanced oxidative stress in the ethylene-insensitive (*ein3-1*) mutant of *Arabidopsis thaliana* exposed to salt stress. *J. Plant Physiol.* 2012, *169*, 360–368.

62. Cheng, M.-C.; Liao, P.-M.; Kuo, W.-W.; Lin, T.-P. The *Arabidopsis* ETHYLENE RESPONSE FACTOR1 regulates abiotic stress-responsive gene expression by binding to different *cis*-acting elements in response to different stress signals. *Plant Physiol.* 2013, *162*, 1566–1582.

63. Omidvar, V.; Abdullah, S.N.A.; Ho, C.L.; Mahmood, M. Isolation

and characterization of an ethylene-responsive element binding protein (EgEREBP) from oil palm (*Elaeis guineensis*). *Aust. J. Crop Sci.* 2013, *7*, 219–226. [Google Scholar]

64. Pan, Y.; Seymour, G.B.; Lu, C.; Hu, Z.; Chen, X.; Chen, G. An ethylene response factor (ERF5) promoting adaptation to drought and salt tolerance in tomato. *Plant Cell Rep.* 2012, *31*, 349–360.
65. Dubois, M.; Skirycz, A.; Claeys, H.; Maleux, K.; Dhondt, S.; de Bodt, S.; Vanden Bossche, R.; de Milde, L.; Yoshizumi, T.; Matsui, M.; *et al.* Ethylene response factor6 acts as a central regulator of leaf growth under water-limiting conditions in *Arabidopsis*. *Plant Physiol.* 2013, *162*, 319–332.
66. Zhang, Z.; Li, F.; Li, D.; Zhang, H.; Huang, R. Expression of ethylene response factor JERF1 in rice improves tolerance to drought. *Planta* 2010, *232*, 765–774.
67. Lin, Z.F.; Zhong, S.L.; Grierson, D. Recent advances in ethylene research. *J. Exp. Bot.* 2009, *60*, 3311–3336.
68. Rodrigues, M.A.; Bianchetti, R.E.; Freschi, L. Shedding light on ethylene metabolism in higher plants. *Front. Plant Sci.* 2014, *5*, 665.
69. Yokoi, S.; Quintero, F.J.; Cubero, B.; Ruiz, M.T.; Bressan, R.A.; Hasegawa, P.M.; Pardo, J.M. Differential expression and function of *Arabidopsis thaliana* NHX $Na^+/H^+$ antiporters in the salt stress response. *Plant J.* 2002, *30*, 529–539.
70. Tanaka, Y.; Sano, T.; Tamaoki, M.; Nakajima, N.; Kondo, N.; Hasezawa, S. Ethylene inhibits abscisic acid-induced stomatal closure in *Arabidopsis*. *Plant Physiol.* 2005, *138*, 2337–2343.
71. Cho, Y.-H.; Yoo, S.-D. Novel connections and gaps in ethylene signaling from the ER membrane to the nucleus. *Front. Plant Sci.* 2015, *5*, 733.
72. Chao, Q.; Rothenberg, M.; Solano, R.; Roman, G.; Terzaghi, W.; Ecker, J.R. Activation of the ethylene gas response pathway in *Arabidopsis* by the nuclear protein ETHYLENE-INSENSITIVE3 and related proteins. *Cell* 1997, *89*, 1133–1144.
73. Charfeddine, M.; Saidi, N.M.; Charfeddine, S.; Hammami, A.; Bouzid, R.G. Genome-wide analysis and expression profiling of the ERF transcription factor family in potato (*Solanum tuberosum* L.).

*Mol. Biotechnol.* 2015, *57*, 348–358.

74. Sharma, M.K.; Kumar, R.; Solanke, A.U.; Sharma, R.; Tyagi, A.K.; Sharma, A.K. Identification, phylogeny, and transcript profiling of ERF family genes during development and abiotic stress treatments in tomato. *Mol. Genet. Genomics* 2010, *284*, 455–475.
75. Burg, M.B.; Ferraris, J.D. Intracellular organic osmolytes: Function and regulation. *J. Biol. Chem.* 2008, *283*, 7309–7313.
76. Zhang, N.; Si, H.-J.; Wen, G.; Du, H.-H.; Liu, B.-L.; Wang, D. Enhanced drought and salintiy tolerance in transgenic potato plants with BADH gene from spinach. *Plant Biotechnol. Rep.* 2011, *5*, 71–77.
77. Sharma, S.; Villamor, J.G.; Verslues, P.E. Essential role of tissue-specific proline synthesis and catabolism in growth and redox balance at low water potential. *Plant Physiol.* 2011, *157*, 292–304.
78. De Ronde, J.A.; Spreeth, M.H.; Cress, W.A. Effect of antisense L-$\Delta^1$-pyrroline-5-carboxylate reductase transgenic soybean plants subjected to osmotic and drought stress. *Plant Growth Regul.* 2000, *32*, 13–26.
79. Stiller, I.; Dulai, S.; Kondrák, M.; Tarnai, R.; Szabó, L.; Toldi, O.; Bánfalvi, Z. Effects of drought on water content and photosynthetic parameters in potato plants expressing the trehalose-6-phosphate synthase gene of *Saccharomyces cerevisiae*. *Planta* 2008, *227*, 299–308.
80. Movahedi, S.; Sayed Tabatabaei, B.E.; Alizade, H.; Ghobadi, C.; Yamchi, A.; Khaksar, G. Constitutive expression of *Arabidopsis DREB1B* in transgenic potato enhances drought and freezing tolerance. *Biol. Plant* 2012, *56*, 37–42.
81. Verbruggen, N.; Hermans, C. Proline accumulation in plants: A review. *Amino Acids* 2008, *35*, 753–759.
82. Ashraf, M.; Foolad, M.R. Roles of glycine betaine and proline in improving plant abiotic stress resistance. *Environ. Exp. Bot.* 2007, *59*, 206–216.
83. Maurel, C.; Reizer, J.; Schroeder, J.I.; Chrispeels, M.J. The vacuolar membrane protein γ-TIP creates water specific channels in *Xenopus* oocytes. *EMBO J.* 1993, *12*, 2241–2247.

84. Danielson, J.A.H.; Johanson, U. Unexpected complexity of the aquaporin gene family in the moss *Physcomitrella patens*. *BMC Plant Biol.* 2008, *8*, 1–15.

85. Ding, L.; Gao, C.; Li, Y.; Li, Y.; Zhu, Y.; Xu, G.; Shen, Q.; Kaldenhoff, R.; Kai, L.; Guo, S. The enhanced drought tolerance of rice plants under ammonium is related to aquaporin (AQP). *Plant Sci.* 2015, *234*, 14–21.

86. Jang, J.Y.; Kim, D.G.; Kim, Y.O.; Kim, J.S.; Kang, H. An expression analysis of a gene family encoding plasma membrane aquaporins in response to abiotic stresses in *Arabidopsis thaliana*. *Plant Mol. Biol.* 2004, *54*, 713–725.

87. Alexandersson, E.; Fraysse, L.; Sjövall-Larsen, S.; Gustavsson, S.; Fellert, M.; Karlsson, M.; Johanson, U.; Kjellbom, P. Whole gene family expression and drought stress regulation of aquaporins. *Plant Mol. Biol.* 2005, *59*, 469–484.

88. Li, J.; Ban, L.; Wen, H.; Wang, Z.; Dzyubenko, N.; Chapurin, V.; Gao, H.; Wang, X. An aquaporin protein is associated with drought stress tolerance. *Biochem. Biophys. Res. Commun.* 2015, *459*, 208–213.

89. Mahdieh, M.; Mostajeran, A.; Horie, T.; Katsuhara, M. Drought stress alters water relations and expression of *PIP*-type aquaporin genes in *Nicotiana tabacum* plants. *Plant Cell Physiol.* 2008, *49*, 801–813.

90. Xu, Y.; Hu, W.; Liu, J.; Zhang, J.; Jia, C.; Miao, H.; Xu, B.; Jin, Z. A banana aquaporin gene, *MaPIP1;1*, is involved in tolerance to drought and salt stresses. *BMC Plant Biol.* 2014, *14*, 59.

91. Lee, H.K.; Cho, S.K.; Son, O.; Xu, Z.; Hwang, I.; Kim, W.T. Drought stress-induced Rma1H1, a RING membrane-anchor E3 ubiquitin ligase homolog, regulates aquaporin levels via ubiquitination in transgenic *Arabidopsis* plants. *Plant Cell* 2009, *21*, 622–641.

92. Moshelion, M.; Halperin, O.; Wallach, R.; Oren, R.; Way, D.A. Role of aquaporins in determining transpiration and photosynthesis in water-stressed plants: Crop water-use efficiency, growth and yield. *Plant Cell Environ.* 2014, 2014.

93. Rodrigues, S.M.; Andrade, M.O.; Gomes, A.P.S.; Damatta, F.M.; Baracat-Pereira, M.C.; Fontes, E.P.B. *Arabidopsis* and tobacco

plants ectopically expressing the soybean antiquitin-like *ALDH7* gene display enhanced tolerance to drought, salinity, and oxidative stress. *J. Exp. Bot.* 2006, *57*, 1909–1918.

94. Badawi, G.H.; Kawano, N.; Yamauchi, Y.; Shimada, E.; Sasaki, R.; Kubo, A.; Tanaka, K. Over-expression of ascorbate peroxidase in tobacco chloroplasts enhances the tolerance to salt stress and water deficit. *Physiol. Plant.* 2004, *121*, 231–238.
95. Gallé, Á.; Csiszár, J.; Benyó, D.; Laskay, G.; Leviczky, T.; Erdei, L.; Tari, I. Isohydric and anisohydric strategies of wheat genotypes under osmotic stress: Biosynthesis and function of ABA in stress responses. *J. Plant Physiol.* 2013, *170*, 1389–1399. [Google Scholar]
96. McKersie, B.D.; Bowley, S.R.; Harjanto, E.; Leprince, O. Water-deficit tolerance and field performance of transgenic alfalfa overexpressing superoxide dismutase. *Plant Physiol.* 1996, *111*, 1177–1181.
97. Fulda, S.; Mikkat, S.; Stegmann, H.; Horn, R. Physiology and proteomics of drought stress acclimation in sunflower (*Helianthus annuus* L.). *Plant Biol.* 2011, *13*, 632–642.
98. Castillejo, M.A.; Maldonado, A.M.; Ogueta, S.; Jorrin, J.V. Proteomic analysis of responses to drought stress in sunflower (*Helianthus annuus*) leaves by 2DE gel electrophoresis and mass spectrometry. *Open Proteomics J.* 2008, *1*, 59–71.
99. Singh, S.; Brocker, C.; Koppaka, V.; Chen, Y.; Jackson, B.C.; Matsumoto, A.; Thompson, D.C.; Vasiliou, V. Aldehyde dehydrogenases in cellular responses to oxidative/electrophilic stress. *Free Radic. Biol. Med.* 2013, *56*, 89–101.
100. Foyer, C.H.; Halliwell, B. Purification and properties of dehydroascorbate reductase from spinach leaves. *Phytochemistry* 1977, *16*, 1347–1350.
101. Shigeoka, S. Regulation and function of ascorbate peroxidase isoenzymes. *J. Exp. Bot.* 2002, *53*, 1305–1319.
102. Gopavajhula, V.R.; Chaitanya, K.V.; Ali Khan, P.A.; Shaik, J.P.; Reddy, P.N.; Alanazi, M. Modeling and analysis of soybean (*Glycine max*. L.) Cu/Zn, Mn and Fe superoxide dismutases. *Genet. Mol. Biol.* 2013, *36*, 225–236.
103. Gregersen, P.L.; Culetic, A.; Boschian, L.; Krupinska, K. Plant

senescence and crop productivity. *Plant Mol. Biol.* 2013, *82*, 603–622.

104. Kusaba, M.; Tanaka, A.; Tanaka, R. Stay-green plants: What do they tell us about the molecular mechanism of leaf senescence. *Photosynth. Res.* 2013, *117*, 221–234.

105. Thomas, H.; Ougham, H. The stay-green trait. *J. Exp. Bot.* 2014, *65*, 3889–3900.

106. Hörtensteiner, S. Stay-green regulates chlorophyll and chlorophyll-binding protein degradation during senescence. *Trends Plant Sci.* 2009, *14*, 155–162.

107. Grbić, V.; Bleecker, A.B. Ethylene regulates the timing of leaf senescence in *Arabidopsis*. *Plant J.* 1995, *8*, 595–602.

108. Oh, S.A.; Park, J.H.; Lee, G.I.; Paek, K.H.; Park, S.K.; Nam, H.G. Identification of three genetic loci controlling leaf senescence in *Arabidopsis thaliana*. *Plant J.* 1997, *12*, 527–535.

109. Zhang, K.; Gan, S.S. An abscisic acid-AtNAP transcription factor-SAG113 protein phosphatase 2C regulatory chain for controlling dehydration in senescing *Arabidopsis* leaves. *Plant Physiol.* 2012, *158*, 961–969.

110. Sharabi-Schwager, M.; Lers, A.; Samach, A.; Guy, C.L.; Porat, R. Overexpression of the *CBF2* transcriptional activator in *Arabidopsis* delays leaf senescence and extends plant longevity. *J. Exp. Bot.* 2010, *61*, 261–273.

111. Rolando, J.L.; Ramírez, D.A.; Yactayo, W.; Monneveux, P.; Quiroz, R. Leaf greenness as a drought tolerance related trait in potato (*Solanum tuberosum* L.). *Environ. Exp. Bot.* 2015, *110*, 27–35.

112. Lenis, J.I.; Calle, F.; Jaramillo, G.; Perez, J.C.; Ceballos, H.; Cock, J.H. Leaf retention and cassava productivity. *Field Crops Res.* 2006, *95*, 126–134.

113. Borras, L.; Slafer, G.A.; Otegui, M.E. Seed dry weight response to source-sink manipulations in wheat, maize and soybean: A quantitative reappraisal. *Field Crops Res.* 2004, *86*, 131–146.

114. Gentinetta, E.; Ceppl, D.; Lepori, C.; Perico, G.; Motto, M.; Salamini, F. A major gene for delayed senescence in maize. Pattern of photosynthates accumulation and inheritance. *Plant Breed.* 1986,

*97*, 193–203.

115. Ding, L.; Wang, K.J.; Jiang, G.M.; Liu, M.Z.; Niu, S.L.; Gao, L.M. Post anthesis changes in photosynthetic traits of maize hybrids released in different years. *Field Crop Res.* 2005, *93*, 108–115.
116. Ding, L.; Wang, K.J.; Jiang, G.M.; Liu, M.Z.; Gao, L.M. Photosynthetic rate and yield formation in different maize hybrids. *Biol. Plant* 2007, *51*, 165–168.
117. Messmer, R.; Fracheboud, Y.; Bänziger, M.; Stamp, P.; Ribaut, J.-M. Drought stress and tropical maize: QTLs for leaf greenness, plant senescence, and root capacitance. *Field Crop Res.* 2011, *124*, 93–103.
118. Borrell, A.K.; Hammer, G.L.; Douglas, A.C.L. Does maintaining green leaf area in sorghum improve yield under drought? I. Leaf growth and senescence. *Crop Sci.* 2000, *40*, 1026–1037.
119. Borrell, A.K.; Hammer, G.L.; Henzell, R.G. Does maintaining green leaf area in sorghum improve yield under drought? II. Dry matter production and yield. *Crop Sci.* 2000, *40*, 1037–1048.
120. Jordan, D.R.; Hunt, C.H.; Cruickshank, A.W.; Borrell, A.K.; Henzell, R.G. The relationship between the stay-green trait and grain yield in elite sorghum hybrids grown in a range of environments. *Crop Sci.* 2011, *52*, 1153–1161.
121. Tian, F.X.; Zhang, M.; Wang, X.; Chen, Y.H.; Wang, W. Antioxidative defence under drought stress in a wheat stay-green mutant. *Biol. Plant* 2015, *59*, 123–130.
122. Kooyers, N.J. The evolution of drought escape and avoidance in natural herbaceous populations. *Plant Sci.* 2015, *234*, 155–162.
123. Monneveux, P.; Ramínez, D.A.; Pino, M.-T. Drought tolerance in potato (*S. tuberosum* L.): Can we learn from drought tolerance research in cereals? *Plant Sci.* 2013, *205*, 76–86.
124. Sprenger, H.; Rudack, K.; Schudoma, C.; Neumann, A.; Seddig, S.; Peters, R.; Zuther, E.; Kopka, J.; Hincha, D.K.; Walther, D.; *et al.* Assessment of drought tolerance and its potential yield penalty in potato. *Funct. Plant Biol.* 2015, *42*, 655–667.
125. Wilkinson, S.; Kudoyarova, G.R.; Veselov, D.S.; Arkhipova, T.N.;

Davies, W.J. Plant hormone interactions: Innovative targets for crop breeding and management. *J. Exp. Bot.* 2012, *63*, 3499–3509.

126. Cominelli, E.; Conit, L.; Tonelli, C.; Galbiati, M. Challenges and perspectives to improve crop drought and salinity tolerance. *New Biotechnol.* 2013, *30*, 355–361.

127. Cabello, J.V.; Loydeyri, A.F.; Zurbriggen, M.D. Novel perspectives for the engineering of abiotic stress tolerance in plants. *Curr. Opin. Plant Biol.* 2014, *26*, 62–70.

128. Dolferus, R. To grow or not to grow: A stressful decision for plants. *Plant Sci.* 2014, *229*, 247–261.

129. Alter, S.; Bader, K.C.; Spannagl, M.; Wang, Y.; Bauer, E.; Schön, C.-C.; Mayer, K.F.X. DroughtDB: An expert-curated compilation of plant drought stress genes and their homologs in nine species. *Database* 2015, *2015*, bav046.

130. Galindo-González, L.; Pinzón-Latorre, D.; Bergen, A.E.; Jensen, C.D.; Deyholos, M.K. Ion torrent sequencing as a tool for mutation discovery in the flax (*Linum usitatissimum* L.) genome. *Plant Methods* 2015, *11*, 19.

131. Yu, S.; Liao, F.; Wang, F.; Wen, W.; Li, J.; Mei, H.; Luo, L. Identification of rice transcription factors associated with drought tolerance using ecotilling method. *PLoS ONE* 2012, *7*, e30765.

132. Ma, X.; Sukiran, N.L.; Ma, H.; Su, Z. Moderate drought causes dramatic floral transcriptomic reprogramming to ensure successful reproductive development in *Arabidopsis*. *BMC Plant Biol.* 2014, *14*, 164.

133. Nakaya, A.; Isobe, S.N. Will genomic selection be a practical method for plant breeding? *Ann. Bot.* 2012, *110*, 13003–13016.

134. Plauborg, F.; Abrahamsen, P.; Gjettermann, B.; Mollerup, M.; Iversen, B.V.; Liu, F.; Andersen, M.N.; Hansen, S. Modelling of root ABA synthesis, stomatal conductance, transpiration and potato production under water saving irrigation regimes. *Agric. Water Manag.* 2010, *98*, 425–439.

CHAPTER

9

# A ROLE FOR BARLEY CALCIUM-DEPENDENT PROTEIN KINASE CPK2A IN THE RESPONSE TO DROUGHT

Agata Cieśla[1], Filip Mituła[1], Lucyna Misztal[1], Olga Fedorowicz-Strońska[2], Sabina Janicka[1], Małgorzata Tajdel-Zielińska[1], Małgorzata Marczak[1], Maciej Janicki[1], Agnieszka Ludwików[1] and Jan Sadowski[1]

[1]Biotechnology Department, Faculty of Biology, Adam Mickiewicz University, Poznań, Poland

[2]Institute of Plant Genetics, Polish Academy of Science, Poznań, Poland

Increasing the drought tolerance of crops is one of the most challenging goals in plant breeding. To improve crop productivity during periods of water deficit, it is essential to understand the complex regulatory pathways that adapt plant metabolism to environmental conditions. Among various plant hormones and second messengers, calcium ions are known to be involved in drought stress perception and signaling. Plants have developed specific calcium-dependent protein kinases that convert calcium signals into phosphorylation events. In this study we attempted to elucidate the role of a calcium-dependent protein kinase in the drought stress response of barley (*Hordeum vulgare* L.), one of the most economically important crops worldwide. The ongoing barley genome project has provided useful information about genes potentially involved in the drought stress response, but information on the role of calcium-dependent kinases is still limited. We found that the gene encoding the calcium-dependent protein kinase *HvCPK2a* was significantly upregulated in response to drought. To better understand the role of HvCPK2a in drought stress signaling, we generated transgenic Arabidopsis plants that overexpressed the corresponding coding sequence. Overexpressing lines displayed drought sensitivity, reduced nitrogen balance index (NBI), an increase in total chlorophyll content and decreased relative water content. In addition, *in vitro* kinase assay experiments combined with mass spectrometry allowed HvCPK2a autophosphorylation sites to be identified. Our results suggest that HvCPK2a is a dual-specificity calcium-dependent protein kinase that functions as a negative regulator of the drought stress response in barley.

**Keywords:** Calcium-dependent protein kinase, *Hordeum vulgare* L., drought stress, mass spectrometry, protein function

## INTRODUCTION

Drought stress is one of the most critical threats to world agriculture. The severity and duration of drought are largely unpredictable, but even short periods of water deficit reduce plant growth and crop productivity. The ensuing reduced crop yields affect certain regional economies by limiting their ability to compete in agricultural markets. Consequently, the improvement of yields under drought conditions is now a major goal of plant breeding programs worldwide. Because the effect of drought on crop yield is determined by multiple factors, understanding the plant response to dry conditions is fundamental to the generation of drought-tolerant cultivars (Zhu, 2002; Rampino et al., 2006).

Plant responses and adaptation to drought are achieved at the molecular, biochemical, and physiological levels (Morari et al., 2015). Abscisic acid (ABA) is recognized as the most critical hormone in plant responses to water deficit. In addition, many functionally diverse second messengers are involved in drought stress tolerance (Zeng et al.,

2015), including calcium and calcium sensors, which are important for signaling and subsequent adaptation to stress conditions (Klimecka and Muszynska, 2007; Kudla et al., 2010; Reddy et al., 2011; Vivek et al., 2013). Plants have evolved several classes of calcium-binding proteins, including calcineurin B-like (CBL), calmodulin (CaM) and calmodulin-related proteins (Cheng et al., 2002; Luan et al., 2002). Calcium-dependent protein kinases (CPKs), which convert calcium signals into phosphorylation events, have been identified as key intracellular protein kinases in the drought stress response. Calcium-binding proteins also play a crucial role in numerous other physiological processes, including hormone signaling and pathogen responses (Sheen, 1996; Romeis et al., 2001; Guo et al., 2002; Fan et al., 2004; Ludwig et al., 2004; Boudsocq and Sheen, 2013).

CPKs are highly conserved proteins involved in plant growth, development, stress signaling, defense responses and proteasome regulation (Sheen, 1996; Romeis et al., 2001; Harper et al., 2004; Boudsocq et al., 2010; Asano et al., 2012; Hubbard et al., 2012). They are monomeric proteins consisting of five domains: a kinase domain, an autoinhibitory domain, a regulatory domain, a variable domain located at the N-terminus, and a short C-terminal domain. CPKs are regulated by $Ca^{2+}$ ions that bind to four EF-hand motifs with distinct calcium affinities (Harmon et al., 2000; Klimecka and Muszynska, 2007; Boudsocq and Sheen, 2013). In addition to activation by calcium binding, CPKs are also regulated by phospholipids, reversible phosphorylation events and adapter proteins (Romeis et al., 2000; Chehab et al., 2004; Harper et al., 2004; Hegeman et al., 2006; Klimecka and Muszynska, 2007; Witte et al., 2010; Oh et al., 2012). Although some CPKs are cytosolic, the great majority of CPKs identified have potential N-myristoylation and palmitylation motifs for membrane association in the N-terminal domain (Harper et al., 2004; Klimecka and Muszynska, 2007; Benetka et al., 2008; Martin et al., 2012; Boudsocq and Sheen, 2013).

Growing evidence indicates that CPKs positively or negatively regulate drought stress responses. Thus, the *Arabidopsis thaliana* CPK, AtCPK6, which controls accumulation of the compatible osmolyte proline (Xu et al., 2010), is a positive regulator of the drought stress response. Other examples of CPKs with a positive regulatory effect include AtCPK10 and OsCPK13 (first named as CDPK7), which are necessary for enhanced drought tolerance in *A. thaliana* and rice (*Oryza sativa*), respectively (Saijo et al., 2000, 2001; Zou et al., 2010). In contrast, AtCPK23, which controls the plasma membrane-bound slow-voltage gated ion channels slow anion channel-associated (SLAC1) and SLAC1 homolog 3 (SLAH3) (Geiger et al., 2010, 2011), negatively regulates drought and salt resistance (Ma and Wu, 2007).

Recently, members of the CPK family were also identified as ABA signaling components that regulate plant responses to abiotic stresses (Ma and Wu, 2007; Zhu et al., 2007; Zhao et al., 2011a,b). For example, AtCPK4, AtCPK11, AtCPK10, AtCPK30, and AtCPK32 all mediate ABA responses by phosphorylating ABA-responsive transcription factors (ABFs), leading to modulation of gene expression (Choi et al., 2005; Zhu et al., 2007. AtCPK12 is involved in ABA signaling in seeds and young seedlings (Zhao et al., 2011b), while AtCPK3, AtCPK6, and AtCPK23 have been identified as players in ABA-regulated stomatal signaling (Mori et al., 2006; Ma and Wu, 2007). Certain AtCPKs, including CPK4, CPK11, and CPK23, which regulate calcium-mediated ABA signaling pathways, are also involved in tolerance to drought and salt stresses (Ma and Wu, 2007; Zhu et al., 2007; Zhao et al., 2011a,b). In addition, 10 out of 14 wheat *CPK* genes appeared to respond to abiotic stresses including drought and salt stress, as well as ABA stimulus (Li et al., 2008). Five CPK family members in wheat (TaCPK1, TaCPK6, TaCPK7, TaCPK9, and TaCPK18) were implicated in abiotic stress response including drought, osmotic stress, salt and temperature (Li et al., 2008; Geng et al., 2011).

Barley (*Hordeum vulgare* L.) is the fourth most significant cereal crop plant after wheat, maize and rice (FAO). The recent availability of the barley genome sequence, together with various classical and large-scale approaches, facilitate the identification of candidate genes involved in drought-stress adaptation (Sreenivasulu et al., 2008; Mochida and Shinozaki, 2013). Despite the importance of a variety of calcium sensors in the response to drought, CPK-dependent signal transduction in barley has not yet been characterized. Here, we investigate the role of HvCPK2a in the barley drought response.

# METHODS

## Plant Growth and Treatments

For liquid chromatography-tandem mass spectrometry (LC-MS/MS) and protoplast isolation, barley cultivar *H. vulgare* L. cv. "Sebastian" was sown in pots filled with a mixture of peat and sand (2:7) with high content of phosphorus and potassium, a standard content of magnesium and a low sulfur content. Pots with 12 barley plants (28 pots for each experimental variant) were grown in a growth room at 21°C/15°C (day/night) with a 16-h photoperiod (photon flux density 800 μmol $m^{-2}$ $s^{-1}$) and 50% air humidity. The soil water content was maintained at 65%. Plants were grown for ~6 weeks (33 growth stage according Zadocs scale). Drought stress was imposed by withholding water, while in control plants the watering regime was unchanged. After 3 days, the soil water content in drought-exposed plants decreased to 35–45%. *A. thaliana* Col-0 was grown and stably transformed using the floral dip method as described in Ludwików et al. (2009). For the survival test, 3-week old plants that overexpressed *HvCPK2a* were subjected to drought stress. After 10 days, plants were rehydrated and assessed after a 4-day recovery period. Plants were scored as survivors if the base of the shoots remained green after the recovery period. Leaf relative water content was calculated according to the following formula: RWC = (fresh weight – dry weight)/(turgid weight – dry weight) × 100.

## Sequence Alignment and Phylogenetic Analysis

Multiple alignments of Arabidopsis, rice and 23 barley CPK amino acid sequences were performed using Clustal W (Thompson et al., 1997). Unrooted phylogenetic trees were constructed with the MEGA6 program (Tamura et al., 2013) using the neighbor joining method. Amino acid sequences were obtained from NCBI. Sequence identity and sequence similarity were calculated using the SIAS tool at http://imed.med.ucm.es/Tools/sias.html, where sequence identity is the number of exactly matching residues (expressed as a percentage) in a sequence alignment between two sequences of the alignment. Percent similarity is used to quantify the amino acid similarity between two sequences of the alignment.

## Immune Complex Kinase Assay

To identify barley calcium protein kinases (CPKs) involved in the drought stress response, total protein was isolated from drought-stressed barley plants using protein isolation buffer (20 mM Hepes pH 7.5, 10 mM $MgCl_2$, 1 mM DTT, 60 mM β-glycerophosphate, 1 mM NaF, 1 mM phenylmethylsulfonyl fluoride, 0.1 mM $NaVO_3$, Roche inhibitors). Cell extracts containing 2 mg total protein were immunoprecipitated overnight at 4°C with 25 μl serum of anti-CPK antibodies AS1 or AS2 (AS13 2754, AS16 3836, respectively; Agrisera); the antibodies were pre-coupled to Dynabeads for 1 h at room temperature in 1x PBS. Alternatively, 2–5 mg cell extracts were centrifuged at 10,000 × g for 20 min at 4°C. The resulting soluble (supernatant) and total microsomal fractions (pellet resuspended in 1 ml isolation buffer) were subjected to immunoprecipitation as indicated above. The immunoprecipitated proteins were washed three times with wash buffer I (20 mM Tris-HCl pH 7.5, 5 mM EDTA, 100 mM NaCl, 1% Triton X-100), once with the same buffer but containing 1 M NaCl, then finally washed with kinase buffer (20 mM Hepes pH 7.5, 10 mM $MgCl_2$, 1 mM DTT, 1 mM $CaCl_2$) and analyzed by *in vitro* kinase assays at 30°C in kinase buffer containing 25 μM ATP, [$\gamma$-$^{32}$P]ATP (2 μCi per reaction) with or without 2 mM EGTA and myelin basic protein (MBP) as a substrate. The reactions were stopped by the addition of SDS-loading buffer after 60 min. SDS-PAGE reaction products were analyzed by autoradiography. For LS-MS/MS analysis, immune complexes were resuspended in Tris pH 7.0 and 150 mM NaCl and analyzed.

## Preparation of Soluble and Membrane Protein Fractions

For membrane fractionation experiments, 5 g tobacco leaves expressing HvCPK2a-GFP-His protein were homogenized with mortar and pestle in homogenization buffer [1xPBS pH 7.4, 0.1 M NaCl, 0.1% NP-40, 1 mM DTT, 1 mM PMSF a cocktail of protease inhibitors (Roche)]. After filtration through two layers of Miracloth (Millipore), the homogenate was centrifuged at 10,000 × g for 20 min at 4°C to remove intact organelles and cell walls. The resulting supernatant was used as a soluble protein fraction. The membrane pellet was resuspended in homogenizing buffer and sonicated (15 cycles, each of 30 s power on/30 s off). The HvCPK2a-His-GFP protein was then purified from the microsomal fraction using

Ni-NTA affinity chromatography and analyzed by immunoblot or by LC-MS/MS in the Laboratory of Mass Spectrometry, Institute of Biochemistry and Biophysics, Polish Academy of Sciences (Warsaw, Poland).

## LC-MS/MS Analysis and Data Processing

For protein identification, independent samples were concentrated and desalted on a RP-C18 pre-column (Waters). Peptide separation was achieved on a nano-Ultra Performance Liquid Chromatography (UPLC) RP-C18 column (Waters, BEH130 C18 column, 75 μm i.d., 250 mm long) of a nanoACQUITY UPLC system using a 45 min linear acetonitrile gradient. Peptides were subjected to electrospray ionization (ESI) using the ion source of an Orbitrap Velos mass spectrometer (Thermo). Higher-energy collisional dissociation (HCD) was used to obtain ion fragments. An electrospray voltage of 1.5 kV was used. Raw data files were pre-processed with Mascot Distiller software (version 2.4.2.0, MatrixScience). The peptide masses and fragmentation spectra obtained were matched against the *A. thaliana* database TAIR10 (70772 sequences/28965710 residues) using the probability based protein identification algorithm Mascot (Mascot Daemon v. 2.4.0, Mascot Server v. 2.4.1, Matrix Science). Search parameters were set as follows: enzyme specificity: semi trypsin; peptide mass tolerance: ±30 ppm; fragment mass tolerance: ±0.1 Da; fixed modifications: carbamidomethylation of cysteine; variable modifications: oxidation of methionine, phosphorylation of serine, threonine, tyrosine and ubiquitination of lysine. The protein mass was left as unrestricted, and mass values as monoisotopic with two missed cleavages allowed. To avoid false positive results, a decoy search was performed. The expected value threshold used for analysis was 0.05, which means that all peptide identifications had less than 1 in 20 chance of being a random match. Unique proteins identified in at least three or more replicates were regarded as putative HvCPK2a protein complex elements. The mass spectrometry proteomics data have been deposited to the ProteomeXchange Consortium via the PRIDE partner repository with the dataset identifier PXD005095 (Vizcaíno et al., 2014, 2016). To identify HvCPK2a phosphorylation sites, tryptic peptides were enriched in phosphopeptides using a magnetic Ti-IMAC method according to the manufacturer›s procedure (MagReSyn, ReSyn Biosciences). The spectra of peptides reported as being phosphorylated were examined manually to confirm the precise sites of phosphorylation.

## qPCR Analysis

The third leaves from 10 barley plants were ground into fine powder in liquid nitrogen and total RNA was extracted using TRIZOL reagent according to the manufacturer's manual (Life Technologies). The isolated RNA was purified with SV Total RNA Isolation System (Promega) and cDNA synthesis was performed in duplicate using a SuperScript cDNA Synthesis Kit (Invitrogen). Quantitative RT-PCR (qPCR) analysis was performed using the Stratagene Mx3000P Cycler system with Brilliant III Ultra-Fast SYBR QPCR MM Supermix (Stratagene) in a total volume of 20 μl. The reactions were performed as technical duplicates using independent cDNA synthesis reactions. For each condition three biological replicates were conducted. Expression values were normalized against ADP-ribosylation factor cDNA, which was amplified using the primers 5′-TGCTGAATGAGGATGAGCTG-3′ and 5′-GTCCCTCGTACAACCCTTCA-3′. Gene-specific primers used for amplification of *HvCPK2a* transcripts were 5′-GGGCAAGCTACACAAAGGAG-3′ and 5′-ACGTAGCCGTGGAGGTTGTA-3′.

## Plasmid Construction

To prepare *HvCPK2a* constructs, cDNA clone 3093 NII, which encodes the barley kinase, was obtained from the National Institute of Agrobiological Sciences, Japan. *HvCPK2a* cDNA was amplified by *Pfu* DNA polymerase using the primers For 5′-CACCATGGGAAACTGCTGCGC-3′ and either N-Rev 5′-TAGCACGACGTCGCGCCGCTT-3′ or C-Rev 5′-CACGACGTCGCGCCGCTTCTT-3′, followed by cloning into the pENTR™/SD/D-TOPO® vector (Thermo Fisher Scientific) as either N- or C- terminal fusion constructs. To generate an N-terminal His-HvCPK2a fusion protein, the *HvCPK2a* clone was recombined with the Gateway® pDEST17™ vector (Invitrogen) using Gateway® LR Clonase® II Enzyme Mix (Invitrogen). To generate variants of *HvCPK2a*, a mutagenesis reaction was performed as described in Mitula et al. (2015) using the following primer pairs: K94M (For 5′-GCCTGCA**T**GACCATCGCCAAG-3′, Rev 5′-CTTGGCGATGGTC**A**TGCAGGC) and D189A

(For 5′-CATCCACCGCGCCCTCAAGCC-3′, Rev 5′-GGCTTGAGGGCGCGGTGGATG-3′). The vector construct *35S:HvCPK2a-His-GFP*(Earley et al., 2006) was used for Arabidopsis protoplast transformations and transient expression assays in *Nicotiana benthamiana*. To localize HvCPK2a and mutated versions of the protein in barley protoplasts, we generated a C-terminal HvCPK2a-GFP fusion in the pS5-DEST-EGFP vector using Gateway® LR Clonase® (Invitrogen).

## Protoplast Isolation, Transformation, and Transient Expression Assays

Transient expression assays in Arabidopsis protoplasts or in tobacco leaves were performed as described in Ludwików et al. (2014) and Mitula et al. (2015). For the preparation of barley protoplasts, the central parts of 20 leaves from 6- to 10-day-old spring barley seedlings were dissected and cut into 0.5 mm strips. Strips were transferred to 10 ml enzyme solution (1.5% cellulase, 0.3% macerozyme, 0.5 M sorbitol, 1 mM $CaCl_2$, 10 mM MES pH 5.7) and incubated at 30°C for 3 h in the dark without shaking. After digestion, protoplasts were transferred to 50 ml tubes, centrifuged for 4 min (50 × g, 4°C), washed twice with wash solution (0.5 M sorbitol, 20 mM KCl, 4 mM MES pH 5.7) and resuspended in MMg solution (0.5 M sorbitol, 15 mM $MgCl_2$,4 mM MES pH 5.7) to a final concentration of $2 \times 10^4$ cells per 100 µl. For protoplast transfections, 5–10 µg plasmid DNA was combined with 100 µl protoplasts and 110 µl PEG solution (40% PEG 4000, 200 mM sorbitol, 10 mM $CaCl_2$). After gentle mixing, the solution was incubated for 15 min at RT in a horizontal position. Then, 450 µl W1 buffer (0.5 M sorbitol, 20 mM KCl, 4 mM MES pH 5.7) was added, mixed and centrifuged for 4 min, 50 × g. Supernatant was removed, protoplasts were resuspended in 300 µl W1 solution and incubated overnight at 22°C in the dark in a horizontal position.

## Chlorophyll Content

Pigment composition was analyzed using a Dualex 4 leaf-clip optical sensor based on the protocols described in Cerovic et al. (2012, 2015). The leaf chlorophyll content, the flavonoid index and the nitrogen balance index (NBI) were quantified for the upper side of 10 leaves/plant by measuring light transmission absorbed by chlorophyll at 710 nm and in the near-infrared at 850 nm. Flavonoid content was quantified by absorption of UV light. The NBI was then calculated as the chlorophyll to flavonoid ratio, which is directly related to nitrogen/carbon allocation in leaves. We analyzed 3-week old Arabidopsis plants, five plants of each line, and performed 10 measurements on five mature leaves of each plant. Measurements were made on the laminae, avoiding midribs.

## His-HvCPK2a Overexpression and Purification

*Escherichia coli* Rosetta 2 (DE3) (Novagen) transformed with pDEST17-*HvCPK2a* plasmid was grown overnight in 20 ml LB medium supplemented with 100 μg/ml ampicillin at 37°C with shaking. Twenty ml fresh LB (plus 100 μg/ml ampicillin) were inoculated with 1 ml overnight culture and grown to an optical density ($OD_{600}$) of 0.4–0.6. The culture was then induced with 0.5 mM IPTG and grown for 3 h, bacteria were collected by centrifugation and the pellet stored at −80°C. The pellet was resuspended in lysis buffer (50 mM Tris-HCl pH 7.5, 150 mM NaCl, 0.5% Nonidet P-40, 1 mM PMSF). His-HvCPK2a protein was purified from membrane fractions using Ni Sepharose resin according to standard procedures recommended by the manufacturer (GE Healthcare).

## Immunoblotting

Proteins were fractionated on 10% resolving gels by SDS-PAGE and then transferred onto Immobilon-P membrane (Millipore) by semi-dry transfer. The membranes were blocked in PBS-T (PBS plus 0.01% Tween 20) with 3% skim milk for 1 h with agitation. Next, membranes were washed three times with PBS-T and incubated with primary antibody AS1 (1:1000, AS13 2754 Agrisera), anti-$H^+$-ATPase (1:1000, Agrisera), anti-GST (1:6000, MoBiTec Molecular Biotechnology) or anti-GFP (1:200; sc-8334, Santa Cruz Biotechnology) for 1 h with agitation. After washing, the membranes were incubated with appropriate secondary antibodies for 1 h. Detection was performed using ECL (Thermo Scientific) and visualized by G:BOX (Syngene).

# RESULTS

## Identification, Sequence Comparison, and Phylogeny of Barley CPKs

Plant CPKs constitute a multigene family, but relatively few CPK family members have been characterized. To identify specific CPK gene copies in barley, we performed a BLASTP search against available databases using full length protein sequences from rice and *Arabidopsis*. A reciprocal analysis was performed to confirm that identified entries indeed gave the best match against the relevant rice and *Arabidopsis* sequences.

To investigate the relatedness of the HvCPKs identified to other known CPKs, a phylogenetic tree was generated including full length protein sequences form *H. vulgare, O. sativa*, and *A. thaliana*. A comparison of kinase domain sequences showed that HvCPKs are distributed very unevenly among the four groups. Each group contains CPKs from all three species investigated, and groups are not dominated by particular species. While the first group contains nine HvCPKs, the next three groups (2nd, 3rd, and 4th) include seven, six and two HvCPKs respectively. Interestingly, of the 24 HvCPKs identified, twenty members have orthologous loci in rice. Importantly, OsCPK9 has neither an orthologous nor a paralogous copy in barley, suggesting that the *OsCPK9* gene has evolved since the two plant species diverged.

## Identification of CPKs Involved in the Drought Stress Response

The structure of the CPK proteins is reported to be highly conserved among plants (Hamel et al., 2014). Our results support this: the sequences of HvCPK kinase domains showed remarkable conservation of a 13 amino acid residue sequence in the active site. We hypothesized that this core sequence is likely to be useful for the identification of barley CPKs involved in the drought stress response. Therefore, we raised polyclonal antisera (AS1 and AS2) against two peptides that correspond various regions of the kinase domain; both peptides contain the conserved 13 amino acid core residue. Alignments of HvCPK proteins indicate that they share sequence identity to the AS1 and AS2 peptides in the ranges 76.9–100% and 69.2–100%, respectively. The AS1 and AS2 antisera were used for the immunoprecipitation (IP) of protein complexes containing CPKs from 6-week-old barley plants subjected to drought stress. AS1 immune complexes contained strong kinase activity, but AS2 did not precipitate active kinases under any conditions used. CPKs are localized in various cell compartments, including the cytosol and plasma membrane; therefore, to identify as many putative HvCPKs as possible, we used the soluble protein fraction as well as membrane-bound proteins recovered from sedimenting (microsome) fractions. To confirm that the AS1 antiserum recognizes CPKs, we performed in vitro kinase assays on the resulting immune complexes using MBP as a substrate. AS1 immune complexes identified strong kinase activity in the membrane protein fraction of drought-stressed barley plants (Figure 1).

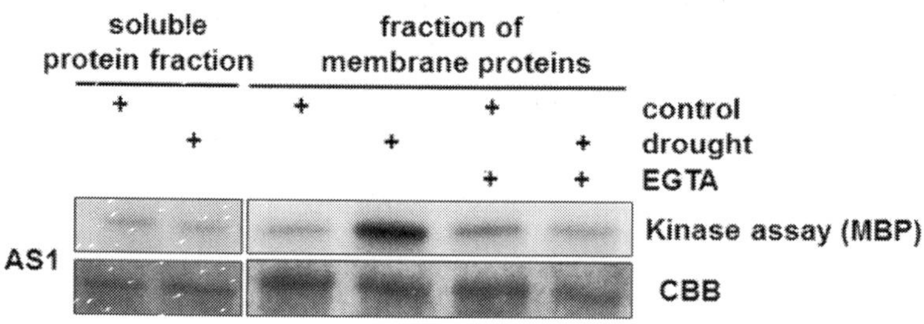


**Figure 1.** Identification of drought-induced CPKs in *Hordeum vulgare*. We used both the soluble protein fraction and membrane-bound proteins isolated from a microsomal fraction from drought-stressed barley to define the subcellular location of AS1-associated kinase activity. Strong kinase activity was present in membrane proteins. Treatment with EGTA ($Ca^{2+}$ chelator) abolished HvCPK2a kinase activity. MBP was used as a substrate; Coomassie Brilliant Blue (CBB) staining confirmed equal loading of MBP.

To identify CPKs within proteins immunoprecipitated using AS1 antibodies, the same immune complexes were analyzed using LC-MS/MS in conjunction with database searching against the TAIR database and the Triticum, Oryza, and Hordeum subsets of the National Center for Biotechnology Information (NCBI). Using this approach, several putative barley CPKs were found in the drought-tolerant cultivar; these included HvCPK24, HvCPK2a, and HvCPK14, which showed 92.7, 95.2, and 92.9% similarity to OsCPK24, OsCPK2, and OsCPK14 from rice, and 77, 85.8, and 84.1% identity to CPK11, CPK34, and CPK17 from Arabidopsis, respectively. Although none of the putative HvCPKs identified have previously been implicated in the drought response, the HvCPK24-orthologous kinases AtCPK11 and OsCPK24 are known to be involved in drought, high-salt or cold stress responses (Wan et al., 2007). In contrast, the HvCPK2a-orthologous kinases AtCPK34, AtCPK17, and OsCPK2, which are almost undetectable in vegetative tissues but are abundant in pollen, flower or seed tissues, are reported to be critical only for male reproduction and have not been linked to any stress response (Valmonte et al., 2014). To confirm the role of HvCPK2a in the response to drought, total RNA was extracted from drought-stressed barley at different time points. The level of *HvCPK2a* transcript abundance was significantly increased after 3 days of drought (Figure 2), showing that *HvCPK2a* is detectable in leaves collected from control and drought-treated plants. Our results suggest functional divergence between HvCPK2a and its orthologs in Arabidopsis and rice. Therefore, our further analyses concentrate on the role of HvCPK2a.

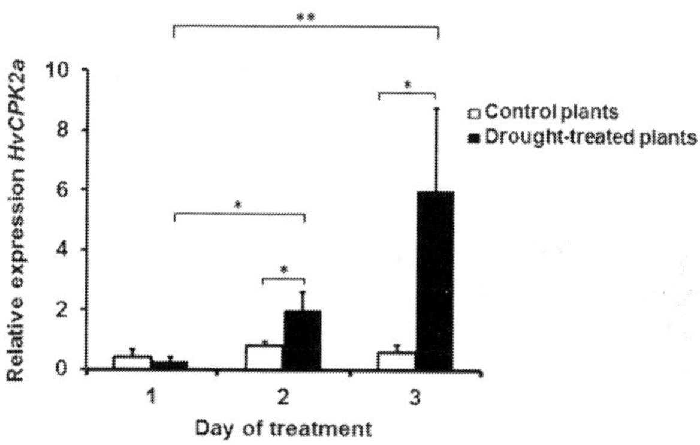


**Figure 2. *HvCPK2a* gene expression profile in *Hordeum vulgare* L. cv. "Sebastian" during drought stress**. Barley *HvCPK2a* gene expression under intensifying drought stress and control conditions: a significant increase in *HvCPK2a* expression occurred on the third day of drought. Relative quantification was determined by qPCR analysis (with ADP-ribosylation factor genes as internal control). Vertical bars correspond to standard errors of the mean. The asterisks indicate significant differences according to Student›s *t*-test; $*P < 0.05$, $**P < 0.005$.

## Subcellular Localization of HvCPK2a

HvCPK2a is a canonical CPK with putative myristoylation/palmitoylation sites suggesting that this kinase might be associated with membranes.

However, it is known that numerous factors can affect CPK localization, including protein activity, calcium and posttranslational modifications (PTMs) (Hegeman et al., 2006; Schulz et al., 2013; Simeunovic et al., 2016). Because CPKs integrate different signaling cascades, rapid CPK inactivation is crucial for proper signal transduction; therefore, we were interested to know whether the kinase activity of HvCPK2a affects its subcellular localization. Accordingly, constructs for the *in vivo* expression of wild type (WT) HvCPK2a, as well as K94M and D189A variants which are reported to have reduced or abolished kinase activity (Patharkar and Cushman, 2000; Franz et al., 2011; Matschi et al., 2013), were generated. The CPK2$^{K94M}$ version of the protein possesses a single amino acid change in its ATP-binding loop, while the CPK2$^{D189A}$ variant carries a single amino acid change within its kinase active site. C-terminal GFP fusions were transiently expressed under the control of the *35S*promoter in *A. thaliana* protoplasts. We immunoprecipitated the WT, K94M, and D189A versions of HvCPK2a from protoplasts extracts and confirmed that both mutants have significantly reduced kinase activity compared to WT. The presence of all three HvCPK2a versions was demonstrated by immunoblotting (Figure 3A).

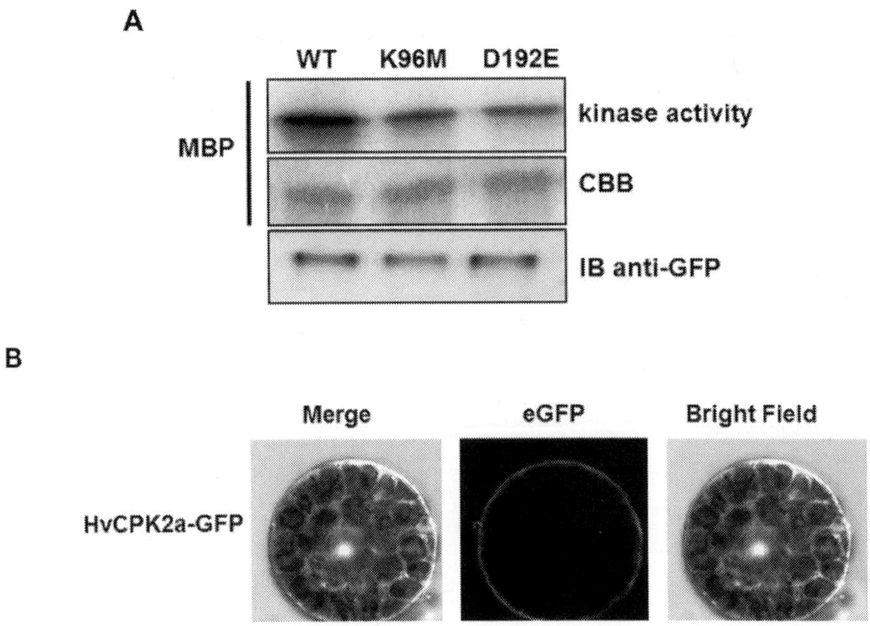

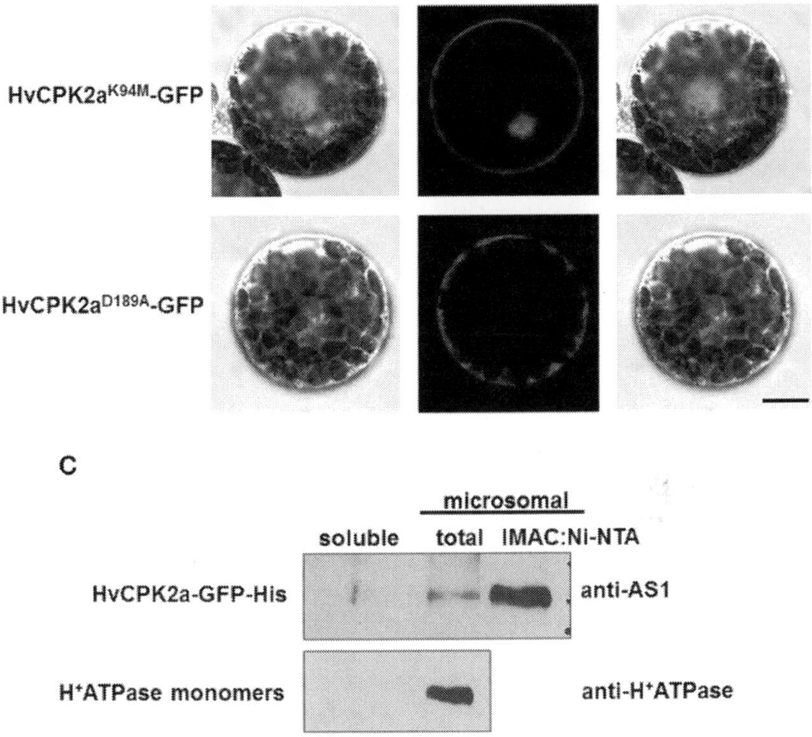


**Figure 3.** Subcellular localization of the HvCPK2a-GFP fusion protein. (A) Microscopy images show that WT HvCPK2a and the D189A variant, visualized as GFP fusion proteins, are associated with the plasma membrane in barley protoplasts. A variant with a single amino acid change within the ATP-binding site (HvCPK2a$^{K94M}$-GFP) showed localization to both membrane and cytoplasm. Scale bar: 10 μm. **(B)** Expression of all three forms of HvCPK2a-GFP in barley protoplasts was confirmed by IB using an anti-GFP antibody. **(C)** HvCPK2a-GFP co-fractionates with the plasma membrane fraction in tobacco plants. Total proteins from tobacco leaves transiently expressing HvCPK2a-GFP-His were separated into soluble and total microsomal fractions. The microsomal fraction, which contained HvCPK2a-GFP-His, was enriched by nickel-affinity chromatography, then separated on a 10% SDS-PAGE gel and transferred to Immobilon-P membrane. $H^+$-ATPase monomers of mass 95 kDa, which are known to be membrane-associated, were detected in the microsomal fraction.

Next, we determined the subcellular localization of HvCPK2a variants in barley protoplasts. Because we were unable to transform barley protoplasts using pEarlyGate vectors, we generated a separate set of vector constructs that facilitated protein localization in a barley transient expression system. Thus, coding sequences for WT HvCPK2a and the two mutant forms were expressed in pS5N3 vectors under control of the UBQ promoter as C-terminal fusions. We observed WT HvCPK2a and both variants localized to the plasma membrane, but the variant with a mutation in the ATP-binding site (K94M) was also found in the nucleus (Figure 3B), suggesting that HvCPK2a is important for the amplification of calcium signals in various cellular compartments. A similar distribution was observed in Arabidopsis as in barley protoplasts, WT HvCPK2 was mostly located along the inner face of the plasma membrane, as was the D189A mutant. Interestingly, however, the K94M variant localized to the nucleus as well as the plasma membrane.

The membrane localization of HvCPK2a was further demonstrated by AS1 immunoblotting of protein extracts from tobacco leaves expressing HvCPK2a-GFP-His (Figure 3C). We identified the HvCPK2a fusion protein in the total microsomal fraction, and also after enrichment of this fraction by nickel-affinity chromatography, but not in the soluble protein fraction. Antibodies against $H^+$-ATPase confirmed the presence of membrane proteins in the microsomal fraction. The localization of the barley kinase is supported by a previous finding showing that the HvCPK2a orthologous kinase AtCPK34 is also associated with the plasma membrane (Myers et al., 2009).

## Identification of HvCPK2a-Interacting Proteins

Despite the importance of CPKs in plants, relatively little is known about their substrates. To improve our understanding of the HvCPK2a signal transduction pathway, we attempted to identify proteins that interact with HvCPK2a using LC-MS/MS. For this purpose, we expressed HvCPK2a-GFP-His in a *Nicotiana benthamiana* transient expression system. Fully developed leaves from 4-week-old tobacco plants were infiltrated with *Agrobacterium* carrying a *35S:HvCPK2a-His-GFP* construct and plant material was collected 5 or 6 days later. As a control, non-infiltrated tobacco leaves were used. Because HvCPK2a is associated with membranes, we recovered HvCPK2a-GFP-His protein from the microsomal fraction using nickel-affinity chromatography and subjected co-purifying proteins to LC-MS/MS. WT tobacco leaves were processed similarly as a negative control.

We initially used the tobacco database with LC-MS/MS data to attempt to identify proteins that interact with HvCPK2a, but this approach mainly highlighted proteins connected with the plant response to *Agrobacterium* infection. Therefore, in further analysis we used the TAIR10 Arabidopis database and as a result identified 29 candidate HvCPK2a-interacting proteins that were not present in control samples. Identification of AtCPK34 validated the method used to purify HvCPK2a. The most abundant proteins found in HvCPK2a complexes were citrate synthase 5 (CSY5), sucrose synthase 4 (SUS4), NAD-dependent malic enzyme 1 (NAD-ME1), selenium binding protein 1 (SBP1), formate dehydrogense (FDH), 40S ribosomal protein SA B (RSAB), chlorophyll A-B binding family protein (NPQ4), GTP-binding elongation factor Tu family protein and glutathione S-transferase TAU23. In four of the seven MS/MS runs we identified pyruvate orthophosphorate dikinase (PPDK), indole-3-butyric acid response 1 (IBR1), BHB domain containing membrane associated protein family, translation elongation factor EF1B, adenine nucleotide alpha hydrolases-like superfamily protein, beta-ureidopropionase and aldehyde dehydrogenase 2B4. Among proteins identified in three or fewer samples were serine hydroxmethyltransferase 2 (SHM2), two members of the family of 90 kDa heat shock proteins (HSP90.2/ERD8 and HSP90.7), phospholipase D alpha 1 (PLDALPHA1) and two isoforms of the $H^+$ATPase family, AHA1 and AHA3.

The candidate proteins listed above were analyzed further with regards to their role in biological processes. The majority of these proteins are known to play essential roles in the stress responses to salt, drought, cold and wounding. These stress-responsive proteins included early response to dehydration (HSP90.2/ERD8), an ER-resident HSP90-like protein (SHD), NAD-ME1, SBP1, UDP-glucose:glycoprotein glucosyltransferases (EBS1, UGGT, PSL2), SUS4 and formate dehydrogense (FDH). Other interesting candidates were proteins involved in ABA signaling ($H^+$-ATPase 1, PLDALPHA1) and development (PLDALPHA1; IBR1; aspartate aminotransferase 3). These results suggest that HvCPK2a is involved in a variety of cellular functions.

## HvCPK2a Overexpression Reduces Tolerance to Drought in Transgenic Arabidopsis

To confirm a role for HvCPK2a in the drought stress response we generated transgenic *A. thaliana* lines that overexpress *HvCPK2a*. Ten independent lines were selected and all were included in further analysis; representative results of three lines are shown. There was no difference in growth between WT and *HvCPK2a*-overexpressing plants, but after exposure to drought, *HvCPK2a*-overexpressing plants (Oex lines #1–3) exhibited lower leaf relative water content (RWC) and lower survival rates than WT plants (Figure 4), indicating that *HvCPK2a*-overexpression reduces tolerance to drought.

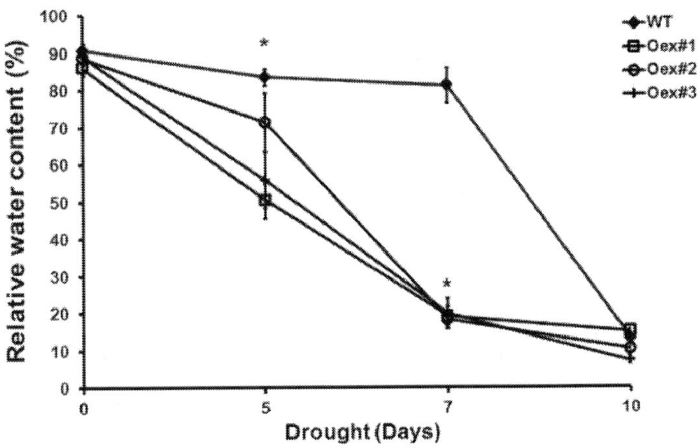


**Figure 4.** Drought-stressed HvCPK2a overexpression lines exhibit reduced water capacity. Relative water content of WT and Oex lines in response to drought stress: drought-stressed transgenic plants lose more water than WT. Data shown are the means of 10 measurements taken from 10 plants ±SE from two independent experiments. The asterisks indicate significant differences in Oex lines compared with WT Col-0 according to Student's *t*-test; $*P < 0.01$.

Chlorophyll and flavonoid levels can be used to monitor senescence and both are good indicators of nitrogen status in plants (Agati et al., 2015, 2016). Therefore, we used a Dualex 4 optical sensor to investigate the influence of drought on chlorophyll (Chl index) and flavonoid (FLAV index) content, which also allowed us to calculate the NBI. As indicated in Figures 5A–C, the Chl index significantly increased during the period of drought in both WT and *HvCPK2a*-overexpressing lines #1–3. This is particularly clear on the 6th and 7th days of drought, when dehydration becomes more pronounced: the Chl index is significantly higher in *HvCPK2a*-overexpressing lines than in WT (Figures 5B,C). Next, we analyzed the FLAV index, and found that, as drought progressed, it also significantly increased in *HvCPK2a*-overexpressing lines compared to WT (Figures 5D–F), becoming three times higher than WT after 7 days (Figure 5F). For the NBI, there was no difference between *HvCPK2a*-overexpressing lines #1–3 and WT under fully hydrated conditions (Figure 5G), but after drought was imposed the NBI decreased significantly for all Oex lines compared to WT (Figures 5H,I). Taken together, these data indicate that the drought-sensitive phenotype of *HvCPK2a*-overexpressing plants is associated with reduced water retention capacity and diversification of specialized secondary metabolism.

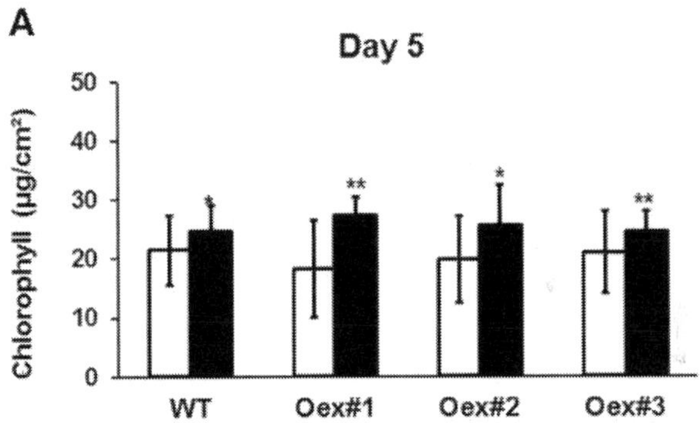


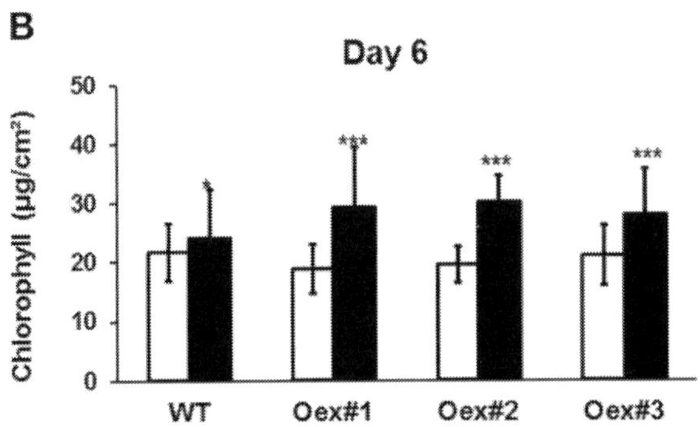

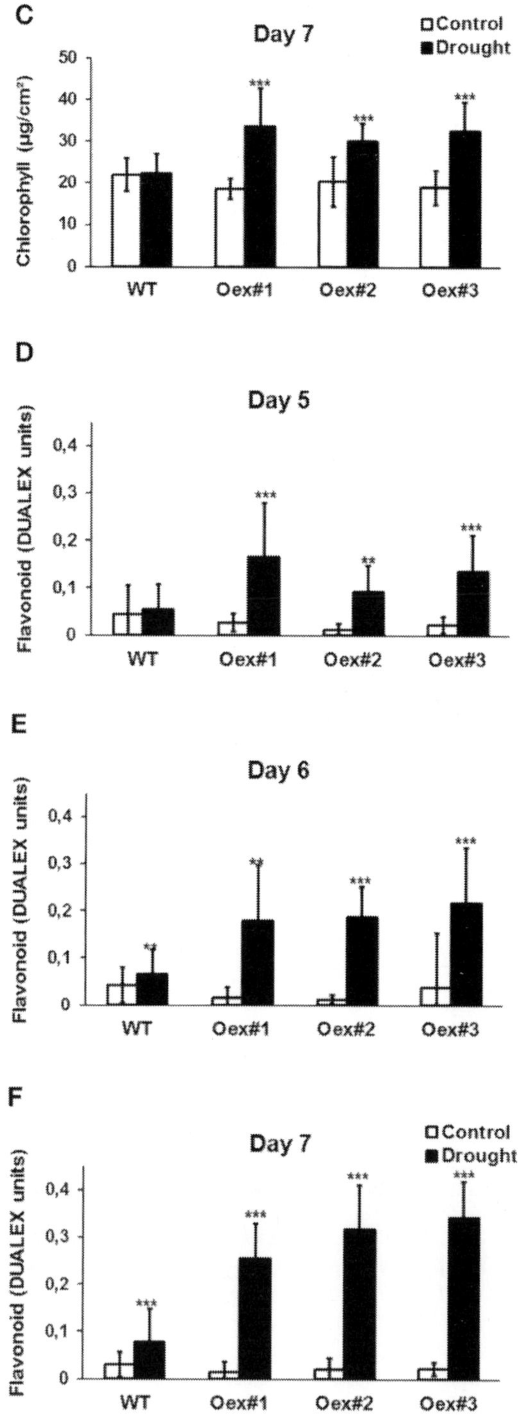
C
Day 7
Control
Drought
Chlorophyll (µg/cm²)
50
40
30
20
10
0
***
***
***
WT
Oex#1
Oex#2
Oex#3
D
Day 5
Flavonoid (DUALEX units)
0,4
0,3
0,2
0,1
0
***
**
***
WT
Oex#1
Oex#2
Oex#3
E
Day 6
Flavonoid (DUALEX units)
0,4
0,3
0,2
0,1
0
**
**
***
***
WT
Oex#1
Oex#2
Oex#3
F
Day 7
Control
Drought
Flavonoid (DUALEX units)
0,4
0,3
0,2
0,1
0
***
***
***
***
WT
Oex#1
Oex#2
Oex#3

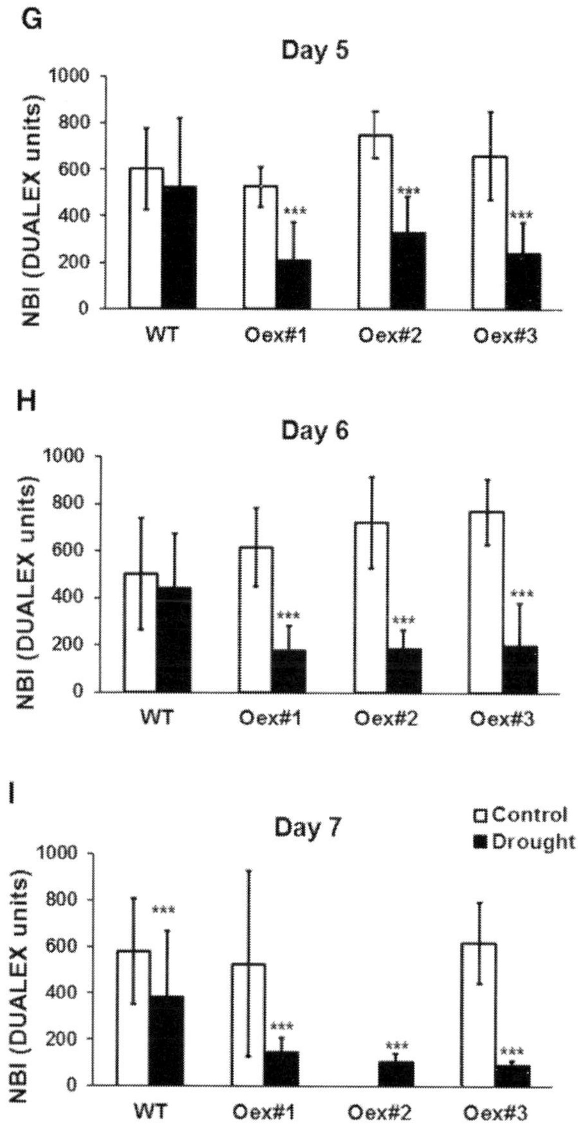


**Figure 5.** Overexpression of *HvCPK2a* **in Arabidopsis reduces plant tolerance to drought**. Drought-stressed *HvCPK2a*-overexpressing plants have higher Chl index (A–C), higher flavonoid content (D–F) and reduced NBI index (G–I) compared to WT plants. Data represent mean ± SE; $n = 10$ plants per line. The asterisks indicate significant differences in Oex lines compared with WT Col-0 according to Student›s *t*-test; $^{*}P < 0.01$; $^{**}P < 0.001$; $^{***}P < 0.0001$.

## Recombinant HvCPK2a is Regulated by Autophosphorylation

Recent studies indicate that the activity of CPK kinases is regulated by autophosphorylation (Chaudhuri et al., 1999; Harmon et al., 2000; Oh et al., 2012). To test whether HvCPK2a kinase is activated by autophosphorylation, His-tagged HvCPK2a protein was affinity purified and used in a kinase assay with MBP as a substrate. As shown in As shown in Figure 6A, at the initial time point His-*Hv*CPK2a was highly active and, accordingly, we did not observe a significant increase in kinase activity (with MBP as substrate) in the presence of $Ca^{2+}$, suggesting that His-HvCPK2a autophosphorylates in bacterial systems. To confirm this observation, we monitored His-HvCPK2a activity in bacterial cultures after IPTG induction in a time course experiment. In the absence of $Ca^{2+}$, we observed phosphorylation of the MBP substrate by His-HvCPK2a as soon as 30 min after IPTG induction. At the same time, de novo autophosphorylation of His-HvCPK2a was barely noticeable. Moreover, in the presence of $Ca^{2+}$ ions there was a substantial increase in MBP substrate phosphorylation, as well as an increase in autophosphorylation of His-HvCPK2a (Figure 6B). When His-HvCPK2a was incubated with both $Ca^{2+}$ and EGTA, a significant attenuation of kinase activity against MBP was observed. In line with this, there was nearly no His-HvCPK2a kinase activity against MBP in the presence of the calcium chelator, suggesting that His-HvCPK2a activity is indeed calcium dependent (Figure 6C). Overall, these results demonstrate that *HvCPK2a* encodes a functional calcium-dependent protein kinase.

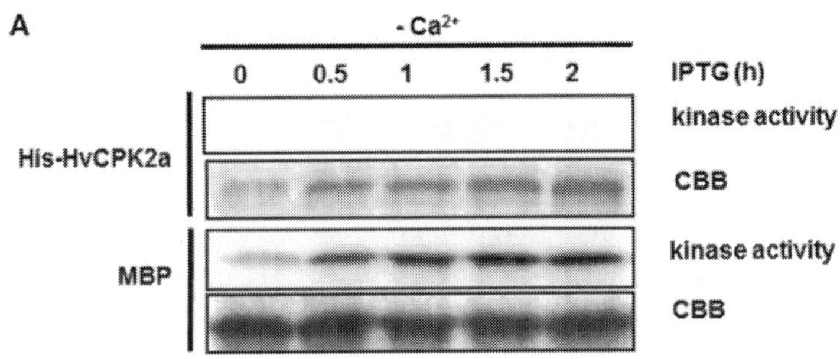

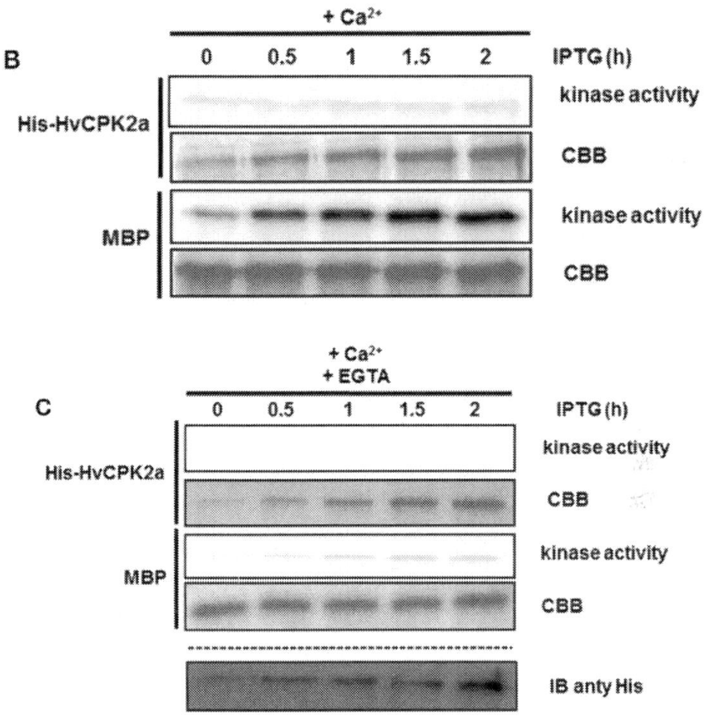


**Figure 6.** Autoactivation of recombinant His-HvCPK2a. Recombinant protein expression was induced by IPTG, collected at the indicated time points, then analyzed for kinase activity using MBP as artificial substrate in the presence (B–C) or absence (A) of $Ca^{2+}$. EGTA was used as a calcium chelator to prevent calcium-dependent phosphorylation. Radioactivity of MBP bands was measured with a phosphoimager. Coomassie Brilliant Blue (CBB) staining confirmed equal loading. His-HvCPK2a was detected using anti-His antibody.

We used mass spectrometry to identify autophosphorylation sites and found nine phosphorylated residues in the activated His-HvCPK2a, comprising three threonine residues (T79, T83, and T95), five serine residues (S102, S199, S214, S229, and S484) and one tyrosine residue (Y231) (Table 1; Data Sheet 1). Seven of the first eight residues listed, namely T79, T83, T95, S102, S199, S214, and S229, were located in the kinase domain. The eighth residue, S484, was identified within the calmodulin-like domain, toward the C-terminus (Figure 7). As shown in Table 1, the most abundant phosphorylation sites were T95 and S229. One of these, T95, was identified in the immediate proximity to K94, a conserved lysine that is crucial for ATP-binding in CPKs (Sheen, 1996). Interestingly, the HvCPK2a homolog AtCPK34 is thought to be a dual specificity kinase (Oh et al., 2012, 2013), which is in accord with the identification of a single autophosphorylated tyrosine residue (Y231) in HvCPK2a, indicating that the barley protein does have tyrosine kinase activity. As well as phosphorylation sites, we also found a single ubiquitination event on K85 (Figure 7), suggesting that HvCPK2a undergoes proteasomal degradation.

**Table 1.** Identification of His-HvCPK2a autophosphorylation sites

| Spectra count | | Sites/ residues | Score | Observed | Mr (expt) | Mr (calc) | Delta | M | Expect | U | Peptide |
|---|---|---|---|---|---|---|---|---|---|---|---|
| *In vivo* autophosphorylation sites | *In vitro* autophosphorylation sites | | | | | | | | | | |
| 1 | 0 | T75 | 37 | 722.8186 | 1443.6226 | 1443.6040 | 0.0186 | 0 | 0.021 | | [illegible] **pT**[illegible] + Phospho (ST) |
| 2 | 0 | T83, K85 | 37.0 | 793.3122 | 1584.6098 | 1584.6354 | −0.0256 | 0 | 0.017 | U | R.SQFGVTYLCpTEubqK.S + GlyGly (K); Phospho (ST) |
| 34 | 56 | T95 | 57.5 | 504.2228 | 1006.4310 | 1006.4381 | −0.0071 | 1 | 0.0041 | | Q.FACpTAK.R + Phospho (ST) |
| 0 | 5 | S102 | 86.4 | 755.8821 | 1509.7096 | 1509.7440 | −0.0343 | 0 | 0.0011 | | R.KLpSKEDVEDVR.R + Phospho (ST) |
| 3 | 13 | S199 | 71 | 748.8919 | 1495.7693 | 1495.7687 | 0.0006 | 0 | 0.00027 | U | R.DLKPENFLLpSK.D + Phospho (ST) |
| 1 | 2 | S213 | 76.7 | 656.3003 | 1310.5981 | 1310.5948 | 0.0033 | 0 | 9.1e-05 | U | K.ATDFGLpSVFFK.P + Phospho (ST) |
| 171 | 260 | S229 | 114 | 859.4178 | 1716.8211 | 1716.8375 | −0.0164 | 0 | 2.3e-08 | U | K.DVGpSAYYIAPEVLK.R + Phospho (ST) |
| 0 | 14 | Y231 | 106 | 859.4266 | 1716.8386 | 1716.8375 | 0.64 | 1 | 1.5e-07 | U | K.DVGSApYYIAPEVLK.R + Phospho (Y) |
| 6 | 9 | S484 | 54.7 | 749.8043 | 1497.5940 | 1497.5984 | −0.0044 | 0 | 0.013 | U | K.ELpSDVDADNDGR.I + Phospho (ST) |

Using LC MS/MS we identified nine autophosphorylation sites within recombinant HvCPK2a. Phosphorylation of residue T83 and ubiquitination of K85 were observed in the same peptide. Spectra count—the number of spectra assigned per peptide; Score—Mascot Score is the log of a probability that the observed match is a random event [−10log10(P), where P is the absolute probability], Observed—experimental m/z value, Mr (expt)—experimental m/z transformed to a relative molecular mass, Mr (calc)—calculated relative molecular mass of the matched peptide, Delta—difference (error) between the experimental and calculated masses, M—the number of missed cleavages, Expect—expectation value, directly equivalent to the E-value in a Blast search (the lower the value, the better), U—unique peptide for the protein hit, Peptide—sequence of identified phosphopeptides. Spectra for each peptide are included in Data Sheet 2.

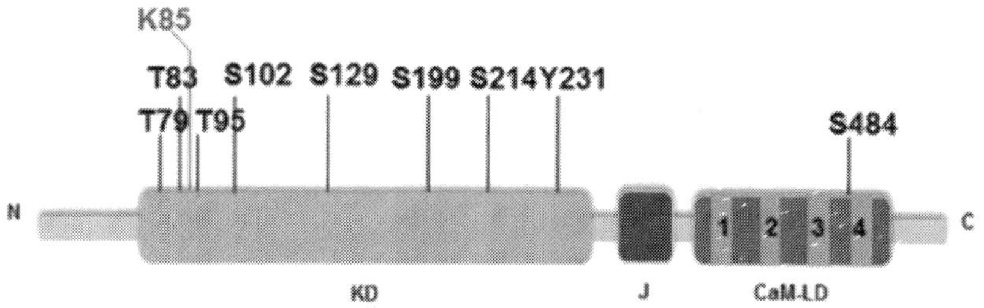


**Figure 7.** HvCPK2a autophosphorylation sites. Schematic of HvCPK2a showing the location of phosphorylation sites identified by LC-MS/MS. The conserved kinase domain (KD) is shown in orange, the junction domain (J) in black and the calmodulin-like domain (CaM-LD) in gray and red. Seven out of the eight phosphorylated residues identified were located within the kinase domain (KD) and one in the calmodulin-like domain in close proximity to the fourth EF-hand structure. Lysine 85 (K85) was identified as a ubiquitination site.

## DISCUSSION

Calcium ions are one of the most important second messenger signaling systems in all eukaryotes. Plants have evolved a multigene family of CPKs, which are involved in development, stress responses and plant hormone signaling (Ludwig et al., 2004; Yoon et al., 2006; Kobayashi et al., 2007; Ma and Wu, 2007; Myers et al., 2009; Mehlmer et al., 2010; Xu et al., 2010; Asano et al., 2011, 2012; Boudsocq and Sheen, 2013). However, in barley, relatively few biological functions have been assigned to CPK family members. In the present study, we explored the possible role of HvCPK2a in the drought stress response. We showed that the *HvCPK2a* transcript level increases in response to drought stress treatment (Figure 2), but that overexpression of HvCPK2a increases drought sensitivity (Figures 4, 5). We also found that HvCPK2a-GFP and HvCPK2a$^{D189A}$-GFP fusion proteins are associated with the plasma membrane (Figure 3A). Moreover, proteins that putatively interact with HvCPK2a are localized in various cellular compartments. Finally, we demonstrated that HvCPK2a is a dual-specificity kinase that autophosphorylates at T79, T83, T95, S102, S199, S214, S229, S484, and Y231 residues (Table 1; Figure 7).

The role of calcium-dependent kinases in the response to drought is best characterized in Arabidopsis and rice (Saijo et al., 2000, 2001;

Ma and Wu, 2007; Zou et al., 2010). There are also examples of CPKs involved in drought signaling in *Vicia faba* (Liu et al., 2006) and wheat (Li et al., 2008). To identify barley CPKs regulated by drought, we used targeted antibodies and identified CPK activity in immune complexes from drought-stressed samples. Using LC-MS/MS, we identified HvCPK2a and HvCPK14 as drought-induced CPKs, although their Arabidopsis homologs are predominantly expressed in mature pollen (Harper et al., 2004; Myers et al., 2009). In contrast to the situation in Arabidopsis, we found that *HvCPK2a* is expressed in leaves in response to drought stress (Figure 2). Interestingly, the sequence identity of HvCPK2a and its rice homologs is rather high (~90%), and only slightly lower (~85%) for HvCPK2a and its Arabidopsis homolog. We conclude therefore that the apparent functional divergence is more likely due to differences in promoter regulatory sequences, resulting in divergent expression patterns between species. However, it cannot be excluded that variation in amino acid sequence may add an additional level of functional diversification, for example, by modifying target protein binding (Dun et al., 2014).

We investigated the role of HvCPK2a in the drought stress response using two approaches. In the first approach, we generated Arabidopsis plants that constitutively overexpress barley *CPK2a*. The transgenic plants were less able to adapt to adverse conditions compared to WT plants, indicating that HvCPK2a is a negative regulator of tolerance to drought stress. It has been reported that Arabidopsis CPK23 negatively regulates drought resistance by affecting ion transport (Ma and Wu, 2007). Importantly, *HvCPK2a*-overexpressing plants have a reduced RWC (Figure 4) and NBI (Figure 5), suggesting an important role for *HvCPK2a* in the regulation of leaf growth and biomass under drought conditions.

In the second approach, we investigated in detail how HvCPK2a contributes to regulation of the drought stress response by looking for proteins that interact with the barley kinase. Using LC-MS/MS, we identified several putative HvCPK2a-interacting partners including a membrane-localized $H^+$-ATPase, which has been shown to be regulated by CPKs (Schaller and Sussman, 1988; Xing et al., 1996; Camoni et al., 1998a,b; Lino et al., 1998; De Nisi et al., 1999; Kinoshita and Shimazaki, 1999; Rutschmann et al., 2002), a glutathione S-transferase (GST), which is crucial in the response to various stress conditions

(Bartlling et al., 1993; Cummins et al., 2011; Kumar et al., 2013), and a selenium binding protein (AtSBP) involved in tolerance to cadmium and selenium (Dutilleul et al., 2008; Hugouvieux et al., 2009; Schild et al., 2014). Another protein identified, a sucrose synthase (SUS), was found to be involved in long-term adaptation to drought stress conditions (Déjardin et al., 1999; Baud et al., 2004). Importantly, SUS protein sequences contain a predicted motif for CPK-mediated phosphorylation (Huber et al., 1996; Nakai et al., 1998; Zhang et al., 1999; Komina et al., 2002; Hardin et al., 2003, 2004; Fedosejevs et al., 2014). For example, maize SUS1 has two conserved phosphorylation sites for CPK (Ser-15 and Ser-170). N-terminal phosphorylation of Ser-15 stimulates SUS1 activity and affects its membrane association. On the other hand, S170 phosphorylation is thought to be involved in the proteolytic turnover of SUS1 (Hardin et al., 2003). N-terminal phosphorylation of SUS1 by CPK has also been demonstrated in *Ricinus communis*. These data suggest that the drought stress response in barley might be controlled in the long term by HvCPK2a phosphorylation of SUS.

Other enzymes identified in HvCPK2a complexes were formate dehydrogenase (FDH), NAD-dependent malic enzyme (NAD-ME) and pyruvate orthophosphate dikinase (PPDK). Recent studies showed that FDH contributes to stress responses; for example, it is involved in the response to iron deficiency in barley roots, and to osmotic stress in potato (Suzuki et al., 1998; Ambard-Bretteville et al., 2003). NAD-ME provides substrates for the tricarboxylic acid cycle and therefore belongs to a key pathway involved in carbon metabolism. In drought-stressed *Amaranthus cruentus*, NAD-ME activity is increased, although the protein level remains stable (Guo et al., 2009; Babayev et al., 2014), suggesting that enzyme activity might be regulated by posttranslational modifications. In rice, PPDK, also known for its role in C4-plant photosynthesis, is regulated by ABA and water deficit (Moons et al., 1998). Overall, these results indicate that HvCPK2a integrates metabolism (formate, malic and pyruvate metabolism) to sustain plant growth and yield during stress conditions.

To understand the role of HvCPK2a in the drought stress response, we examined the mechanism of its activation (Figure 6). CPKs are classified as serine/threonine protein kinases, but there are data demonstrating that CPKs are dual-specificity kinases, which autophosphorylate their own Ser/

Thr and Tyr residues (Oh et al., 2012). Our data demonstrate that HvCPK2a is a $Ca^{2+}$-dependent dual-specificity kinase that autophosphorylates in bacteria on T, S and Y residues localized within the catalytic (T79, T83, T95, S102, 199, S214, S229, Y231) and a calmodulin-like domain (S484) (Figure 7; Data Sheet 2). Previous studies have demonstrated that phosphorylation of the catalytic domain CPK is critical for kinase activity (Chaudhuri et al., 1999; Chehab et al., 2004; Oh et al., 2012). As eight out of nine HvCPK2a autophosphorylation sites are within the catalytic domain, it will be important to determine in future studies which specific phosphosites are kinase-activating. The function of phosphorylation within the calmodulin-like domain is currently unknown and likewise deserves investigation, particularly as corresponding phosphosites were found in tomato LeCPK1 (Ser-439) and ice plant MeCPK1 (Ser-420) (Rutschmann et al., 2002; Chehab et al., 2004).

In summary, this study provides new insights into the role of barley CPKs in the drought stress response. Compared to its Arabidopsis homologs, HvCPK2a has apparently acquired different, but key functions in the regulation of drought tolerance. The results presented here provide the basis for further analysis of CPK2a functions in barley.

## AUTHOR CONTRIBUTIONS

AC, FM, LM, OF, SJ, MT, MM, and AL conceived the study, carried out the experiments and analyzed the data. MT and MJ assisted in manuscript data acquisition and analysis. AC and MJ were involved in manuscript data deposition. AL and JS participated in study design and coordination. FM, LM and OF participated in drafting the article. AC and AL wrote the manuscript. All authors revised and approved the final manuscript.

## CONFLICT OF INTEREST STATEMENT

The authors declare that the research was conducted in the absence of any commercial or financial relationships that could be construed as a potential conflict of interest.

## ACKNOWLEDGMENTS

We thank Hanna Korcz-Szatkowska for excellent technical assistance. This research was supported by the European Regional Development Fund through the Innovative Economy for Poland 2007–2013, POLAPGEN-BD "Biotechnological tools for breeding cereals with increased resistance to drought" (project No. WND-POIG.01.03.01-00-101/08) to JS and KNOW Poznan RNA Centre, 01/KNOW2/2014.

## REFERENCES

1. Agati, G., Foschi, L., Grossi, N., and Volterrani, M. (2015). In field non-invasive sensing of the nitrogen status in hybrid bermudagrass (*Cynodon dactylon* × *C. transvaalensis* Burtt Davy) by a fluorescence-based method. *Eur. J. Agron.* 63, 89–96. doi: 10.1016/j.eja.2014.11.007
2. Agati, G., Tuccio, L., Kusznierewicz, B., Chmiel, T., Bartoszek, A., Kowalski, A., et al. (2016). Nondestructive optical sensing of flavonols and chlorophyll in white head cabbage (*Brassica oleracea* L. var. *capitata* subvar. *alba*) grown under different nitrogen regimens. *J. Agric. Food Chem.* 64, 85–94. doi: 10.1021/acs.jafc.5b04962
3. Ambard-Bretteville, F., Sorin, C., Rébeillé, F., Hourton-Cabassa, C., and des Francs-Small, C. C. (2003). Repression of formate dehydrogenase in *Solanum tuberosum* increases steady-state levels of formate and accelerates the accumulation of proline in response to osmotic stress. *Plant Mol. Biol.* 52, 1153–1168. doi: 10.1023/B:PLAN.0000004306.96945.ef
4. Asano, T., Hakata, M., Nakamura, H., Aoki, N., Komatsu, S., Ichikawa, H., et al. (2011). Functional characterisation of OsCPK21, a calcium-dependent protein kinase that confers salt tolerance in rice. *Plant Mol. Biol.* 75, 179–191. doi: 10.1007/s11103-010-9717-1
5. Asano, T., Hayashi, N., Kikuchi, S., and Ohsugi, R. (2012). CDPK-mediated abiotic stress signaling. *Plant Signal Behav.* 7, 817–821. doi: 10.4161/psb.20351
6. Babayev, H., Mehvaliyeva, U., Aliyeva, M., Feyziyev, Y., and Guliyev, N. (2014). The study of NAD-malic enzyme in *Amaranthus cruentus* L. under drought. *Plant Physiol. Biochem.* 81, 84–89. doi: 10.1016/j.plaphy.2013.12.022
7. Bartlling, D., Radzio, R., Steiner, U., and Weiler, E. W. (1993). A

glutathione S-transferase with glutathione-peroxidase activity from *Arabidopsis thaliana*. *Eur. J. Biochem.* 216, 579–586. doi: 10.1111/j.1432-1033.1993.tb18177.x

8. Baud, S., Vaultier, M.-N., and Rochat, C. (2004). Structure and expression profile of the sucrose synthase multigene family in Arabidopsis. *J. Exp. Bot.* 55, 397–409. doi: 10.1093/jxb/erh047
9. Benetka, W., Mehlmer, N., Maurer-Stroh, S., Sammer, M., Koranda, M., Neumüller, R., et al. (2008). Experimental testing of predicted myristoylation targets involved in asymmetric cell division and calcium-dependent signalling. *Cell Cycle* 7, 3709–3719. doi: 10.4161/cc.7.23.7176
10. Boudsocq, M., and Sheen, J. (2013). CDPKs in immune and stress signaling. *Trends Plant Sci.* 18, 30–40. doi: 10.1016/j.tplants.2012.08.008
11. Boudsocq, M., Willmann, M. R., McCormack, M., Lee, H., Shan, L., He, P., et al. (2010). Differential innate immune signalling via $Ca^{2+}$ sensor protein kinases. *Nature* 464, 418–422. doi: 10.1038/nature08794
12. Camoni, L., Fullone, M. R., Marra, M., and Aducci, P. (1998a). The plasma membrane H+-ATPase from maize roots is phosphorylated in the C-terminal domain by a calcium-dependent protein kinase. *Physiol. Plant* 104, 549–555. doi: 10.1034/j.1399-3054.1998.1040405.x
13. Camoni, L., Harper, J. F., and Palmgren, M. G. (1998b). 14-3-3 proteins activate a plant calcium-dependent protein kinase (CDPK). *FEBS Lett.* 430, 381–384. doi: 10.1016/S0014-5793(98)00696-6
14. Cerovic, Z. G., Ghozlen, N. B., Milhade, C., Obert, M., Debuisson, S., and Le Moigne, M. (2015). Nondestructive diagnostic test for nitrogen nutrition of grapevine (*Vitis vinifera* L.) based on dualex leaf-clip measurements in the field. *J. Agric. Food Chem.* 63, 3669–3680. doi: 10.1021/acs.jafc.5b00304
15. Cerovic, Z. G., Masdoumier, G., Ghozlen, N. B., and Latouche, G. (2012). A new optical leaf-clip meter for simultaneous non-destructive assessment of leaf chlorophyll and epidermal flavonoids. *Physiol. Plant.* 146, 251–260. doi: 10.1111/j.1399-3054.2012.01639.x

16. Chaudhuri, S., Seal, A., and Gupta, M. D. (1999). Autophosphorylation-dependent activation of a calcium-dependent protein kinase from groundnut. *Plant Physiol.* 120, 859–866. doi: 10.1104/pp.120.3.859
17. Chehab, E. W., Patharkar, O. R., Hegeman, A. D., Taybi, T., and Cushman, J. C. (2004). Autophosphorylation and subcellular localization dynamics of a salt-and water deficit-induced calcium-dependent protein kinase from ice plant. *Plant Physiol.* 135, 1430–1446. doi: 10.1104/pp.103.035238
18. Cheng, S.-H., Willmann, M. R., Chen, H.-C., and Sheen, J. (2002). Calcium signaling through protein kinases. the Arabidopsis calcium-dependent protein kinase gene family. *Plant Physiol.* 129, 469–485. doi: 10.1104/pp.005645
19. Choi, H., Park, H.-J., Park, J. H., Kim, S., Im, M.-Y., Seo, H.-H., et al. (2005). Arabidopsis calcium-dependent protein kinase AtCPK32 interacts with ABF4, a transcriptional regulator of abscisic acid-responsive gene expression, and modulates its activity. *Plant Physiol.* 139, 1750–1761. doi: 10.1104/pp.105.069757
20. Cummins, I., Dixon, D. P., Freitag-Pohl, S., Skipsey, M., and Edwards, R. (2011). Multiple roles for plant glutathione transferases in xenobiotic detoxification. *Drug Metab. Rev.* 43, 266–280. doi: 10.3109/03602532.2011.552910
21. Déjardin, A., Sokolov, L. N., and Kleczkowski, L. A. (1999). Sugar/osmoticum levels modulate differential abscisic acid-independent expression of two stress-responsive sucrose synthase genes in Arabidopsis. *Biochem. J.* 344(Pt 2), 503–509. doi: 10.1042/0264-6021:3440503
22. De Nisi, P., Dell›Orto, M., Pirovano, L., and Zocchi, G. (1999). Calcium-dependent phosphorylation regulates the plasma-membrane $H^+$-ATPase activity of maize (*Zea mays* L.) roots. *Planta* 209, 187–194.
23. Dun, X., Shen, W., Hu, K., Zhou, Z., Xia, S., Wen, J., et al. (2014). Neofunctionalization of duplicated Tic40 genes caused a gain-of-function variation related to male fertility in *Brassica oleracea* lineages. *Plant Physiol.* 166, 1403–1419. doi: 10.1104/pp.114.246470

24. Dutilleul, C., Jourdain, A., Bourguignon, J., and Hugouvieux, V. (2008). The Arabidopsis putative selenium-binding protein family: expression Study and characterization of SBP1 as a potential new player in cadmium detoxification processes. *Plant Physiol.* 147, 239–251. doi: 10.1104/pp.107.114033

25. Earley, K. W., Haag, J. R., Pontes, O., Opper, K., Juehne, T., Song, K., et al. (2006). Gateway-compatible vectors for plant functional genomics and proteomics. *Plant J.* 45, 616–629. doi: 10.1111/j.1365-313X.2005.02617.x

26. Fan, L.-M., Zhao, Z., and Assmann, S. M. (2004). Guard cells: a dynamic signaling model. *Curr. Opin. Plant Biol.* 7, 537–546. doi: 10.1016/j.pbi.2004.07.009

27. Fedosejevs, E. T., Ying, S., Park, J., Anderson, E. M., Mullen, R. T., She, Y.-M., et al. (2014). Biochemical and molecular characterization of RcSUS1, a cytosolic sucrose synthase phosphorylated *in vivo* at serine 11 in developing castor oil seeds. *J. Biol. Chem.* 289, 33412–33424. doi: 10.1074/jbc.M114.585554

28. Franz, S., Ehlert, B., Liese, A., Kurth, J., Cazalé, A. C., and Romeis, T. (2011). Calcium-dependent protein kinase CPK21 functions in abiotic stress response in *Arabidopsis thaliana*. *Mol. Plant* 4, 83–96. doi: 10.1093/mp/ssq064

29. Geiger, D., Maierhofer, T., Al-Rasheid, K. A., Scherzer, S., Mumm, P., Liese, A., et al. (2011). Stomatal closure by fast abscisic acid signaling is mediated by the guard cell anion channel SLAH3 and the receptor RCAR1. *Sci. Signal.* 4, ra32–ra32. doi: 10.1126/scisignal.2001346

30. Geiger, D., Scherzer, S., Mumm, P., Marten, I., Ache, P., Matschi, S., et al. (2010). Guard cell anion channel SLAC1 is regulated by CDPK protein kinases with distinct $Ca^{2+}$ affinities. *Proc. Natl. Acad. Sci. U.S.A.* 107, 8023–8028. doi: 10.1073/pnas.0912030107

31. Geng, S., Zhao, Y., Tang, L., Zhang, R., Sun, M., Guo, H., et al. (2011). Molecular evolution of two duplicated CDPK genes CPK7 and CPK12 in grass species: a case study in wheat (*Triticum aestivum* L.). *Gene* 475, 94–103. doi: 10.1016/j.gene.2010.12.015

32. Guo, P., Baum, M., Grando, S., Ceccarelli, S., Bai, G., Li, R., et al. (2009). Differentially expressed genes between drought-tolerant

and drought-sensitive barley genotypes in response to drought stress during the reproductive stage. *J. Exp. Bot.* 60, 3531–3544. doi: 10.1093/jxb/erp194

33. Guo, Y., Xiong, L., Song, C.-P., Gong, D., Halfter, U., and Zhu, J.-K. (2002). A calcium sensor and its interacting protein kinase are global regulators of abscisic acid signaling in Arabidopsis. *Dev. Cell* 3, 233–244. doi: 10.1016/S1534-5807(02)00229-0
34. Hamel, L.-P., Sheen, J., and Séguin, A. (2014). Ancient signals: comparative genomics of green plant CDPKs. *Trends Plant Sci.* 19, 79–89. doi: 10.1016/j.tplants.2013.10.009
35. Hardin, S. C., Tang, G.-Q., Scholz, A., Holtgraewe, D., Winter, H., and Huber, S. C. (2003). Phosphorylation of sucrose synthase at serine 170: occurrence and possible role as a signal for proteolysis. *Plant J.* 35, 588–603. doi: 10.1046/j.1365-313X.2003.01831.x
36. Hardin, S. C., Winter, H., and Huber, S. C. (2004). Phosphorylation of the amino terminus of maize sucrose synthase in relation to membrane association and enzyme activity. *Plant Physiol.* 134, 1427–1438. doi: 10.1104/pp.103.036780
37. Harmon, A. C., Gribskov, M., and Harper, J. F. (2000). CDPKs–a kinase for every $Ca^{2+}$ signal? *Trends Plant Sci.* 5, 154–159. doi: 10.1016/S1360-1385(00)01577-6
38. Harper, J. F., Breton, G., and Harmon, A. C. (2004). Decoding $Ca^{2+}$ signals through plant protein kinases. *Annu. Rev. Plant Biol.* 55, 263–288. doi: 10.1146/annurev.arplant.55.031903.141627
39. Hegeman, A. D., Rodriguez, M., Han, B. W., Uno, Y., Phillips, G. N., Hrabak, E. M., et al. (2006). A phyloproteomic characterization of *in vitro* autophosphorylation in calcium-dependent protein kinases. *Proteomics* 6, 3649–3664. doi: 10.1002/pmic.200500926
40. Hubbard, K. E., Siegel, R. S., Valerio, G., Brandt, B., and Schroeder, J. I. (2012). Abscisic acid and $CO_2$ signalling via calcium sensitivity priming in guard cells, new CDPK mutant phenotypes and a method for improved resolution of stomatal stimulus–response analyses. *Ann. Bot.* 109, 5–17. doi: 10.1093/aob/mcr252
41. Huber, S. C., Huber, J. L., Liao, P. C., Gage, D. A. Jr., McMichael, R. W., Chourey, P.S., et al. (1996). Phosphorylation of serine-15 of maize leaf sucrose synthase (occurrence *in vivo* and possible

regulatory significance). *Plant Physiol.* 112, 793–802. doi: 10.1104/pp.112.2.793

42. Hugouvieux, V., Dutilleul, C., Jourdain, A., Reynaud, F., Lopez, V., and Bourguignon, J. (2009). Arabidopsis putative selenium-binding protein1 expression is tightly linked to cellular sulfur demand and can reduce sensitivity to stresses requiring glutathione for tolerance. *Plant Physiol.* 151, 768–781. doi: 10.1104/pp.109.144808
43. Kinoshita, T., and Shimazaki, K. (1999). Blue light activates the plasma membrane $H^+$-ATPase by phosphorylation of the C-terminus in stomatal guard cells. *EMBO J.* 18, 5548–5558. doi: 10.1093/emboj/18.20.5548
44. Klimecka, M., and Muszynska, G. (2007). Structure and functions of plant calcium-dependent protein kinases. *Acta Biochim. Pol.* 54, 219–233.
45. Kobayashi, M., Ohura, I., Kawakita, K., Yokota, N., Fujiwara, M., Shimamoto, K., et al. (2007). Calcium-dependent protein kinases regulate the production of reactive oxygen species by potato NADPH oxidase. *Plant Cell* 19, 1065–1080. doi: 10.1105/tpc.106.048884
46. Komina, O., Zhou, Y., Sarath, G., and Chollet, R. (2002). *In vivo and in vitro* phosphorylation of membrane and soluble forms of soybean nodule sucrose synthase. *Plant Physiol.* 129, 1664–1673. doi: 10.1104/pp.002360
47. Kudla, J., Batistič, O., and Hashimoto, K. (2010). Calcium signals: the lead currency of plant information processing. *Plant Cell* 22, 541–563. doi: 10.1105/tpc.109.072686
48. Kumar, S., Asif, M. H., Chakrabarty, D., Tripathi, R. D., Dubey, R. S., and Trivedi, P. K. (2013). Expression of a rice Lambda class of glutathione S-transferase, OsGSTL2, in Arabidopsis provides tolerance to heavy metal and other abiotic stresses. *J. Hazard. Mater.* 248–249, 228–237. doi: 10.1016/j.jhazmat.2013.01.004
49. Li, A.-L., Zhu, Y.-F., Tan, X.-M., Wang, X., Wei, B., Guo, H.-Z., et al. (2008). Evolutionary and functional study of the CDPK gene family in wheat (*Triticum aestivum* L.). *Plant Mol. Biol.*66, 429–443. doi: 10.1007/s11103-007-9281-5
50. Lino, B., Baizabal-Aguirre, V. M., and González de la Vara, L. E. (1998). The plasma-membrane $H^+$-ATPase from beet root is inhibited

by a calcium-dependent phosphorylation. *Planta*204, 352–359. doi: 10.1007/s004250050266

51. Liu, G., Chen, J., and Wang, X. (2006). VfCPK1, a gene encoding calcium-dependent protein kinase from *Vicia faba*, is induced by drought and abscisic acid. *Plant Cell Environ.* 29, 2091–2099. doi: 10.1111/j.1365-3040.2006.01582.x

52. Luan, S., Kudla, J., Rodriguez-Concepcion, M., Yalovsky, S., and Gruissem, W. (2002). Calmodulins and calcineurin B–like proteins calcium sensors for specific signal response coupling in plants. *Plant Cell* 14, S389–S400. doi: 10.1105/tpc.001115

53. Ludwig, A. A., Romeis, T., and Jones, J. D. (2004). CDPK-mediated signalling pathways: specificity and cross-talk. *J. Exp. Bot.* 55, 181–188. doi: 10.1093/jxb/erh008

54. Ludwików, A., Cieśla, A., Kasprowicz-Maluśki, A., Mituła, F., Tajdel, M., Gałgański, ł., et al. (2014). Arabidopsis protein phosphatase 2C ABI1 interacts with type I ACC synthases and is involved in the regulation of ozone-induced ethylene biosynthesis. *Mol. Plant* 7, 960–976. doi: 10.1093/mp/ssu025

55. Ludwików, A., Kierzek, D., Gallois, P., Zeef, L., and Sadowski, J. (2009). Gene expression profiling of ozone-treated *Arabidopsis* abi1td insertional mutant: protein phosphatase 2C ABI1 modulates biosynthesis ratio of ABA and ethylene. *Planta* 230, 1003–1017. doi: 10.1007/s00425-009-1001-8

56. Ma, S.-Y., and Wu, W.-H. (2007). AtCPK23 functions in Arabidopsis responses to drought and salt stresses. *Plant Mol. Biol.* 65, 511–518. doi: 10.1007/s11103-007-9187-2

57. Martin, B. R., Wang, C., Adibekian, A., Tully, S. E., and Cravatt, B. F. (2012). Global profiling of dynamic protein palmitoylation. *Nat. Methods* 9, 84–89. doi: 10.1038/nmeth.1769

58. Matschi, S., Werner, S., Schulze, W. X., Legen, J., Hilger, H. H., and Romeis, T. (2013). Function of calcium-dependent protein kinase CPK28 of *Arabidopsis thaliana* in plant stem elongation and vascular development. *Plant J.* 73, 883–896. doi: 10.1111/tpj.12090

59. Mehlmer, N., Wurzinger, B., Stael, S., Hofmann-Rodrigues, D., Csaszar, E., Pfister, B., et al. (2010). The $Ca^{2+}$-dependent

protein kinase CPK3 is required for MAPK-independent salt-stress acclimation in Arabidopsis. *Plant J.* 63, 484–498. doi: 10.1111/j.1365-313X.2010.04257.x

60. Mitula, F., Tajdel, M., Ciesla, A., Kasprowicz-Maluśki, A., Kulik, A., Babula-Skowrońska, D., et al. (2015). Arabidopsis ABA-activated kinase MAPKKK18 is regulated by protein phosphatase 2C ABI1 and the ubiquitin–proteasome pathway. *Plant Cell Physiol.* 56, 2351–2367. doi: 10.1093/pcp/pcv146
61. Mochida, K., and Shinozaki, K. (2013). Unlocking Triticeae genomics to sustainably feed the future. *Plant Cell Physiol.* 54, 1931–1950. doi: 10.1093/pcp/pct163
62. Moons, A., Valcke, R., and Van Montagu, M. (1998). Low-oxygen stress and water deficit induce cytosolic pyruvate orthophosphate dikinase (PPDK) expression in roots of rice, a C3 plant. *Plant J.* 15, 89–98. doi: 10.1046/j.1365-313X.1998.00185.x
63. Morari, F., Meggio, F., Lunardon, A., Scudiero, E., Forestan, C., Farinati, S., et al. (2015). Time course of biochemical, physiological, and molecular responses to field-mimicked conditions of drought, salinity, and recovery in two maize lines. *Front. Plant Sci.* 6:314. doi: 10.3389/fpls.2015.00314
64. Mori, I. C., Murata, Y., Yang, Y., Munemasa, S., Wang, Y.-F., Andreoli, S., et al. (2006). CDPKs CPK6 and CPK3 function in ABA regulation of guard cell S-type anion-and $Ca^{2+}$-permeable channels and stomatal closure. *PLoS Biol.* 4:e327. doi: 10.1371/journal.pbio.0040327
65. Myers, C., Romanowsky, S. M., Barron, Y. D., Garg, S., Azuse, C. L., Curran, A., et al. (2009). Calcium-dependent protein kinases regulate polarized tip growth in pollen tubes. *Plant J.* 59, 528–539. doi: 10.1111/j.1365-313X.2009.03894.x
66. Nakai, T., Konishi, T., Zhang, X.-Q., Chollet, R., Tonouchi, N., Tsuchida, T., et al. (1998). An increase in apparent affinity for sucrose of mung bean sucrose synthase is caused by *in vitro*phosphorylation or directed mutagenesis of Ser11. *Plant Cell Physiol.* 39, 1337–1341. doi: 10.1093/oxfordjournals.pcp.a029339
67. Oh, M.-H., Wu, X., and Huber, S. C. (2013). Impact of $Ca^{2+}$ on structure of soybean CDPKβ and accessibility of the Tyr-24

autophosphorylation site. *Plant Signal. Behav.* 8:e27671. doi: 10.4161/psb.27671

68. Oh, M.-H., Wu, X., Kim, H. S., Harper, J. F., Zielinski, R. E., Clouse, S. D., et al. (2012). CDPKs are dual-specificity protein kinases and tyrosine autophosphorylation attenuates kinase activity. *FEBS Lett.* 586, 4070–4075. doi: 10.1016/j.febslet.2012.09.040
69. Patharkar, O. R., and Cushman, J. C. (2000). A stress-induced calcium-dependent protein kinase from *Mesembryanthemum crystallinum* phosphorylates a two-component pseudo-response regulator. *Plant J.* 24, 679–691. doi: 10.1046/j.1365-313x.2000.00912.x
70. Rampino, P., Pataleo, S., Gerardi, C., Mita, G., and Perrotta, C. (2006). Drought stress response in wheat: physiological and molecular analysis of resistant and sensitive genotypes. *Plant Cell Environ.* 29, 2143–2152. doi: 10.1111/j.1365-3040.2006.01588.x
71. Reddy, A. S., Ali, G. S., Celesnik, H., and Day, I. S. (2011). Coping with stresses: roles of calcium-and calcium/calmodulin-regulated gene expression. *Plant Cell* 23, 2010–2032. doi: 10.1105/tpc.111.084988
72. Romeis, T., Ludwig, A. A., Martin, R., and Jones, J. D. G. (2001). Calcium-dependent protein kinases play an essential role in a plant defence response. *EMBO J.* 20:5556–5567. doi: 10.1093/emboj/20.20.5556
73. Romeis, T., Piedras, P., and Jones, J. D. G. (2000). Resistance gene-dependent activation of a calcium-dependent protein kinase in the plant defense response. *Plant Cell* 12, 803–815. doi: 10.1105/tpc.12.5.803
74. Rutschmann, F., Stalder, U., Piotrowski, M., Oecking, C., and Schaller, A. (2002). LeCPK1, a calcium-dependent protein kinase from tomato. Plasma membrane targeting and biochemical characterization. *Plant Physiol.* 129, 156–168. doi: 10.1104/pp.000869
75. Saijo, Y., Hata, S., Kyozuka, J., Shimamoto, K., and Izui, K. (2000). Over-expression of a single $Ca2^{+}$-dependent protein kinase confers both cold and salt/drought tolerance on rice plants. *Plant J.* 23, 319–327. doi: 10.1046/j.1365-313x.2000.00787.x
76. Saijo, Y., Kinoshita, N., Ishiyama, K., Hata, S., Kyozuka, J.,

Hayakawa, T., et al. (2001). A $Ca^{2+}$-dependent protein kinase that endows rice plants with cold-and salt-stress tolerance functions in vascular bundles. *Plant Cell Physiol.* 42, 1228–1233. doi: 10.1093/pcp/pce158

77. Schaller, G. E., and Sussman, M. R. (1988). Phosphorylation of the plasma-membrane $H^+$-ATPase of oat roots by a calcium-stimulated protein kinase. *Planta* 173, 509–518. doi: 10.1007/BF00958964
78. Schild, F., Kieffer-Jaquinod, S., Palencia, A., Cobessi, D., Sarret, G., Zubieta, C., et al. (2014). Biochemical and biophysical characterization of the selenium-binding and reducing site in *Arabidopsis thaliana* homologue to mammals selenium-binding protein 1. *J. Biol. Chem.* 289, 31765–31776. doi: 10.1074/jbc.M114.571208
79. Schulz, P., Herde, M., and Romeis, T. (2013). Calcium-dependent protein kinases: hubs in plant stress signaling and development. *Plant Physiol.* 163, 523–530. doi: 10.1104/pp.113.222539
80. Sheen, J. (1996). $Ca^{2+-}$dependent protein kinases and stress signal transduction in plants. *Science* 274, 1900–1902. doi: 10.1126/science.274.5294.1900
81. Simeunovic, A., Mair, A., Wurzinger, B., and Teige, M. (2016). Know where your clients are: subcellular localization and targets of calcium-dependent protein kinases. *J. Exp. Bot.* 67, 3855–3872. doi: 10.1093/jxb/erw157
82. Sreenivasulu, N., Usadel, B., Winter, A., Radchuk, V., Scholz, U., Stein, N., et al. (2008). Barley grain maturation and germination: metabolic pathway and regulatory network commonalities and differences highlighted by new MapMan/PageMan profiling tools. *Plant Physiol.* 146, 1738–1758. doi: 10.1104/pp.107.111781
83. Suzuki, K., Itai, R., Suzuki, K., Nakanishi, H., Nishizawa, N.-K., Yoshimura, E., et al. (1998). Formate dehydrogenase, an enzyme of anaerobic metabolism, is induced by iron deficiency in barley roots. *Plant Physiol.* 116, 725–732. doi: 10.1104/pp.116.2.725
84. Tamura, K., Stecher, G., Peterson, D., Filipski, A., and Kumar, S. (2013). MEGA6: molecular evolutionary genetics analysis version 6.0. *Mol. Biol. Evol.* 30, 2725–2729. doi: 10.1093/molbev/mst197
85. Thompson, J. D., Gibson, T. J., and Plewniak, F. (1997). The

CLUSTAL_X windows interface: flexible strategies for multiple sequence alignment aided by quality analysis tools. *Nucleic Acids Res.* 25, 4876–4882. doi: 10.1093/nar/25.24.4876

86. Valmonte, G. R., Arthur, K., Higgins, C. M., and MacDiarmid, R. M. (2014). Calcium-dependent protein kinases in plants: evolution, expression and function. *Plant Cell Physiol.* 55, 551–569. doi: 10.1093/pcp/pct200
87. Vivek, P. J., Tuteja, N., and Soniya, E. V. (2013). CDPK1 from ginger promotes salinity and drought stress tolerance without yield penalty by improving growth and photosynthesis in Nicotiana tabacum. *PLoS ONE* 8:e76392. doi: 10.1371/journal.pone.0076392
88. Vizcaíno, J. A., Csordas, A., del-Toro, N., Dianes, J. A., Griss, J., Lavidas, I., et al. (2016). 2016 update of the PRIDE database and its related tools. *Nucleic Acids Res.* 44, D447–D456. doi: 10.1093/nar/gkv1145
89. Vizcaíno, J, A., Deutsch, E. W., Wang, R., Csordas, A., Reisinger, F., Ríos, D., et al. (2014). ProteomeXchange provides globally coordinated proteomics data submission and dissemination. *Nat. Biotechnol.* 32, 223–226. doi: 10.1038/nbt.2839
90. Wan, B., Lin, Y., and Mou, T. (2007). Expression of rice $Ca^{2+}$-dependent protein kinases (CDPKs) genes under different environmental stresses. *FEBS Lett.* 581, 1179–1189. doi: 10.1016/j.febslet.2007.02.030
91. Witte, C.-P., Keinath, N., Dubiella, U., Demoulière, R., Seal, A., and Romeis, T. (2010). Tobacco calcium-dependent protein kinases are differentially phosphorylated *in vivo* as part of a kinase cascade that regulates stress response. *J. Biol. Chem.* 285, 9740–9748. doi: 10.1074/jbc.M109.052126
92. Xing, T., Higgins, V. J., and Blumwald, E. (1996). Regulation of plant defense response to fungal pathogens: two types of protein kinases in the reversible phosphorylation of the host plasma membrane $H^+$-ATPase. *Plant Cell* 8, 555–564.
93. Xu, J., Tian, Y.-S., Peng, R.-H., Xiong, A.-S., Zhu, B., Jin, X.-F., et al. (2010). AtCPK6, a functionally redundant and positive regulator involved in salt/drought stress tolerance in Arabidopsis. *Planta* 231, 1251–1260. doi: 10.1007/s00425-010-1122-0

94. Yoon, G. M., Dowd, P. E., Gilroy, S., and McCubbin, A. G. (2006). Calcium-dependent protein kinase isoforms in Petunia have distinct functions in pollen tube growth, including regulating polarity. *Plant Cell* 18, 867–878. doi: 10.1105/tpc.105.037135

95. Zeng, H., Xu, L., Singh, A., Wang, H., Du, L., and Poovaiah, B. W. (2015). Involvement of calmodulin and calmodulin-like proteins in plant responses to abiotic stresses. *Front. Plant Sci.* 6:600. doi: 10.3389/fpls.2015.00600

96. Zhang, X.-Q., Lund, A. A., Sarath, G., Cerny, R. L., Roberts, D. M., and Chollet, R. (1999). Soybean nodule sucrose synthase (nodulin-100): further analysis of its phosphorylation using recombinant and authentic root-nodule enzymes. *Arch. Biochem. Biophys.* 371, 70–82. doi: 10.1006/abbi.1999.1415

97. Zhao, R., Sun, H.-L., Mei, C., Wang, X.-J., Yan, L., Liu, R., et al. (2011a). The Arabidopsis $Ca^{2+}$-dependent protein kinase CPK12 negatively regulates abscisic acid signaling in seed germination and post-germination growth. *New Phytol.* 192, 61–73. doi: 10.1111/j.1469-8137.2011.03793.x

98. Zhao, R., Wang, X.-F., and Zhang, D.-P. (2011b). CPK12: a $Ca^{2+}$-dependent protein kinase balancer in abscisic acid signaling. *Plant Signal. Behav.* 6, 1687–1690. doi: 10.4161/psb.6.11.17954

99. Zhu, J.-K. (2002). Salt and drought stress signal transduction in plants. *Annu. Rev. Plant Biol.* 53, 247. doi: 10.1146/annurev.arplant.53.091401.143329

100. Zhu, S.-Y., Yu, X.-C., Wang, X.-J., Zhao, R., Li, Y., Fan, R.-C., et al. (2007). Two calcium-dependent protein kinases, CPK4 and CPK11, regulate abscisic acid signal transduction in Arabidopsis. *Plant Cell* 19, 3019–3036. doi: 10.1105/tpc.107.050666

101. Zou, J.-J., Wei, F.-J., Wang, C., Wu, J.-J., Ratnasekera, D., Liu, W.-X., et al. (2010). Arabidopsis calcium-dependent protein kinase CPK10 functions in abscisic acid-and $Ca^{2+}$-mediated stomatal regulation in response to drought stress. *Plant Physiol.* 154, 1232–1243. doi: 10.1104/pp.110.157545

CHAPTER

# 10

# CHLORIDE REGULATES LEAF CELL SIZE AND WATER RELATIONS IN TOBACCO PLANTS

Juan D. Franco-Navarro[1], Javier Brumós[2,3], Miguel A. Rosales[1,4], Paloma Cubero-Font[1], Manuel Talón[2] and José M. Colmenero-Flores[1]

[1] Institute of Natural Resources and Agrobiology, CSIC, Avda Reina Mercedes 10, 41012-Sevilla, Spain

[2] Valencian Institute of Agrarian Research, Center of Genomics, Ctra Moncada-Náquera Km 4.6, 46113-Moncada, Valencia, Spain

[3] Department of Plant and Microbial Biology, North Carolina State University, Raleigh, NC 27695-7614, USA.

[4] Biochemistry and Molecular Physiology of Plants (BPMP), UMR 5004 (CNRS/INRA/SupAgro/UM), 2 Place Viala, F-34060 Montpellier Cedex 2, France

## ABSTRACT

Chloride ($Cl^-$) is a micronutrient that accumulates to macronutrient levels since it is normally available in nature and actively taken up by higher plants. Besides a role as an unspecific cell osmoticum, no clear biological roles have been explicitly associated with $Cl^-$ when accumulated to macronutrient concentrations. To address this question, the glycophyte tobacco (*Nicotiana tabacum* L. var. Habana) has been treated with a basal nutrient solution supplemented with one of three salt combinations containing the same cationic balance: $Cl^-$-based (CL), nitrate-based (N), and sulphate+phosphate-based (SP) treatments. Under non-saline conditions (up to 5mM $Cl^-$) and no water limitation, $Cl^-$ specifically stimulated higher leaf cell size and led to a moderate increase of plant fresh and dry biomass mainly due to higher shoot expansion. When applied in the 1–5mM range, $Cl^-$ played specific roles in regulating leaf osmotic potential and turgor, allowing plants to improve leaf water balance parameters. In addition, $Cl^-$ also altered water relations at the whole-plant level through reduction of plant transpiration. This was a consequence of a lower stomatal conductance, which resulted in lower water loss and greater photosynthetic and integrated water-use efficiency. In contrast to $Cl^-$, these effects were not observed for essential anionic macronutrients such as nitrate, sulphate, and phosphate. We propose that the abundant uptake and accumulation of $Cl^-$ responds to adaptive functions improving water homeostasis in higher plants.

**Key words:** Beneficial nutrient, chloride nutrition, growth, osmotic potential, turgor, water potential, water balance, water relations, water-use efficiency, WUE

## INTRODUCTION

Chloride ($Cl^-$) is one of the 16 elements essential for plant growth. Because it is supposedly needed in small quantities for healthy growth of plants (<50–100 μM in the nutrient media), $Cl^-$ is classified as a micronutrient (Johnson *et al.*, 1957; Whitehead, 1985). Usually, in non-halophytic plants, the critical deficiency concentration is <0.2mg $g^{-1}$ shoot DW (reviewed in Flowers, 1988; Xu *et al.*, 2000; White and Broadley, 2001; Broadley *et al.*, 2012a). Under this critical threshold, plants tend to show a significant decrease in leaf area as a result of a reduction in cell division rates (Terry, 1977). As an essential micronutrient, $Cl^-$ is involved in the stabilization of the water splitting system of photosystem II (PSII) and the regulation of enzyme activities such as the asparagine synthethase, amylases, and the tonoplast $H^+$-ATPase. Important $Cl^-$ functions are also related to the electrical charge balance of essential cations such as $K^+$ and $H^+$, playing main roles in the stabilization of the electric potential of cell membranes and the regulation of pH gradients (reviewed in Xu *et al.*,

2000; White and Broadley, 2001; Hänsch and Mendel, 2009; Broadley *et al.*, 2012a). In addition, $Cl^-$ is an osmotically active solute in the vacuole. Participation of $Cl^-$ in cell osmotic regulation has been debated largely through its involvement in the regulation of cell turgor processes such as stomatal movement and the activity of motor cells controlling nastic movements (reviewed in Flowers, 1988; White and Broadley, 2001; Broadley *et al.*, 2012a). Preferential accumulation of $Cl^-$ has been reported in epidermal cells of barley (Leigh and Tomos, 1993) and epidermal cells from elongating internodes of *Pisum sativum* (Yamagami *et al.*, 2004). Since higher plants are thought to perform normally with low $Cl^-$ content, it is generally accepted that specific osmoregulatory functions of $Cl^-$ are confined to these tissues, where $Cl^-$ content should be higher than the average of the bulk tissues (Flowers, 1988).

However $Cl^-$ does not appear to be a typical micronutrient since the actual $Cl^-$concentration in plants, in the range of 2–20mg $g^{-1}$ DW (Xu *et al.*, 2000; Brumós *et al.*, 2010; Broadley *et al.*, 2012a), is 10–100 times higher than the concentration required as an essential micronutrient. This is relevant provided that all other mineral micronutrients (B, Cu, Fe, Mn, Mo, Ni, and Zn) are present at much lower concentrations in plant tissues ($10^{-4}$–$10^{-1}$ mg $g^{-1}$DW), while the accumulation to higher levels results in plant toxicity (Hänsch and Mendel, 2009). $Cl^-$ transport occurs primarily via the symplastic pathway (Pitman, 1982; Brumós *et al.*, 2010), and $Cl^-$ uptake under non-saline conditions is an electrogenic $Cl^-/2H^+$ symport mechanism requiring metabolic energy (Felle, 1994; White and Broadley, 2001; Britto and Kronzucker, 2006). As $Cl^-$ accumulation to macronutrient concentrations requires a considerable use of energy (Brumós *et al.*, 2010), it is feasible to think that $Cl^-$ plays a broader and poorly understood biological role, one that is certainly not critical under normal growth conditions. Since $Cl^-$ appears to be particularly well suited to accomplishing osmoregulatory roles, we wondered whether $Cl^-$accumulation at macronutrient concentrations is specifically involved in the regulation of water relations in plants at both the cell and the whole-plant level. It is generally assumed that, when accumulated in large quantities, $Cl^-$serves a non-specific osmotic function in the vacuole, where it is interchangeable with other solutes such as nitrate ($NO_3^-$; Fricke *et al.*, 1994). This is a controversial issue for several reasons: (i) usually the role of $Cl^-$ is not adequately differentiated from that of their accompanying cations (reviewed in Flowers, 1988); (ii) or, very frequently, some physiological and molecular responses have been entirely attributed to the cation (e.g. $K^+$) in experimental approaches where the effect of the accompanying anion ($Cl^-$) has not been controlled (see, for instance, Armengaud *et al.*, 2004; Benlloch-Gonzalez *et al.*, 2008); (iii) it is unclear to what extent $Cl^-$ is preferred by plants to fulfil osmoregulatory roles or whether other inorganic anions, such as the macronutrients $NO_3^-$, sulphate ($SO_4^{2-}$), or phosphate ($PO_4^{3-}$) can replace $Cl^-$ in such functions; and (iv) the idea linking $Cl^-$ homeostasis with osmotic/turgor regulation has been frequently discussed in the context of salt stress and halophyte species (Yeo and Flowers, 1980, 1986; Flowers and Yeo, 1986; Flowers, 1988; Perez-Perez *et al.*, 2007). For instance, $Cl^-$ was frequently assumed to move passively in plants, resulting in toxic effects in leaves of some woody species such as Citrus under high or moderate salt stress (Moya *et al.*, 1999; Storey and Walker, 1999; Moya *et al.*, 2003). We have reported that under non-saline conditions $Cl^-$ is actively taken up and accumulated into leaf tissues of Citrus plants to levels that exceed the critical content requirement by one order of magnitude in the so-called $Cl^-$-excluder rootstocks and by two orders in $Cl^-$-includer (salt-sensitive) rootstocks, through symplastically regulated transport mechanisms (Tadeo *et al.*, 2008; Brumos *et al.*, 2010). This is indicative that Citrus (and probably other glycophyte plants) handle $Cl^-$ homeostasis similarly to a macronutrient such as $K^+$ rather than a toxic ion such as $Na^+$.

In this work, it is our aim to establish the specific role of $Cl^-$ in glycophyte plants when accumulated to macronutrient levels, and it is shown here that $Cl^-$ nutrition in the millimolar range (1–5mM) specifically stimulates growth and increased leaf cell size, regulating leaf tissue water balance and water relations in tobacco plants.

# MATERIALS AND METHODS

## Experimental Design

Tobacco plants (*Nicotiana tabacum* L. var. Habana) were grown under greenhouse conditions at 24±2 °C/17±2 °C (day/night), a relative humidity of 60±10% (EL-1-USB Data-logger, Lascar Electronics Inc., Erie, PA, USA), and a 16h/8h photoperiod with a photosynthetic photon flux density (average PAR) of 300–350 μmol $m^{-2}$ $s^{-1}$ (quantum sensor, LI-6 400; Li-COR, Lincoln, NE, USA) and a luminous emittance of 9000–10 000 lux (Digital Lux Meter, LX1010B; Carson Electronics, Valemount, Canada). Tobacco seeds were sown in flat trays (cell size 4 cm×4 cm×10cm) containing peat that had been previously washed with the corresponding nutrient solution. After vernalization for 2 d in a cold chamber (4 °C), seeds were transferred to the greenhouse. After 3 weeks, seedlings were transplanted to 7.5 litre pots (pot size 20 cm×17 cm×25cm), containing a mix of perlite:coarse sand:vermiculite (2:3:5), where plants were watered through a semi-hydroponics regime with the different nutrient solutions described below.

In Johnson *et al.* (1957), 50 μM $Cl^-$ was established as the treatment ensuring $Cl^-$ micronutrient requirements in different plant species, whereas deficiency symptoms were obtained in plants with no $Cl^-$ addition. In this study, 75 μM $Cl^-$ (added as $CoCl_2$ 11 μM+KCl 53 μM) was chosen to be always present in the basal nutrient solution (BS) to fulfil plant requirements in low $Cl^-$ treatments, but not to allow significant $Cl^-$ accumulation in plant tissues. This $Cl^-$concentration was confirmed through direct measurement of the $Cl^-$concentration in the BS medium using a chloridometer (see below). Other nutrients present in the BS solution were: $KNO_3$ 1.25mM, $KH_2PO_4$ 0.725mM, $K_2HPO_4$ 0.073mM, $Ca(NO_3)_2$ 2mM, $MgSO_4$ 1mM, FeNa-EDTA 0.1mM, $H_3BO_3$0.1mM, $MnSO_4$ 0.1mM, $ZnSO_4$ 29 μM, $CoCl_2$ 0.11 μM, KCl 53 μM, $CuSO_4$ 0.1 μM, $Na_2MoO_4$ 1 μM, and 5 μM KI.

This work contains results from five independent experiments: January–February 2012 (JF2012); March–April 2012 (MA2012); November–December 2012 (ND2012); April–May 2013; (AM2013); and September–October 2013 (SO2013). Most experiments (JF2012, MA2012, ND2012, and SO2013) were performed with the application of 5mM $Cl^-$ (CL treatment), which included the following salt mixtures supplemented in the BS solution: 2.5mM KCl, 0.625mM $MgCl_2$, and 0.625mM $CaCl_2$. In order to evaluate the specificity of $Cl^-$ in the studied phenomena, two additional treatments were used: 5mM nitrate ($NO_3^-$) treatment (N) containing 2.5mM $KNO_3$, 0.625mM $Mg(NO_3)_2$, and 0.625mM $Ca(NO_3)_2$; and sulphate+phosphate ($SO_4^{2-}$+$PO_4^{3-}$) treatment (SP) containing 1.25mM $KH_2PO_4$, 0.625mM $K_2SO_4$, 0.625mM $MgSO_4$, and 0.625mM $CaSO_4$. For the AM2013 experiment, increasing concentrations of anions were used in CL treatments: 0mM $Cl^-$ (BS treatment containing 0.075mM $Cl^-$), 0.15mM $Cl^-$, 0.30mM $Cl^-$, 1.0mM $Cl^-$, 2.5mM $Cl^-$, or 5.0mM $Cl^-$; and the equivalent SP treatments ensuring the same cationic balance as in the different CL treatments. All experimental solutions were adjusted to pH 5.7 with KOH.

Pots were irrigated up to field capacity (3.5ml $g^{-1}$ substrate) throughout the experiment. Pots were weighed each week at field capacity to estimate indirectly the increase of plant FW over time. After 6–7 weeks (65–72 d after sowing, DAS), shoots were harvested, FW values were obtained, and samples were dried in a forced-air oven at 75 °C for 48h to obtain the DW. Roots were rinsed with tap water and subsequently with distilled water and, after removing the excess water with filter paper, FW measurements were obtained. DW was obtained as for shoot harvesting. Dry tissues were ground to powder using a homogenizer (Taurus, 25 790 Barcelona, Spain), for subsequent analyses.

## Nutrient Content Determination

Fully photosynthetic and expanded mature leaves (non-senescent) from plants of 65–72 DAS were ground to powder as previously described. For $Cl^-$content determination, powdered leaf tissue was incubated overnight in a 0.1M $HNO_3$ and 10% glacial acetic acid solution. After centrifuging, 0.5ml of the extract was used for the determination of $Cl^-$ concentration in a Corning 926 chloridometer (Sherwood Scientific Ltd, Cambridge, UK) by silver ion titration according to the manufacturer›s instructions and to Gilliam (1971). Other soluble macro- and micronutrients were extracted in water from powdered dry tissues: the $NO_3^-$ concentration was measured through a multiparameter 'Bran+Luebbe' autoanalyser (Bran+Luebbe Analytics, Norderstedt, Germany); $SO_4^{2-}$ and $PO_4^{3-}$ concentrations were measured as described in Novozamsky and Van Eck (1977), and Hogue *et al.* (1970), respectively; and cation content was determined through inductively coupled plasma optical emission spectrometry (ICP-OES) using a Varian ICP 720-ES spectrometer (Varian Inc., Palo Alto, CA, USA).

## Leaf Area

Detached leaves were scanned in an Epson Stylus DX4 000 multifunction printer (Seiko Epson Corp., Owa, Japan). Scanning settings were defined as: colorless b/w and a very low resolution (72 ppp). Leaf area was measured through pixel quantification with the ADN Software 'Medición de Hojas v1.0' (Developed at the Department of Ecology, University of Seville, Spain; Taguas and Rivero, 1989). Data were obtained in $cm^2$. Specific leaf area (SLA) was calculated as follows (Marcelis *et al.*, 1998):

$$SLA = (\text{Total leaf area})(\text{Total leaf DW})^{-1} \quad (1)$$

## Water Parameters and Photosynthesis

Water content (WC, Equation 2), relative water content (RWC, Equation 3), and succulence (Equation 4) were determined in leaves obtained from 4–6 plants, using 3–4 leaves per plant and 10 discs per leaf (1cm diameter), and calculated as follows (Barrs and Weatherley, 1962):

$$WC\,(\%) = [(100 \times (FW - DW)](FW)^{-1} \quad (2)$$

$$RWC\,(\%) = (FW - DW)(TW - DW)^{-1} \quad (3)$$

$$\text{Succulence}\,(g\,cm^{-2}) = (FW - DW)(\text{leaf area})^{-1} \quad (4)$$

where DW is the dry weight, FW is the fresh weight, and TW is the fully hydrated weight of the leaf. For TW calculation, the petiole of the whole leaf was imbibed in water for 4h before measuring the fresh weight of the leaf at full turgor.

Leaf osmotic potential ($\Psi_\pi$) was calculated from the leaf sap obtained from 360 leaf discs: 20 leaf discs per sample×three samples per plant×six plants per treatment. Leaf sap was extracted from leaf discs by transferring the samples, placed in 0.5ml microcentrifuge tubes, from a block heated to 90 °C to liquid nitrogen. Tube caps were sealed with parafilm to avoid water evaporation. This thermal shock was repeated five times and leaf sap was collected in 1.5ml microcentrifuge tubes by centrifugation and filtration of tissue debris. Leaf water potential ($\Psi_w$) was directly obtained from a fresh leaf disc (6–8 disc measurements per plant and six plants per treatment). Both parameters, $\Psi_\pi$ and $\Psi_w$, were recorded using the dew-point microvoltimeter (model HR-33T, Wescor, UT, USA) and the C-52 sample chamber as previously described in Cochrane (1994) and Colmenero-Flores *et al.* (1999). Before measuring, samples were incubated in the chambers for 40min to reach water vapor equilibration. Leaf turgor (or pressure) potential ($\Psi_p$) was calculated from the $\Psi_w$ and $\Psi_\pi$ values (experimentally obtained) according to Equation 5:

$\Psi_p(MPa)=|\Psi_\pi|-|\Psi_w|$ (5)

Water consumption was quantified by measuring the volume of the semi-hydroponics nutrient solution consumed by the plants. Integrated water-use efficiency (WUE) was calculated as the increase of plant FW over time related to the accumulated water consumption (g FW ml $H_2O^{-1}$), as well as the final FW obtained after harvesting related to total water consumption (g FW ml $H_2O^{-1}$) (Abbate *et al.*, 2004).

Leaf gas exchange measurements were conducted between 12:00h and 14:00h using a gas-exchange system (LI-6 400: LI-COR, Lincoln, NE, USA). For each treatment, three photosynthetically active and fully expanded intermediate leaves from 4–6 plants (52–62 DAS) were used. Photosynthesis was induced with ambient light and 400 μmol $mol^{-1}$ $CO_2$ surrounding the leaf. Leaf temperature was maintained at 25 °C, and the leaf-to-air vapor pressure deficit was kept between 1 kPa and 1.3 kPa. These conditions were kept constant for the determinations of the net photosynthetic rate ($A_N$; μmol $CO_2 m^{-2} s^{-1}$) and stomatal conductance ($g_s$; mol $H_2O$ $m^{-2} s^{-1}$). The intrinsic water-use efficiency (WUE) was calculated as the ratio between the photosynthetic rate and stomatal conductance ($AN/g_s$; μmol $CO_2$ $mol^{-1}$ $H_2O$; Rosales *et al.*, 2012). For the AM2013 experiment, which consisted of growing plants in anion concentration treatments, $g_s$ was measured using the Decagon Leaf Porometer (Decagon Devices Inc., Pullman, WA, USA).

Assay of the leaf transpiration in an individual detached leaf, also called fresh weight loss (FWL), was carried out as previously described (Fernández *et al.*, 1997; Cheong *et al.*, 2007; Li *et al.*, 2014) with some modifications. FWL was calculated according to Equation 6 as the percentage of fresh weight loss over time upon leaf detachment:

$FWL(\%)=(100\times LWC_{t=x})(LWC_{t=0})^{-1}$ (6)

where LWC*t* is the leaf water content at time '*t*=*x*' or at time zero '*t*=0'.

Leaves of plants at 65 DAS were detached and dehydrated over a filter paper at 24±2 °C, a relative humidity of 60±10%, and a photosynthetic photon flux density of 300 μmol $m^{-2}$ $s^{-1}$. Measurements of FW were recorded 0, 0.5, 1, 1.5, 2.5, 3, and 6h after leaf separation. Six plants per treatment and three detached leaves per plant were used.

Using the above FWL values, the percentage of saved water in treated plants (SP, N, and CL) was obtained according to Equation 7 and Equation 8 by subtracting the percentage of water loss of BS plants 6h after detachment. Water saving at the whole-plant level was calculated using integrated WUE values obtained by gravimetric measurement as described previously and according to Equation 9 and Equation 10 by subtracting the percentage of water loss of treated plants from the percentage of water loss of BS plants.

$$\text{Water loss (Transpiration, \%)} = 100 - \text{FWL (\%)} \quad (7)$$

$$\text{Water saving (Transpiration, \%)} = \text{Water loss BS(\%)} - \text{Water loss treated (\%)} \quad (8)$$

$$\text{Water loss (Consume, \%)} = (100 \times \text{WUE treated, \%})(\text{WUE BS, \%})^{-1} \quad (9)$$

$$\text{Water saving (Consume, \%)} = 100 - \text{Water loss (Consume, \%)} \quad (10)$$

The PSII quantum yield (QY) is a plant stress marker that quantifies the PSII efficiency. Chlorophyll fluorescence in light-adapted plants was measured using a portable fluorometer (FluorPen FP-100; Photon System Instruments, Brno, Czech Republic). The PSII QY in light-adapted plants (Equation 11) was calculated according to Maxwell and Johnson (2000) through the measurement of the following variable parameters (Equation 12). For each determination, four readings were taken from each leaf and averaged; three leaves from each plant and six plants per treatment were monitored every day.

$$QY = \Phi_{PSII} = F_m'(F_v')^{-1} \quad (11)$$

$$F_v' = (F'_m - F_t) \quad (12)$$

where $F_v'$ is the difference between $F_m'$ (the maximum fluorescence in the light-adapted state) and $F_t$ (the basal fluorescence in the light-adapted state).

## Anatomical Parameters

To measure leaf thickness, histological preparations of tobacco leaves were produced as previously described in Scafaro *et al.* (2011). Analysis of abaxial leaf cells was carried out in epidermal peels and epidermal impressions. Epidermal impressions were performed as described in Gitz and Baker (2009). Peelings were carried out as described in He *et al.* (2013) and Allen *et al.*(1999), but with slight differences: peels were obtained from the abaxial surface by gently ripping the epidermis with a scalpel and removing it with tweezers. Peels were transferred to incubation buffer (10mM MES-KOH pH 6.5, 10mM KCl, and 50 μM $CaCl_2$) before analysis by light microscopy. Histological preparations of leaf tissue were made to analyse mesophyll cells (spongy and palisade cells). Leaf tissues were fixed overnight in FAA (4% formaldehyde, 5% acetic acid, and 50% ethanol), dehydrated through a graded ethanol series, and then embedded in Paraplast Plus (Sigma-Aldrich, http://www.sigmaaldrich.com) as described previously (Langdale, 1994). Microtome sections (8 μm) were stained with Safranin/Fast Green (Sigma-Aldrich). Examination of epidermal peels, epidermal impressions, and histological preparations was performed on a Zeiss Axioskop microscope equipped with Nomarski optics, AxioCam MRc5, and the Zeiss AxioVision software (Freeware 'Carl Zeiss AxioVision Rel.4.9.1.0' available at the Zeiss Homepage http://www.zeiss.com/, Carl Zeiss Microscopy GmbH, Jena, Germany). Cell size was determined as either the cross-sectional area of mesophyll cells or the cell surface measured from epidermal impressions in epidermal cells. Cell count was performed using the Counterall® software (www.counterall.com, Bioscripts.net-IRNAS-CSIC). The total number of epidermal cells per leaf was estimated according to the Equation 13:

$$\text{Total no. of epidermal cells per leaf} = (\text{Epidermal cell frequency}) \times \text{Total leaf area (cm}^2) \quad (13)$$

Where epidermal cell frequency is the total number of epidermal cells per $cm^2$. Measurement of cell size was performed using the outline tool of the AxioVision Software.

## Statistical Analysis

Statistical analysis was performed using the STATGRAPHICS Centurion XVI software (http://www.statgraphics.com; StatPoint Technologies, Warrenton, VA, USA). Shapiro–Wilk test was used to verify the normality of the data sets. One-way ANOVA and multivariate analysis of variance (MANOVA) were performed to determine significant differences between groups of samples, and levels of significance are indicated in the figures by asterisks: *$P \leq 0.05$; **$P \leq 0.01$; ***$P \leq 0.001$.

Non-significant (ns) differences were indicated when $P>0.05$. Multiple comparisons of means were determined by the Tukey's HSD (honestly significant difference) and MRT (multiple range test) statistical tests included in the above-mentioned software. Values represent the mean of 4–6 tobacco plants in each treatment. Correlations between logarithm of anion concentration and physiological parameters were calculated through the Pearson's product–moment correlation coefficient ($R^2$).

## RESULTS

We initially studied the effects of $Cl^-$ application in the millimolar range (CL treatment) on plant $Cl^-$ content and growth, and we also verified that plants subjected to low $Cl^-$ treatments (BS, N, and SP) were not experiencing $Cl^-$deficiency. The leaf cation concentration ($K^+$, $Ca^{2+}$, $Mg^{2+}$) was similar in plants treated with CL, N, and SP supplements, whereas anions were differentially accumulated according to the respective treatment (Table 1). The $Cl^-$ concentration in CL-treated plants was similar to the concentration of the macronutrient $K^+$, and significantly higher than the $NO_3^-$ concentration in N-treated plants or the $SO_4^{2-}+PO_4^{3-}$ concentration in SP-treated plants. Comparing the anion concentration in CL versus N and SP treatments, it was observed that the molar excess of $Cl^-$ over $NO_3^-$ was 2.30, and that of $Cl^-$ over $SO_4^{2-}+PO_4^{3-}$ was 3.05, indicating that $Cl^-$ (not assimilated) was preferentially accumulated over other anionic macronutrients (which are assimilated) (Table 1). $Cl^-$ was preferentially accumulated in leaves, with a molar excess of 4.2-fold over root $Cl^-$ concentration and 3.2-fold over stem $Cl^-$concentration. Consistent with the intracellular accumulation of negative charges of biomolecules, a higher concentration of inorganic cations was observed in all treatments (Tables 2, 3). However, the contribution of inorganic anions to equimolarity with inorganic cations (Table 2) or to electrical neutrality (Table 3) was higher in CL-treated plants. This indicates that CL plants require less accumulation of organic molecules (e.g. organic acids) involved in osmoregulation or electric neutrality processes (Tables 2, 3). The $Cl^-$ concentration in leaves of plants treated with low $Cl^-$(BS, SP, and N treatments) exceeded by 3–5 times the threshold of $Cl^-$deficiency, and no deficiency symptoms such as wilting, chlorosis, or bronzing were observed (not shown). $Cl^-$ is an essential cofactor in photosynthesis, although the primary aspect involved in the reduced growth under $Cl^-$ deficiency was a lower rate of cell division in the leaves (Terry, 1977). We verified that photosynthesis and leaf cell division rates were not impaired in plants subjected to low $Cl^-$ treatments (BS, SP, or N). The highest photosynthetic activity (Figure 1A) and the highest number of epidermal leaf cells (Figure 1B) were observed in N-treated plants. Since the lowest $Cl^-$ content was always observed in N and SP plants, we concluded that low $Cl^-$ treatments covered essential $Cl^-$ requirements. Although $Cl^-$ is thought to be a micronutrient, tobacco plants responded positively to 5mM $Cl^-$ application in terms of FW (Figure 1C) and DW (Figure 1D) when compared with BS or SP treatments. As expected, N-treated plants showed the highest growth and biomass because of the important role of nitrogen in plant growth and development (Hawkesford *et al.*, 2012). Although biomass varied depending on the season in which the experiments were conducted, the CL treatment always determined a significant biomass increase compared with BS and SP treatments. The biomass increase determined by the CL treatment required $Cl^-$ application at >1mM external concentration (Figure 1E, F), ~50–100 times the external concentration required as a micronutrient. Taken together, these results show that $Cl^-$ in the millimolar range stimulates plant growth, and ruled out the possibility that plants subjected to low $Cl^-$ treatments were experiencing $Cl^-$ deficiency.

**Table 1.** Ion concentration in leaves subjected to different treatments

| Treatment | Ion concentration (mmol $g^{-1}$ DW) | | | | | | |
|---|---|---|---|---|---|---|---|
| | $K^+$ | $Ca^{2+}$ | $Mg^{2+}$ | $Cl^-$ | $NO_3^-$ | $PO_4^{3-}$ | $SO_4^{2-}$ |
| BS | 0.905±0.047 b | 0.391±0.029 | 0.252±0.021 | 0.028±0.003 b | 0.255±0.061 b | 0.128±0.002 a | 0.126±0.022 b |
| SP | 1.238±0.059 a | 0.355±0.033 | 0.336±0.046 | 0.015±0.002 b | 0.144±0.028 bc | 0.136±0.080 a | 0.336±0.039 a |
| N | 1.259±0.060 a | 0.465±0.016 | 0.305±0.033 | 0.017±0.002 b | 0.627±0.102 a | 0.081±0.006 b | 0.105±0.005 b |
| CL | 1.268±0.047 a | 0.381 ±0.054 | 0.317 ±0.032 | 1.441±0.061 a | 0.039±0.007 c | 0.104±0.004 b | 0.111±0.008 b |
| *P*-value | *** | ns | ns | *** | *** | *** | *** |

Treatment consisted of the basal nutrient solution (BS) supplemented with 5mM chloride (CL), 5mM nitrate (N), or the sulphate+phosphate (SP) salt mixture containing the same cationic balance as in the CL and N treatments.

Mean values ± SE, $n$=6. Levels of significance: $P$>0.05 ('ns', non-significant differences); ***$P \leq 0.001$). 'Homogeneous group' statistics were calculated through ANOVA test.

**Table 2.** Contribution of inorganic anions to equimolarity with cations in the leaf

| | Ionic concentration (mmol $g^{-1}$ DW) | | Equimolarity (%) | | Contribution to equimolarity (%) | | |
|---|---|---|---|---|---|---|---|
| | Cations | Anions | Inorganic anions | Organic molecules*a* | $SO_4^{2-}+PO_4^{3-}$ | $NO^-$ | $Cl^-$ |
| BS | 1.76±0.05 b | 0.58±0.09 b | 32.6±4.4 b | 67.4±4.4 a | 14.3±0.9 b | 17.1±3.4 a | 1.1±0.2 b |
| SP | 1.96±0.08 ab | 0.65±0.08 b | 32.7±3.6 b | 67.3±3.6 a | 24.3±1.9 a | 7.7±2.2 b | 0.7±0.05 b |
| N | 2.10±0.05 a | 0.72±0.06 b | 34.4±3.0 b | 65.6±3.0 a | 8.9±0.3 b | 25.0±2.9 a | 0.5±0.02 b |

| CL | 2.02±0.07 a | 1.49±0.04 a | 74.3±2.0 a | 25.7±2.0 b | 10.6±0.4 b | 0.8±0.1 b | 62.9±2.0 a |
|---|---|---|---|---|---|---|---|
| *P*-value | * | *** | *** | *** | *** | *** | *** |

Treatments consisted of the basal nutrient solution (BS) alone or supplemented with 5mM nitrate (N) or the sulphate+phosphate (SP) salt mixture containing the same cationic balance as in the CL and N treatments.

Anions and cations measured to calculate the data in the respective columns were: $Cl^-$, $NO_3^-$, $SO_4^{2-}$, and $PO_4^{3-}$ anions; and $K^+$, $Ca^{2+}$, and $Mg^{2+}$ cations.

Mean values ± SE, $n$=4–6. Levels of significance: *$P \leq 0.05$ and ***$P<0.001$. 'Homogeneous group' statistics was calculated through ANOVA test.

*a* Not measured: deduced from quantification of inorganic ions.

**Table 3.** Contribution of inorganic anions to electrical neutrality with cations in the leaf

| | **Ionic concentration (mEq l$^{-1}$)** | | **Contribution to electrical neutrality (%)** | | **Contribution of individual anions to electrical neutrality (%)** | | |
|---|---|---|---|---|---|---|---|
| | **Positive charges of inorganic cations** | **Negative charges of inorganic anions** | **Inorganic anions** | **Organic molecules*a*** | **$SO_4^{2-}$ +$PO_4^{3-}$** | **$NO_3^-$** | **$Cl^-$** |
| BS | 205.83±7.12 | 81.48±8.53 b | 40.6±5.0 b | 59.3±5.0 a | 26.6±2.0 b | 13.1±3.0 ab | 0.9±0.2 b |
| SP | 216.10±8.33 | 108.66±13.48 b | 49.1±4.1 b | 50.9±4.1 a | 42.6±2.8 a | 6.0±1.7 bc | 0.5±0.04 b |
| N | 208.51±1.75 | 75.11±5.33 b | 35.9±2.3 b | 64.0±2.3 a | 16.5±0.6 c | 19.0±2.0 a | 0.4±0.02 b |
| CL | 211.47±6.60 | 146.69±4.71 a | 69.9±2.8 a | 30.1±2.8 b | 20.3±0.9 bc | 1.0±0.3 c | 48.5±2.0 a |
| *P*-value | ns | *** | *** | *** | *** | *** | *** |

Treatments consisted of the basal nutrient solution (BS) alone or supplemented with 5mM nitrate (N) or the sulphate+phosphate (SP) salt mixture containing the same cationic balance as in the CL and N treatments.

Anions and cations measured to calculate the data in the respective columns were: $Cl^-$, $NO_3^-$, $SO_4^{2-}$, and $PO_4^{3-}$anions; and $K^+$, $Ca^{2+}$, and $Mg^{2+}$ cations.

Mean values ± SE, $n$=4–6. Levels of significance: $P$>0.05 ('ns', non-significant differences) and ***$P$<0.001. 'Homogeneous group' statistics were calculated through ANOVA test.

*a* Not measured: deduced from quantification of inorganic ions.

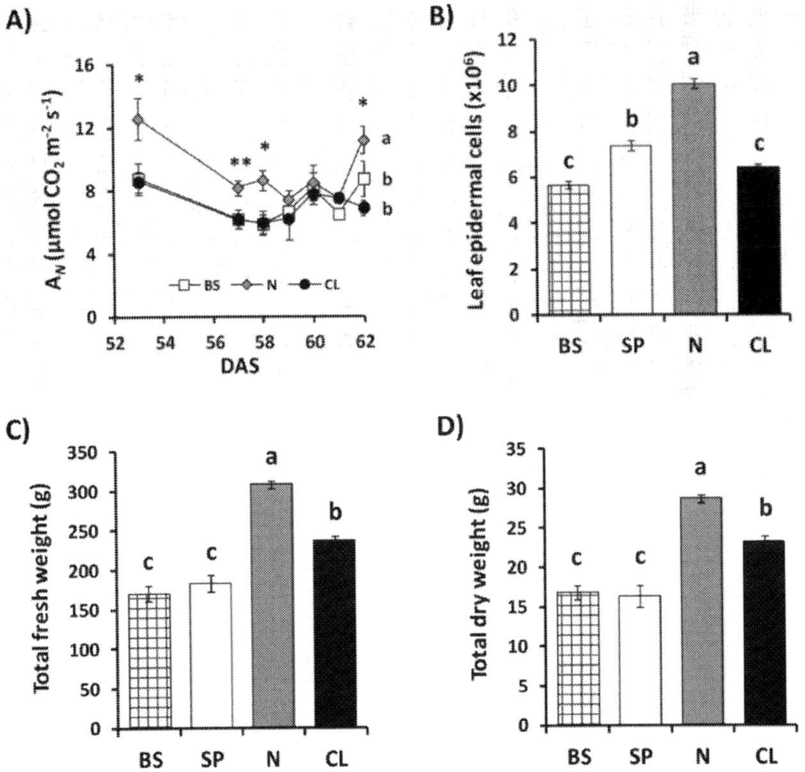

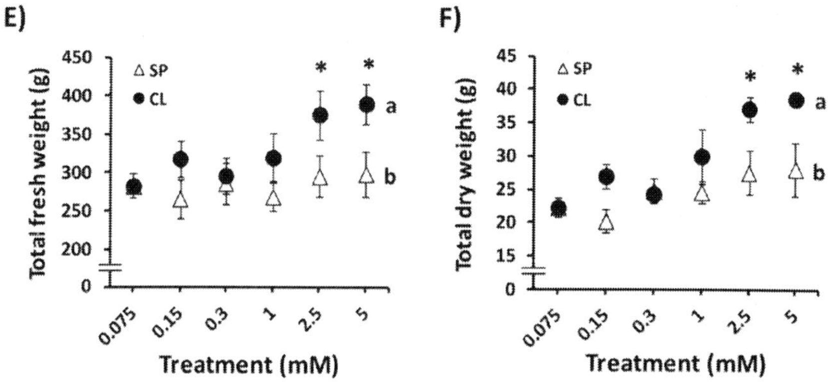


**Figure 1.** Effect of $Cl^-$ nutrition on photosynthesis, cell division, and growth. In (A–D), treatments consisted of the application of the basal nutrient solution (BS) alone or supplemented with 5mM chloride (CL), 5mM nitrate (N), or the sulphate+phosphate (SP) salt mixture containing the same cationic balance as in the CL and N treatments. (A) Effect on net photosynthetic rate ($A_N$) measured in fully expanded, photosynthetically active leaves from tobacco plants between 53 d and 62 d after sowing (DAS). (B) Effect on cell division rate, quantified as the number of epidermal cells per leaf (abaxial side). (C, D) Effect on total (shoot and roots) FW and DW at the end of the experiment. (E, F) Effect of increasing concentrations of $Cl^-$ (CL) or sulphate+phosphate (SP) treatments, maintaining the same cationic balance, on total (shoot and roots) FW and DW, respectively. A non-linear scale is used on the horizontal axis for dose–response experiments (E and F). Mean values ± SE, $n$=6. Levels of significance: *$P \leq 0.05$; **$P \leq 0.01$. 'Homogeneous group' statistics were calculated through ANOVA (A–F) and MANOVA (A, E, F) tests.

Compared with the BS and SP treatments, the CL treatment stimulated growth of shoot organs (stems and leaves), but not of the root. The CL treatment stimulated total leaf FW (Figure 2A), DW (Figure 2B), and total leaf area (Figure 2C), but not the number of leaves, which was stimulated only with the N treatment (Figure 2D). Again, these growth parameters were increased with $Cl^-$ treatments >1mM concentration (Figure 2E–G), whereas no differences between SP and CL treatments were observed in root growth (Figure 2H). The CL treatment specifically stimulated cell size of different leaf cell types including epidermal cells, mesophyll cells, guard cells, and trichomes (Figure 3). This was particularly evident in epidermal cells (Figure 3A, G) and mesophyll cells (Figure 3B, C), which were significantly more expanded in $Cl^-$-treated plants.

In addition, the CL treatment determined higher leaf thickness (Figure 3F). Compared with the basal solution, containing 75 μM $Cl^-$ (giving rise to 2.49mM leaf $Cl^-$ concentration), strong cell growth stimulation was observed with a $Cl^-$treatment as low as 150 μM (giving rise to 6.55mM leaf $Cl^-$ concentration), and this effect was progressively enhanced with increasing concentrations of $Cl^-$ (Figure 3H).

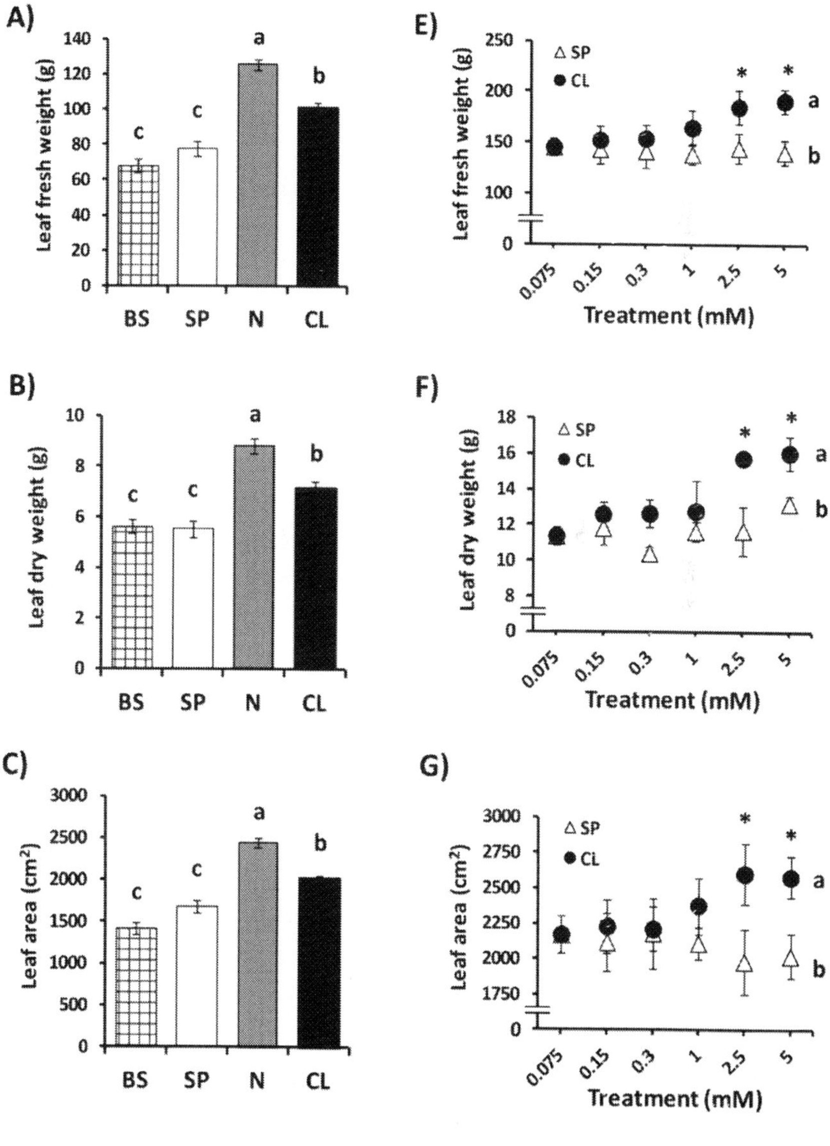

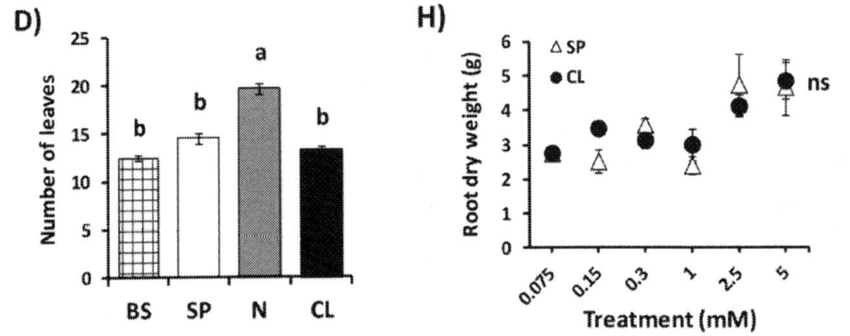


**Figure 2.** Effect of $Cl^-$ nutrition on leaf growth parameters. In (A–D), treatments consisted of the application of the basal nutrient solution (BS) alone or supplemented with 5mM chloride (CL), 5mM nitrate (N), or the sulphate+phosphate (SP) salt mixture containing the same cationic balance as in the CL and N treatments. In (E–H), treatments consisted of increasing concentrations of $Cl^-$ (CL) or sulphate+phosphate (SP) salts maintaining the same cationic balance. (A and E) Effect on total leaf FW. (B and F) Effect on total leaf DW. (C and G) Effect on total leaf area. (D) Effect on number of leaves. (H) Effect on root DW. A non-linear scale used on the horizontal axis for dose–response experiments (E–H). Mean values ± SE, $n$=6. Levels of significance: $P > 0.05$ ('ns', non-significant differences); $*P \leq 0.05$; and 'homogeneous group' statistics were calculated through ANOVA (A–H) and MANOVA (E–H) tests.

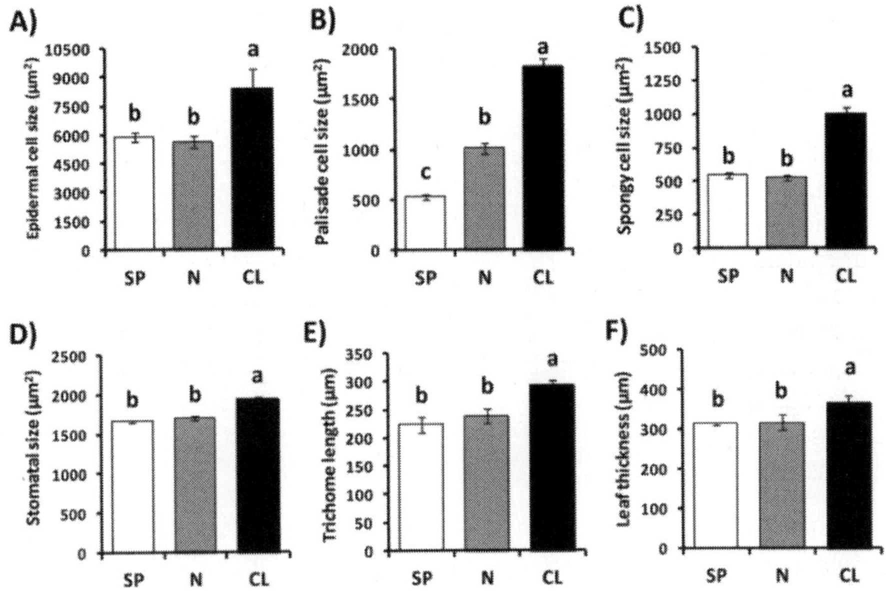

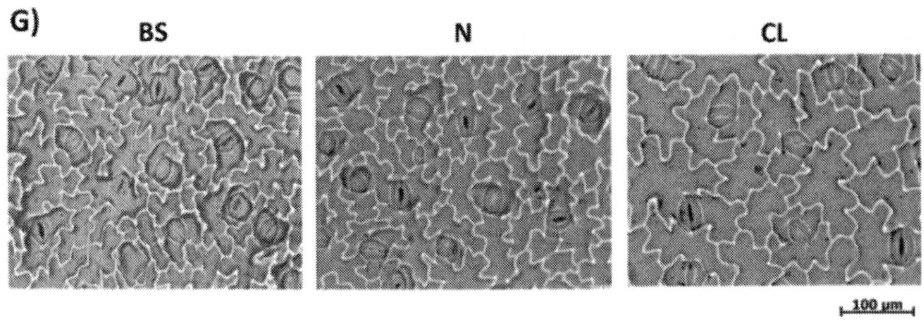


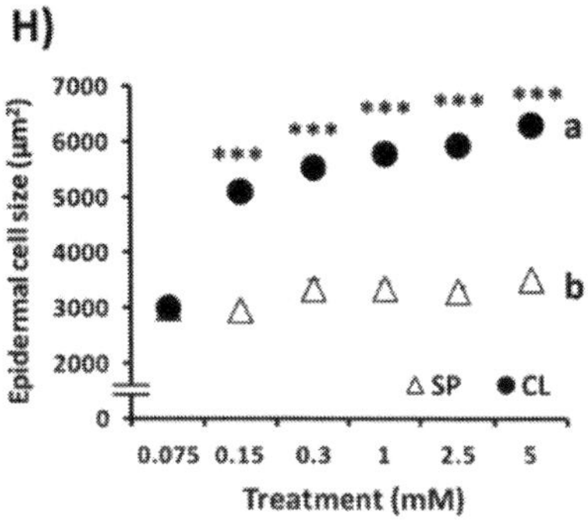


| | | Treatment (mM) | | | | | |
|---|---|---|---|---|---|---|---|
| | *P* | 0.075 | 0.15 | 0.3 | 1 | 2.5 | 5 |
| SP | *0.0035* | b | b | ab | ab | ab | a |
| CL | *<0.0001* | d | c | bc | ab | ab | a |

**Figure 3.** Effect of $Cl^-$ nutrition on leaf cell size. In (A–G), treatments consisted of the application of the basal nutrient solution (BS) alone or supplemented with 5mM chloride (CL), 5mM nitrate (N), or the sulphate + phosphate (SP) salt mixture containing the same cationic balance as in the CL and N treatments. In (H), treatments consisted of increasing concentrations of $Cl^-$ (CL) or sulphate+phosphate (SP), maintaining the same cationic balance. (A, H) Effect on epidermal cell size. (B) Effect on palisade mesophyll cell size. (C) Effect on spongy mesophyll cell size. (D) Effect on stomatal size (combined surface area of the two guard cells and the pore space between). (E) Effect on trichome length. (F) Effect on leaf thickness. (G) Effect of the treatments BS, N, and CL on epidermal cell size observed in microscopy images obtained from abaxial leaf epidermal impressions. Epidermal and stomatal size values were obtained from abaxial epidermal imprints; mesophyll cells size and leaf thickness values were obtained from transversal sections; and trichome length was directly measured from epidermal peel preparations. A non-linear scale is used on the horizontal axis for the dose–response experiment (H). Mean values ± SE, $n$=6. Levels of significance: ***$P \leq 0.001$; and 'homogeneous group' statistics were calculated through ANOVA (A–F, H) and MANOVA (H) tests.

Our working hypothesis was that $Cl^-$ nutrition in the millimolar range might specifically improve the hydric state of plant tissues and water parameters at the whole-plant level. In the leaf, water content (Figure 4A), RWC (Figure 4C), and succulence (Figure 4D) were specifically stimulated by $Cl^-$ over SP and N treatments. An increase of leaf water content required $Cl^-$ application in the millimolar range (Figure 4B). To determine whether higher water content was associated with higher osmotic capacity of $Cl^-$-treated plants, leaf osmotic potential ($\Psi_\pi$) and leaf water potential ($\Psi_w$) values were measured. We confirmed that $\Psi_\pi$ was more negative in both mature and young leaves of plants treated with $Cl^-$, indicating a higher amount of osmotically active solutes in their tissues (Figure 5A, D). Interestingly, leaf $\Psi_w$ of CL-treated plants was less negative than leaf $\Psi_w$ measured in plants subjected to BS, SP, and N treatments (Figure 6B, E). Leaf turgor ($\Psi_p$) of CL plants was significantly higher than $\Psi_p$ of BS, SP, and N plants (Figure 5C, F). Treatments with increasing concentrations of both CL and SP salt supplements determined progressive reductions of leaf $\Psi_\pi$, although the effect was stronger and the differences were significant with CL treatments in the millimolar range (Figure 5G). The observation that leaf $\Psi_w$ was less negative in CL plants compared with BS, SP, and N treatments indicated that besides the osmotic effect, other events occurred in leaves of CL plants, determining higher water accumulation. Water potential values of well-watered tobacco plants in the range of –0.9MPa to –1.3MPa may seem quite negative for a cultivated plant. The dew-point method to determine leaf $\Psi_w$ is a well-accepted method yet is also well documented to lead easily to incorrect values in the lower MPa range. In any case, these values are within the range of $\Psi_w$ values measured by others in well-irrigated *Solanaceae* species (see, for example, Vos and Oyarzún, 1987; Yelenosky and Guy, 1989; Chartzoulakis and Drosos, 1995; Smith and Hare, 2004). In addition, these measurements were made in plants with leaf water content values in the range of 90–93% (Figure 4A) and RWC values in the range of 95–98% (Figure 4C), which correspond to non-stressed plants with well-hydrated leaf tissues.

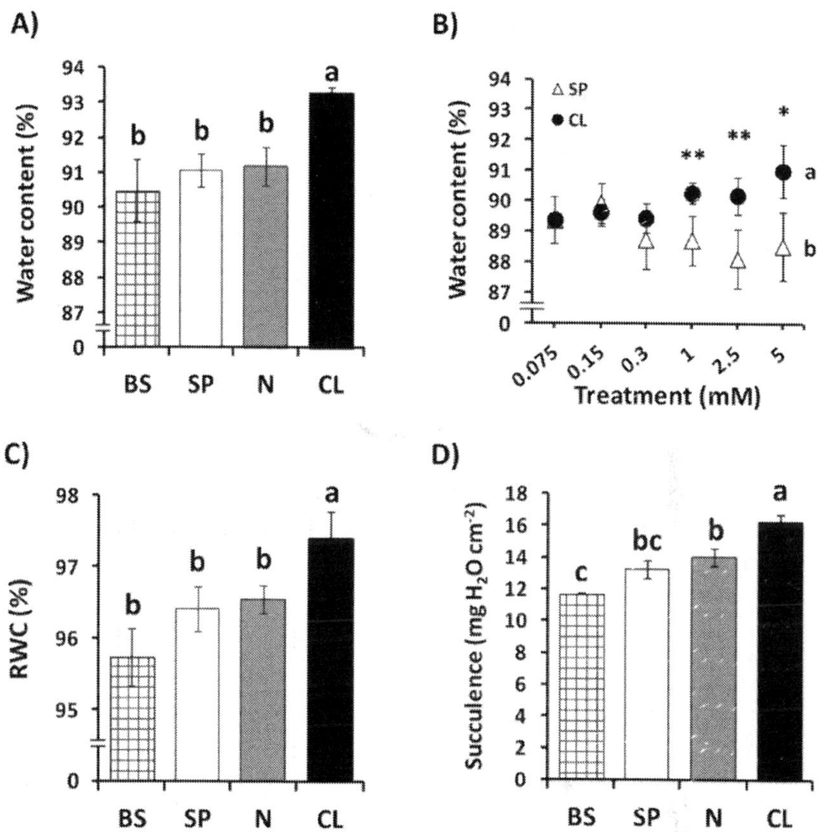


**Figure 4.** Effect of $Cl^-$ nutrition on leaf water parameters. In (A and C–E), treatments consisted of the basal nutrient solution (BS) alone or supplemented with 5mM chloride (CL), 5mM nitrate (N), or the sulphate+phosphate (SP) salt mixture containing the same cationic balance as in the CL and N treatments. In (B), treatments consisted of increasing concentrations of $Cl^-$ (CL) or sulphate+phosphate (SP), maintaining the same cationic balance. (A, B) Effect on leaf water content. (C) Effect on leaf relative water content (RWC). (D) Effect on leaf succulence. A non-linear scale is used on the horizontal axis for the dose–response experiment (B). Mean values ± SE, $n$=6. Levels of significance: $*P \leq 0.05$; $**P \leq 0.01$; and 'homogeneous group' statistics were calculated through ANOVA (A-–E) and MANOVA (B) tests.

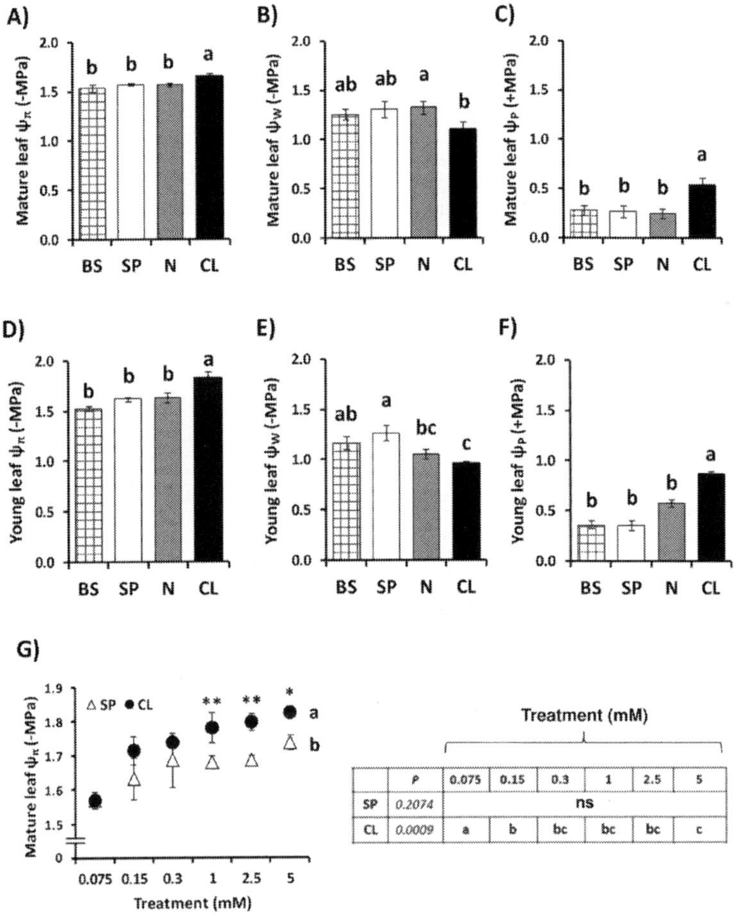


| | P | 0.075 | 0.15 | 0.3 | 1 | 2.5 | 5 |
|---|---|---|---|---|---|---|---|
| SP | *0.2074* | ns | | | | | |
| CL | *0.0009* | a | b | bc | bc | bc | c |

**Figure 5.** Effect of $Cl^-$ nutrition on leaf osmotic potential, water potential, and estimated turgor. In (A–F), treatments consisted of the basal nutrient solution (BS) alone or supplemented with 5mM chloride (CL), 5mM nitrate (N), or the sulphate+phosphate (SP) salt mixture containing the same cationic balance as in the CL and N treatments. In (G), treatments consisted of increasing concentrations of $Cl^-$(CL) or sulphate+phosphate (SP) salts maintaining the same cationic balance. (A–G) Effect on leaf osmotic potential ($\psi_\pi$), water potential ($\psi_w$), and turgor ($\psi_p$) measured using discs from mature (fully expanded) and young leaves (still expanding), harvested before dawn. A non-linear scale is used on the horizontal axis for the dose–response experiment (G). Values ± SE, *n*=6. Levels of significance: $P>0.05$ ('ns', non-significant differences); $^*P\leq0.05$; $^{**}P\leq0.01$; and 'homogeneous group' statistics were calculated through ANOVA (A–G) and MANOVA (G) tests.

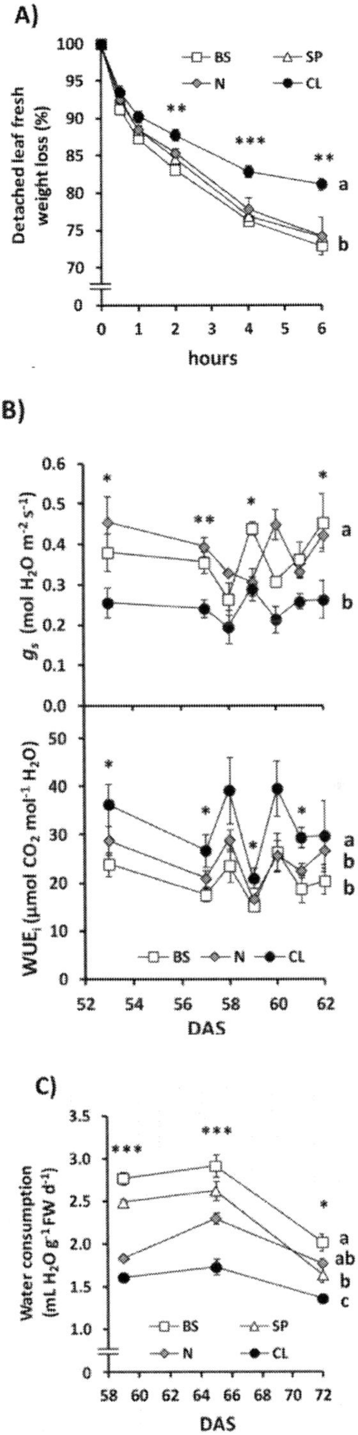
A)
Detached leaf fresh weight loss (%)
hours
BS
SP
N
CL
B)
$g_s$ (mol $H_2O$ $m^{-2}$ $s^{-1}$)
WUE$_i$ (µmol $CO_2$ $mol^{-1}$ $H_2O$)
DAS
C)
Water consumption (mL $H_2O$ $g^{-1}$ FW $d^{-1}$)
DAS

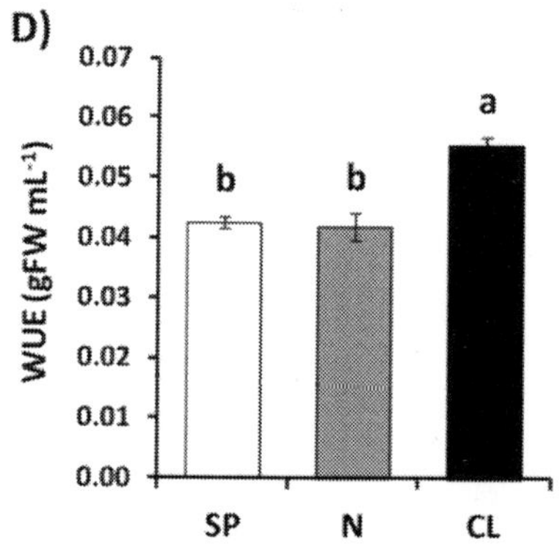


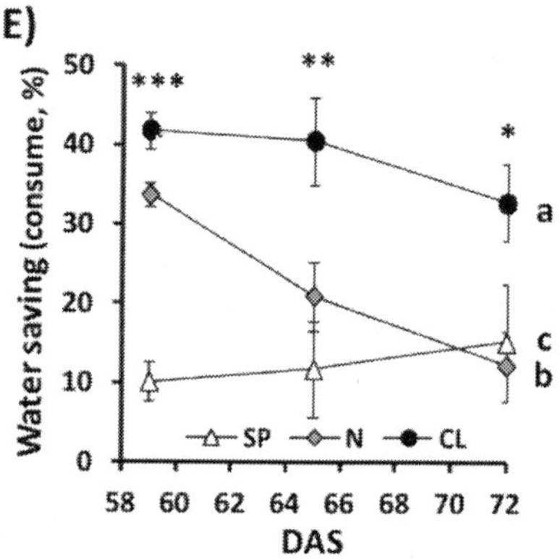


**Figure 6.** Effect of Cl⁻ nutrition on stomatal conductance, leaf-level water loss, whole-plant water use, and water-use efficiency. Treatments consisted of the basal nutrient solution (BS) alone or supplemented with 5mM chloride (CL), 5mM nitrate (N), or the sulphate+phosphate (SP) salt mixture containing the same cationic balance as in the CL and N treatments. (A) Effect on leaf transpiration measured as the FW loss of detached leaves over time. (B) Effect on stomatal conductance ($g_s$) and photosynthetic or instantaneous water-use efficiency ($WUE_i$) measured from plants between 53 d and 62 d after sowing (DAS). WUEi was calculated from both $A_N$ data presented in Figure 1A and $g_s$ data presented in Figure 6B. (C) Effect on total water consumed relative to the plant FW measured at different times after sowing and on each occasion over a 24h period. (D) Effect on integrated WUE obtained from total biomass produced in relation to total water consumed. (E) Effect on plant water saving capacity calculated as the water consumption (mL $H_2O$ $g^{-1}$ FW $d^{-1}$) of treated plants relative to the water consumption of BS plants (from data of (C) measured from plants between 58 and 72 DAS. Mean values ± SE, $n$ = 4-6. Levels of significance: *$P \leq 0.05$; **$P \leq 0.01$; ***$P \leq 0.001$; and 'homogeneous group' statistics were calculated through ANOVA (A–E) and MANOVA (A–C, E) tests.

In plants, the most important factor regulating water content is leaf transpiration. Interestingly, CL plants showed lower transpiration than SP, N, and BS plants, quantified as the relative loss of FW measured in detached leaves (Figure 6A). Reduced transpiration of CL plants was a consequence of the lower stomatal conductance ($g_s$) observed in intact Cl⁻-treated plants (Figure 6B), which in turn determined a higher photosynthetic WUE (Figure 6B). As a consequence, CL plants consumed less water (Figure 6C) and had higher integrated WUE, measured as total plant weight relative to total water consumed (Figure 6D). The basal treatment (BS) always resulted in higher water consumption, which was reduced with the application of salt supplements (Figure 6C). We defined the parameter water saving as the percentage of water saved by the SP, N, or CL treatments in relation to the water consumed in BS plants. We distinguished two different water saving parameters: water saving from whole-plant water consumption, and water saving from leaf transpiration (obtained from the FWL assay). The CL treatment showed the greatest ability to promote water saving from whole-plant water consumption (~40% of BS consumption), significantly higher than the water saving promoted by SP and N treatments (Figure 6E). Furthermore, the water saving determined by the 2.5–5mM Cl⁻ treatments compared with the SP treatment (~25% water saving; Figure 7D) was quite similar to the $g_s$ reductions quantified in plants treated with 2.5–5mM Cl⁻relative to the corresponding SP treatments (~20–36% $g_s$ reduction; Figure 7A). Reduction of $g_s$ and water consumption, the increased WUE, and plant water saving required the application of Cl⁻ in the millimolar range (Figure 7).

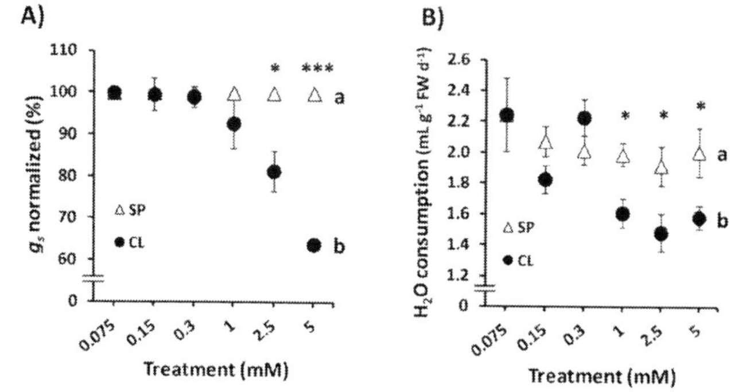

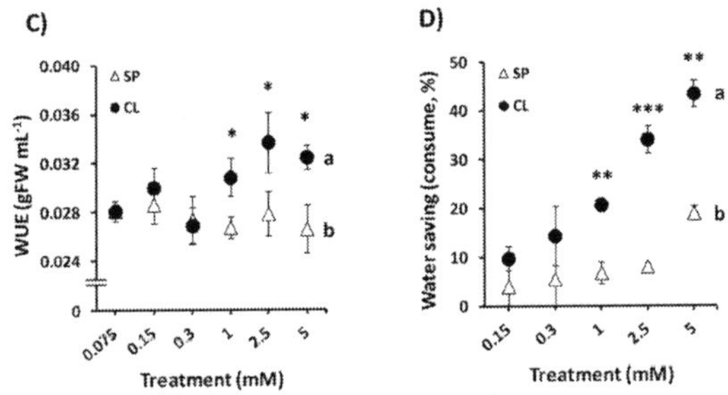


**Figure 7.** Effect of increasing concentrations of anions on whole-plant water relations. Treatments consisted of increasing concentrations of $Cl^-$ (CL) or sulphate+phosphate (SP) salts maintaining the same cationic balance. The results were obtained on days 41, 42, and 47 after sowing (DAS). (A) Effect on stomatal conductance ($g_s$) normalized to the $g_s$ activity of SP plants. (B) Effect on total water consumption relative to the FW of plants. (C) Effect on integrated water-use efficiency (WUE) obtained as total biomass produced in relation to total water consumed. (D) Effect on plant water saving capacity calculated as the water consumption (mL $H_2O$ $g^{-1}$ FW $d^{-1}$) of treated plants relative to the water consumption of BS plants. A non-linear scale is used on the horizontal axis for dose–response experiments. Mean values ± SE, $n$=6. Levels of significance: *$P \leq 0.05$, **$P \leq 0.01$, and ***$P \leq 0.001$; and 'homogeneous group' statistics were calculated through ANOVA and MANOVA tests.

There are three important parameters identified in this work specifically associated with $Cl^-$ treatments determining improved water homeostasis: leaf cell size, leaf osmotic potential, and water saving, obtained from both leaf transpiration and whole-plant water consumption data. These parameters were correlated to the internal leaf concentrations of $Cl^-$ or $SO_4^{2-}$+$PO_4^{3-}$obtained from plants treated with increasing concentrations of anions in CL and SP treatments, respectively. We found a positive correlation of leaf $Cl^-$concentration with these parameters, whereas no correlations were observed with $SO_4^{2-}$+$PO_4^{3-}$ concentration (Figure 8).

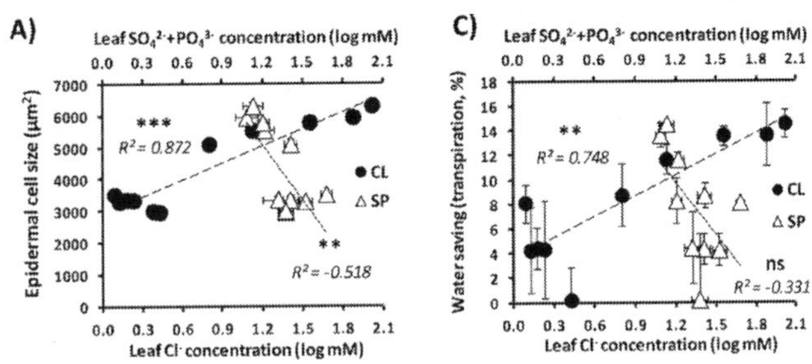

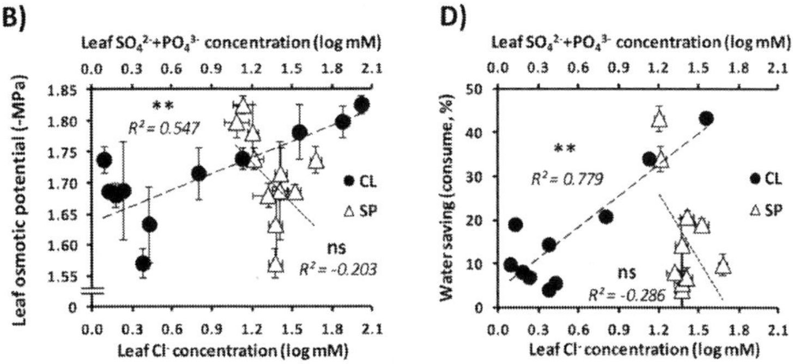


**Figure 8.** Correlations of leaf anion concentration with anatomical and plant water parameters. Treatments consisted of increasing concentrations of $Cl^-$ (CL) or sulphate+phosphate (SP) salts maintaining the same cationic balance. Correlations with chloride concentration measured in both CL and SP plants are given with filled circles and correlations with sulphate+phosphate concentration measured in both CL and SP plants are given with open triangles. (A) Correlations with epidermal cell size. (B) Correlations with leaf osmotic potential. (C) Correlations with water saving capacity based on leaf transpiration measured as FW loss 6h after leaf detachment. (D) Correlations with water saving capacity calculated as the accumulated water consumption (mL $H_2O$ $g^{-1}$ FW $d^{-1}$) of treated plants relative to the accumulated water consumption of BS plants. Logarithms are in base 10. Mean values ± SE, $n$=6. Levels of significance represented by the Pearson's $R^2$ linear correlation test and $P$>0.05 ('ns', non-significant differences); **$P \leq 0.01$; ***$P \leq 0.001$.

Finally, to compare SP and CL treatments in plants with a similar internal leaf concentration of $Cl^-$ and $SO_4^{2-}$+$PO_4^{3-}$ anions, different combinations of CL and SP treatments have been tested. Using 1mM CL and 2.5mM SP treatments, we obtained comparable internal concentrations of $Cl^-$ (35.82±1.52mM) and $SO_4^{2-}$+$PO_4^{3-}$ (33.65±3.72mM) anions, respectively. Also, using 1.83mM CL and 5mM SP treatments, we obtained comparable internal concentrations of $Cl^-$ (55.46±7.50mM) and $SO_4^{2-}$+$PO_4^{3-}$ (48.01±4.25mM) anions, respectively. When these treatments were compared, we still observed increased capacity of $Cl^-$ to promote the effects previously described herein (Tables 4, 5).

**Table 4.** Comparison of different water parameters in SP and CL plants containing similar concentrations of $SO_4^{2-}$+$PO_4^{3-}$ or $Cl^-$ anions. respectively (~33mM)

| | Leaf $\Psi_\pi$ (–MPa) | Leaf WC (%) | Leaf succulence (mg $H_2O$ $cm^{-2}$) | WUE (g FW $ml^{-1}$) | Water consumed (mL $H_2O$ g $FW^{-1}$ $d^{-1}$) | Water saving (transpiration, %) | Water saving (consume, %) |
|---|---|---|---|---|---|---|---|
| SP 2.5 mM | 1.68±0.02 b | 88.1±0.9 b | 14.34±0.68 | 0.026±0.001 b | 1.91±0.13 a | 4.6±1.7 b | 8.1±0.8 b |
| CL 1 mM | 1.79±0.02 a | 90.2±0.3 a | 15.47±0.66 | 0.032±0.001 a | 1.54±0.07 b | 12.4±1.6 a | 19.2±2.0 a |
| *P*-value | * | * | ns | * | * | * | ** |

Different external concentrations of $Cl^-$ (CL) or sulphate+phosphate (SP) salt supplements were applied to obtain plants with comparable leaf internal concentrations of $Cl^-$ and $SO_4^{2-}$+$PO_4^{3-}$anions. We distinguished two different water saving parameters: water saving calculated from whole-plant water consumption (measured gravimetrically) and water saving from leaf transpiration (obtained from the 'detached fresh weight loss' assay shown in Figure 6A). Both parameters were calculated as the water loss of BS plants minus the water loss of treated plants.

$\Psi_\pi$, leaf osmotic potential; WC, water content; WUE, water-use efficiency.

Mean values ± SE, $n$=4–6. Levels of significance: $P>0.05$ ('ns', non-significant differences), *$P\leq0.05$ and **$P<0.01$. 'Homogeneous group' statistics were calculated through ANOVA test.

**Table 5.** Comparison of different water parameters in SP and CL plants containing similar concentration of $SO_4^{2-}$+$PO_4^{3-}$ or $Cl^-$ anions respectively (~50mM)

| | **Leaf $\Psi_\pi$(–MPa)** | **Leaf WC (%)** | **Leaf succulence (mg $H_2O$ cm$^{-2}$)** | **WUE (g FW ml$^{-1}$)** | **Water consumed (mL $H_2O$ g FW$^{-1}$ d$^{-1}$)** | **Water saving (transpiration, %)** | **Water saving (consume, %)** |
|---|---|---|---|---|---|---|---|
| SP 5 mM | 1.73±0.02 b | 88.1±0.9 b | 14.59±0.52 b | 0.025±0.001 b | 2.11±0.17 a | 5.9±1.0 b | 19.0±1.5 b |
| CL 1.83 mM | 1.82±0.02 a | 90.7±0.3 a | 15.71±0.50 a | 0.032±0.002 a | 1.51±0.07 b | 14.5±1.5 a | 38.7±2.6 a |
| *P*-value | * | *** | * | * | ** | * | * |

Different external concentrations of $Cl^-$ (CL) or sulphate+phosphate (SP) salt supplements were applied to obtain plants with comparable leaf internal concentrations of $Cl^-$ and $SO_4^{2-}$+$PO_4^{3-}$anions. We distinguished two different water saving parameters: water saving calculated from whole-plant water consumption (measured gravimetrically) and water saving from leaf transpiration (obtained from the 'detached fresh weight loss' assay shown in Figure 6A). Both parameters were calculated as the water loss of BS plants minus the water loss of treated plants.

$\Psi_\pi$, leaf osmotic potential; WC, water content; WUE, water-use efficiency.

Mean values ± SE, $n$=4–6. Levels of significance: *$P\leq0.05$, **$P\leq0.01$, and ***$P\leq0.001$. 'Homogeneous group' statistics were calculated through ANOVA test.

# DISCUSSION

$Cl^-$ is ubiquitous in nature (typically found at concentrations of 1–5mM; Eaton, 1966) and easily available for plants (Broadley *et al.*, 2012a). Although it is described as a micronutrient, we show in this work that tobacco plants take up $Cl^-$ to levels that are typical of the concentration of a macronutrient, playing specific biological roles that cannot be induced by *bona fide* macronutrient anions such as $NO_3^-$, $SO_4^{2-}$, and $PO_4^{3-}$. The most significant findings of this work are summarized in Figure 9. When fed with $Cl^-$ levels in the millimolar range (1–5mM), tobacco plants accumulate this anion in leaf tissues to high concentrations (40–110mM in the bulk leaf extract), promoting growth, leaf expansion, a better hydration state, reduced transpiration, higher WUE, and water saving. These acquired properties were associated with anatomical and physiological alterations in leaf tissues, which were also specifically triggered by $Cl^-$, such as induction of larger cell size, lower osmotic potential, and higher cell turgor.

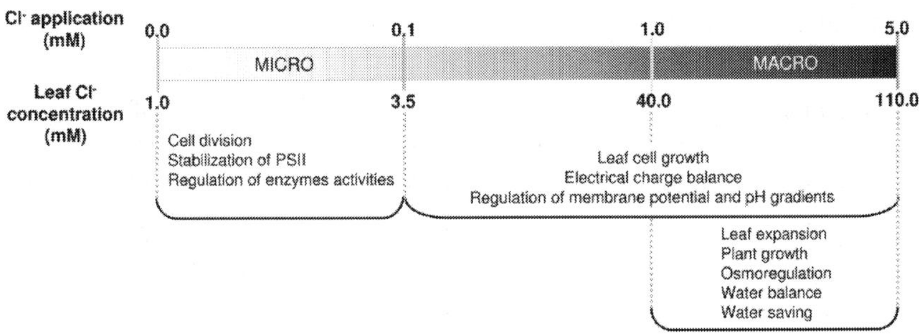


**Figure 9.** Schematic representation of $Cl^-$ functions according to availability in the micro- or macronutrient range. Different biological functions are defined according to the external $Cl^-$ application (0–5mM $Cl^-$) or the resulting leaf bulk tissue concentration (1–110mM $Cl^-$). Functions indicated in the 0.0–0.1mM $Cl^-$ treatment range summarize previous knowledge of $Cl^-$ roles as an essential micronutrient (see the Introduction); biological functions indicated in the 0.1–5.0mM $Cl^-$ treatment range summarize the results obtained in this work.

## Far from Stress and Deficiency Conditions

In the context of plant biology, $Cl^-$ has been generally associated with two extreme situations: its role as a micronutrient, required at low concentrations, and its effect on plants in salt stress conditions, due to excessive accumulation. It is important to note that the experimental conditions used in this work do not lead to any of these situations. In tobacco plants treated with 5mM $Cl^-$, leaf $Cl^-$ concentration was 51.08mg $g^{-1}$ DW (1.44 mmol $g^{-1}$ DW), comparable with the 49.56mg $g^{-1}$ DW (1.27 mmol $g^{-1}$ DW) potassium concentration measured in the same plants (Table 1). This value is similar to the concentration of $Cl^-$ accumulated in leaves of $Cl^-$-includer Citrus rootstocks treated with 4.5mM $Cl^-$ for several months (Brumós *et al.*, 2010). These $Cl^-$ treatments did not represent a stressful situation for plants in view of their normal appearance (not shown), normal growth (Figs 1, 2), and the normal performance of highly stress-sensitive parameters such as the stability of PSII (QY; $F_m{}' F_v{}'^{-1}$) and the plant photosynthetic rate. To draw firm conclusions, it was very important to demonstrate also that under low $Cl^-$ treatments no nutritional deficiency (as an essential micronutrient; Broadley *et al.*, 2012a) occurred in our control treatments. Unless elaborate precautions are taken to exclude all traces of chlorine/$Cl^-$ from the moisture in the air and from the nutrient media, it is usually very difficult to induce $Cl^-$ deficiency in plants (Johnson *et al.*, 1957; Whitehead, 1985). In this work, the minimal $Cl^-$treatment used was 75 μM, present in the basal nutrient solution, giving rise to leaf $Cl^-$ concentrations >0.5mg $g^{-1}$ DW, clearly exceeding the deficiency threshold of <0.2mg $g^{-1}$ DW (Xu *et al.*, 2000; Broadley *et al.*, 2012a). Furthermore, the possibility that plants subjected to low $Cl^-$ treatments were experiencing nutritional $Cl^-$ deficiency was ruled out since: (i) BS, SP, and N plants, treated with low $Cl^-$, did not exhibit visible deficiency symptoms; and (ii) N plants, normally containing half the concentration of $Cl^-$ of BS plants, showed the highest performance in terms of photosynthetic activity, leaf cell division rates, and biomass (Figure 1), which rules out a phenotype of $Cl^-$deficiency.

## Effect of $Cl^-$ on Plant Biomass

As expected, plants supplemented with $NO_3^-$ exhibited the highest biomass as a consequence of the extra nitrogen fertilization giving rise to higher $CO_2$assimilation (Figure 1A) and strong stimulation of cell division and growth of all vegetative organs (Figure 1B–D). However, in contrast to the stimulation of cell division observed in N plants, the biomass increase induced by $Cl^-$ was associated with the stimulation of higher cell size and leaf growth (Figs 2, 3). Agronomic studies have reported a substantial increase in yield for many crops in response to $Cl^-$ fertilization (reviewed in Xu *et al.*, 2000). It was not clear, however, to what extent this effect was due to the accompanying cations, or whether other anions could replace $Cl^-$ in such a growth-promoting effect. The increase of biomass observed in this work (CL versus SP; Figs 1E, F, 2E–H) must be entirely attributed to $Cl^-$ since the resulting internal concentration of cations in leaf tissues remained unchanged among the different treatments (Table 1). The difference in biomass was not due only to the increased water content of CL plants as its DW was also higher (Figs 1D–F, 2B–F), indicating that $Cl^-$ treatments in the millimolar range also improved plant metabolism and growth.

The positive growth response to $Cl^-$ was possibly related to a better hydration state of CL plants (Figs 4, 5 and 6), to more light interception of plants with greater leaf area (Fig 2C, G), and/or the idea previously suggested by Flowers (1988), indicating that high $Cl^-$ accumulation could prevent the diversion of important nutrients such as $NO_3^-$ or malate. In conditions of low $Cl^-$, osmoregulatory and charge balance functions, more importantly occurring at the vacuole, should be compensated by compartmentalization of other molecules (also involved in osmoregulatory functions in plant cells), typically $NO_3^-$ and malate. Once stored in the vacuole, these important sources of N and C are less available for cell metabolism. This suggestion is clearly supported by our results, showing that leaf $NO_3^-$ concentration was reduced >30-fold in plants when $Cl^-$ was increased from the basal (BS) concentration to 5mM $Cl^-$, giving rise to a reduction from 49.30mM to 1.52mM $NO_3^-$. Leaf $NO_3^-$ concentration due to the SP treatment was reduced only 3.49-fold when $SO_4^{2-}+PO_4^{3-}$ was increased from the basal concentration to 5mM $SO_4^{2-}+PO_4^{3-}$, giving rise to a reduction from 49.30mM to 14.21mM $NO_3^-$. Therefore, the reduction in leaf $NO_3^-$ concentration caused by the CL treatment was 9.35-fold greater than the reduction in leaf $NO_3^-$ concentration caused by the SP treatment. The antagonism between $Cl^-$ and $NO_3^-$ accumulation has been widely described and has been attributed to $Cl^-$ and $NO_3^-$ competition for plant root uptake (Xu *et al.*, 2000, and references therein). We cannot accept, however, that $NO_3^-$ uptake is reduced by 10-fold in CL plants compared with SP plants given that $NO_3^-$ is the only source of nitrogen used in the nutrient media. This postulate is inconsistent with the increased biomass of CL plants relative to SP plants. Our data clearly support the $NO_3^-$ diversion hypothesis by which macronutrient accumulation of $Cl^-$ prevents the use of $NO_3^-$ as an osmolyte or charge-balancing molecule, facilitating its assimilation and increasing plant biomass. Although $Cl^-$ has been reported to be interchangeable by $NO_3^-$ (Fricke *et al.*, 1994), we propose that plants preferentially use $Cl^-$ for osmoregulatory purposes, while $NO_3^-$, an essential nitrogen source for land plants, is used as an osmolyte when $Cl^-$ is not sufficiently available in the soil, or in response to high external $NO_3^-$concentrations (Siddiq *et al.*, 1991; Radcliffe *et al.*, 2005). This would represent a radical change in the perception of $Cl^-$, from an $NO_3^-$ antagonist to a nutrient that promotes a more efficient use of nitrogen. We are currently quantifying the role of $Cl^-$ in regulating nitrogen-use efficiency in plants.

Similarly, enhanced $Cl^-$ concentrations in leaves could also reduce accumulation of organic anions such as malate, as suggested by the asymmetry between the concentration of inorganic cations and anions (Tables 2, 3). For example, the CL treatment contributes twice as much as the N treatment to achieve electrical balance between cations and inorganic ions (Table 3). From our results we have estimated that the contribution of organic anions to balance the positive charges of inorganic cations is ~65% in N-treated plants, 67% in SP-treated plants, and 25% in CL-treated plants (Table 2). Considering the concentration of the most abundant inorganic ions measured in bulk-mature leaf tissue of plants treated with $Cl^-$, an estimation of osmotic potential is obtained, which corresponds to –1.10MPa. This value is less than the –1.65MPa measured for the plants in the CL treatment (Figure 5A). This difference of –0.55MPa, accounting for 33% of the osmotic potential, could have been made up by organic and neutral solutes as estimated previously (Tables 2, 3). In other plant species, the contribution of inorganic/organic solutes quantified experimentally in non-stressed plants was similar: 63%/37% in potato (Steward, 1986); 65%/35% in cotton (Meloni *et al.*, 2001); and 77%/23% in wheat (Kameli and Lösel, 1995). Supporting the role of $Cl^-$ in preventing diversion of important metabolites, it has been reported that halophyte plants growing under $Cl^-$-deficient conditions accumulate higher concentrations of $NO_3^-$ (Neales and Sharkey, 1981; Yeo and Flowers, 1986) and malate (Flowers and Hall, 1978) than those plants growing in optimal NaCl concentrations.

Regarding the photosynthetic metabolism, it should be expected that a significant reduction of $g_s$ in CL plants (Figure 6B) gave rise to a reduction of $A_N$. However, $A_N$ of CL plants was maintained similar to values of BS (Figure 1A) and SP plants. This could be a consequence of a higher diffusion conductance to carbon dioxide ($CO_2$) in the leaf mesophyll of CL plants, probably as a consequence of anatomical alterations induced by $Cl^-$ in leaf tissues (Figure 3). We are presently studying this phenomenon.

## Effect of $Cl^-$ on Osmotic Properties and Leaf Cell Size

It is generally assumed that $Cl^-$ is serving a non-specific osmotic function and that other inorganic anions can provide osmolarity to the plant vacuoles or balance positive charges. $NO_3^-$, for example, can certainly do so in the stomata (Geiger *et al.*, 2011). However, this work shows evidence indicating that $Cl^-$ is quantitatively and qualitatively a superior osmolyte in plants. Quantitatively, Table 1 shows that $Cl^-$, which is not assimilated throughout anabolic metabolism, is accumulated in leaf tissues to a higher concentration than anionic macronutrients such as $NO_3^-$, $SO_4^{2-}$, or $PO_4^{3-}$ (Table 1). Calculating the accumulation efficiency of ions according to the molar concentration accumulated in the bulk leaf extract versus the molar concentration applied, it is observed that the accumulation efficiency of $Cl^-$ is comparable with that of potassium, while $NO_3^-$, $SO_4^{2-}$, and $PO_4^{3-}$ exhibited lower values. These latter three anionic macronutrients are assimilated in plant metabolism, reducing their internal concentration and therefore their accumulation efficiency. Compared with $Cl^-$, $NO_3^-$ accumulation efficiency was about four times lower and that of $SO_4^{2-}+PO_4^{3-}$ was about three times lower.

$Cl^-$ was specifically required to stimulate leaf cell size (Figure 3). Stimulation of cell size occurred with $Cl^-$ treatments as low as 150 μM (Figure 3H), which represents an increase in the leaf $Cl^-$ concentration from 2.49mM (in BS plants) to 6.55mM (in CL plants treated with 150 μM $Cl^-$). This concentration increment is not important enough to play a relevant osmoregulatory role. In the same tissues, $SO_4^{2-}+PO_4^{3-}$ and $NO_3^-$ concentrations were much higher (26.23mM and 16.30mM, respectively), suggesting that $Cl^-$ plays a specific signalling role in the stimulation of leaf cell growth. With increasing $Cl^-$ treatments, a growing response was further observed (Figure 3H). At least in the millimolar treatment range, this response could be due to a progressive increase in leaf turgor, significantly higher in plants treated with 5mM $Cl^-$ (Figure 5C, F). This could be a consequence of the specific stimulation by $Cl^-$ of the tonoplast ATPase (Churchill and Sze, 1984), which favours the vacuolar compartmentalization of $Cl^-$, resulting in a more negative osmotic potential and, consequently, a higher turgor than the equivalent SP and N treatments (Figure 5). Taking both factors together, larger and more turgid cells give rise to leaf tissues with higher capacity for water accumulation (Figure 4), which is also evidenced by greater leaf thickness (Figure 3F) and succulence (Figure 4D) of CL-treated plants. Previous studies have shown that WUE is related to the morphological characteristics of leaves. Wright *et al.* (1994) proposed that there is a close negative relationship between WUE and specific leaf area (SLA), indicating a direct correlation between WUE and leaf thickness (Liu and Stützel, 2004). This correlation was observed in CL-treated plants, but not in SP-treated plants, indicating a relationship between leaf morphological changes and the higher WUE. In addition, the stability of the interaction of water molecules is atypically high in the solvation shell of halogen anions (Kropman and Bakker, 2001), making $Cl^-$ an osmolyte with uncommon physical properties, very suitable to stimulate the retention of water in leaf tissue. As a result of the above, $Cl^-$ was clearly more efficient in providing leaf osmotic potential, turgor, and water accumulation capacity (Figure 5), leading to a greater hydration of leaves (Figure 4).

Specific roles for $Cl^-$ in osmotic and cell volume regulation in plants have been previously reported for specialized motor organs or cell types such as the coleoptile of grass seedlings (Babourina *et al.*, 1998), the stigma of grasses at the onset of flower anthesis (Heslop-Harrison and Reger, 1986), the pulvini of *Mimosa pudica* and *Phaseolus vulgaris* during seismonastic leaf movement (Leigh and Wyn Jones, 1985; Fromm and Eschrich, 1988; Iino *et al.*, 2001), epidermal cells from elongating internodes of *Pisum sativum* (Yamagami *et al.*, 2004), and guard cells (Broadley *et al.*, 2012a). Most of these processes have been proved to be responsive to auxin, which stimulates cell $Cl^-$ uptake (Babourina *et al.*, 1998; Iino *et al.*, 2001; Yamagami *et al.*, 2004) as a possible prerequisite for cell elongation. We have characterized a $Cl^-$ transporter (*At*CCC) strongly expressed in organs involved in primary auxin production

(Colmenero-Flores *et al.*, 2007). Lack of function of the *At*CCC gene produces frequent collapse of the elongation zone of the inflorescence stem, possibly due to a failure in the synchronization of physical cell growth and osmolyte (or water) supply during the very fast elongation process of Arabidopsis flowering stems (Colmenero-Flores *et al.*, 2007). These earlier results and the sensitive response of leaf cell size to $Cl^-$ shown in this study (Figure 3) suggest a cross-talk between $Cl^-$ nutrition and auxin activity. In marine organisms, animals, and halophyte plants, $Cl^-$ plays key roles in cell osmotic, hydric, and turgor (or cell volume) regulation (Flowers, 1988; Sardini *et al.*, 2006; Colmenero-Flores *et al.*, 2007). Through changes in the rate of shoot $Cl^-$buildup, plants might sense $Cl^-$ availability and therefore stimulate cell growth to favour more effective compartmentalization of ions and higher water accumulation capacity in plants when such 'high quality' electrolyte is accessible. This hypothesis requires a positive feedback regulation mechanism by which $Cl^-$ could stimulate the synthesis or stability of auxin. Such a regulatory mechanism has been described in some legumes, where the covalent interaction of chlorine/$Cl^-$ with auxin [e.g. indole acetic acid (IAA)] produces a chlorinated auxin (e.g. 4-Cl-IAA) much more active than the non-chlorinated molecule (Böttger *et al.*, 1978; Reinecke, 1999). Under this hypothetical regulatory model, $Cl^-$ first induces higher auxin activity, which in turn stimulates both $Cl^-$ uptake and cell elongation; two processes that we believe are interconnected. Larger leaf cells with higher ability for ion compartmentalization result in increased water storage capacity. In addition, larger cells also resulted in greater leaf area (Figure 2C, G), which is expected to determine more light interception and growth. This may also explain why plants treated with $Cl^-$ have greater dry biomass (Figure 1D, F).

## Effect of $Cl^-$ on Whole-Plant Water Relations

An additional and unexpected effect of $Cl^-$ nutrition on the physiology of tobacco plants was a reduction of leaf transpiration (Figure 6A) as a consequence of the lower stomatal conductance of CL-treated plants (Figure 6B). This phenomenon could be a consequence of: (i) lower stomatal opening; (ii) lower stomatal density; of (iii) lower stomatal index (percentage of stomata out of the total number of epidermal cells plus stomata). The role of anion transport in the regulation of stomatal function is well known. In particular, SLAC1 and SLAH3 channels, which mediate $Cl^-$ and $NO_3^-$ release from guard cells, are the components that connect the signalling of environmental cues with the physical events that trigger stomatal closure (Negi *et al.*, 2008; Vahisalu *et al.*, 2008; Geiger *et al.*, 2011). Elevated $Cl^-$ content in leaf tissues might alter the flow of anions of epidermal cells, for example due to high extracellular accumulation of $Cl^-$, thus modifying the regulatory properties of stomatal activity. On the other hand, specific induction by $Cl^-$ of larger epidermal cell size may alter the distribution of the stomata on the leaf surface. Stomatal density, or the number of stomata per unit area, is a function of cell size (Salisbury, 1927). It is expected that an increase in size of epidermal cells, including guard cells, determines a lower stomatal density, which may reduce stomatal conductance per unit leaf area. Although the stomatal index is independent of cell size, an effect of $Cl^-$ on the stomatal index cannot be ruled out. We are currently quantifying these anatomical parameters to determine the link between leaf $Cl^-$ content and the reduction of stomatal conductance.

## $Cl^-$ as a Beneficial Macronutrient

$Cl^-$ is an essential micronutrient (Broyer *et al.*, 1954). The average concentration of $Cl^-$ required to meet essential functions as a micronutrient and to determine adequate plant growth is 3.0 μmol $g^{-1}$ DW (Kirkby and Marschner, 2012). We have demonstrated here that $Cl^-$ further stimulates tobacco plant growth when accumulated at a concentration 500 times higher (~1,500 μmol $g^{-1}$ DW; Table 1), tissue levels typical of a macronutrient such as potassium. This accumulation specifically determines an increase of plant biomass associated with improved leaf water status, greater WUE, and a remarkable water saving capacity. Effects on leaf water balance and plant water relations are a consequence of anatomical and physiological changes including stimulation of leaf cell size, reduction of stomatal conductance, and induction of additional osmolarity and turgor, which are specifically triggered by $Cl^-$ and not by essential anionic macronutrients such as $NO_3^-$, $SO_4^{2-}$, and $PO_4^{3-}$. Beneficial elements are defined as those elements that stimulate growth, but are not essential or are essential in certain plant species, or under specific conditions (Broadley *et al.*, 2012b). Since $Cl^-$ is not an essential macronutrient but it stimulates growth under such conditions, we propose that $Cl^-$ be defined as an essential micronutrient and a beneficial macronutrient. We expect that to a greater or lesser extent $Cl^-$ plays the same role in other glycophyte plants. Moreover, we suggest that the biological functions regulated by $Cl^-$ availability will probably influence adaptive mechanisms that regulate water homeostasis and the ability of plants to withstand water deficit, a hypothesis that we are currently exploring.

## ACKNOWLEDGEMENTS

This work was supported by the Spanish Ministry of Science and Innovation-FEDER grant AGL2009-08339/AGR. The help, expertise, and technical assistance of C. Rivero, A. Vázquez, S. Luque, B.J. Sañudo, F.J. Durán, Y. Pinto, and J. Espartero are gratefully acknowledged. We would like to extend our gratitude to the valuable reviews and contributions by the anonymous referees and the editor, Timothy Colmer, which helped us to improve the manuscript substantially.

## REFERENCES

1. Abbate PE Dardanelli JL Cantarero MG Maturano M Melchiori RJM Suero EE . 2004. Climatic and water availability effects on water-use efficiency in wheat. Crop Science 44, 474–483.
2. Allen GJ Kuchitsu K Chu SP Murata Y Schroeder JI . 1999. Arabidopsis abi1-1 and abi2-1 phosphatase mutations reduce abscisic acid-induced cytoplasmic calcium rises in guard cells. The Plant Cell 11, 1785–1798.
3. Armengaud P Breitling R Amtmann A . 2004. The potassium-dependent transcriptome of Arabidopsis reveals a prominent role of jasmonic acid in nutrient signaling. Plant Physiology 136, 2556–2576.
4. Babourina O Shabala S Newman I . 1998. Auxin stimulates Cl- uptake by oat coleoptiles. Annals of Botany 82, 331–336.
5. Barrs HD Weatherley PE . 1962. A re-examination of relative turgidity technique for estimating water deficits in leaves. Australian Journal of Biological Sciences 15, 413–428.
6. Benlloch-Gonzalez M Arquero O Maria Fournier J Barranco D

Benlloch M . 2008. K+ starvation inhibits water-stress-induced stomatal closure. Journal of Plant Physiology 165, 623–630.

7. Böttger M Engvild KC Soll H . 1978. Growth of Avena coleoptiles and pH drop of protoplast suspensions induced by chlorinated indoleacetic acids. Planta 140, 89–92.
8. Britto DT Kronzucker HJ . 2006. Futile cycling at the plasma membrane: a hallmark of low-affinity nutrient transport. Trends in Plant Science 11, 529–534.
9. Broadley M Brown P Cakmak I Ma JF Rengel Z Zhao F Marschner P . 2012b. Beneficial elements. In: Marschner P, ed. Marschner's mineral nutrition of higher plants, 3rd edn . San Diego: Academic Press, 249–269.
10. Broadley M Brown P Cakmak I Rengel Z Zhao F . 2012a. Function of nutrients: micronutrients. In: Marschner P, ed. Marschner's mineral nutrition of higher plants, 3rd edn . San Diego: Academic Press, 191–248.
11. Broyer TC Carlton AB Johnson CM Stout PR . 1954. Chlorine—a micronutrient element for higher plants. Plant Physiology 29, 526.
12. Brumós J Talón M Bouhlal RYM Colmenero-Flores JM . 2010. Cl– homeostasis in includer and excluder citrus rootstocks: transport mechanisms and identification of candidate genes. Plant, Cell and Environment 33, 2012–2027.
13. Chartzoulakis K Drosos N . 1995. Water use and yield of greenhouse grown eggplant under drip irrigation. Agricultural Water Management 28, 113–120.
14. Cheong YH Pandey GK Grant JJ Batistic O Li L Kim BG Lee S-C Kudla J Luan S . 2007. Two calcineurin B-like calcium sensors, interacting with protein kinase CIPK23, regulate leaf transpiration and root potassium uptake in Arabidopsis. The Plant Journal 52, 223–239.
15. Churchill KA Sze H . 1984. Anion-sensitive, H+-pumping ATPase of oat roots: direct effects of Cl–, NO3–, and disulfonic stilbene. Plant Physiology 76, 490–497.
16. Cochrane TT . 1994. A new equation for calculating osmotic potential. Plant, Cell and Environment 17, 427–433.

17. Colmenero-Flores JM Moreno LP Smith CE Covarrubias AA . 1999. Pvlea-18, a member of a new late-embryogenesis-abundant protein family that accumulates during water stress and in the growing regions of well-irrigated bean seedlings. Plant Physiology 120, 93–104.
18. Colmenero-Flores JM Martinez G Gamba G Vazquez N Iglesias DJ Brumos J Talon M . 2007. Identification and functional characterization of cation–chloride cotransporters in plants. The Plant Journal 50, 278–292.
19. Eaton FM . 1966. Chlorine. In: Chapman HD, ed. Diagnostic criteria for plant and soils . Riverside: University of California Division of Agricultural Sciences, 98–135.
20. Felle HH . 1994. The H+/Cl– symporter in root-hair cells of Sinapis alba (an electrophysiological study using ion-selective microelectrodes). Plant Physiology 106, 1131–1136.
21. Fernández JE Moreno F Giron IF Blazquez OM . 1997. Stomatal control of water use in olive tree leaves. Plant and Soil 190, 179–192.
22. Flowers TJ . 1988. Chloride as a nutrient and as an osmoticum. In: Tinker PBLaüchli A, eds. Advances in plant nutrition , Vol. 3. New York: Praeger, 55–78.
23. Flowers TJ Hall JL . 1978. Salt tolerance in the halophyte, Suaeda maritima (L.) Dum.: the influence of the salinity of the culture solution on the content of various organic compounds. Annals of Botany 42, 1057–1063.
24. Flowers TJ Yeo AR . 1986. Ion relations of plants under drought and salinity. Functional Plant Biology 13, 75–91.
25. Fricke W Leigh RA Tomos AD . 1994. Epidermal solute concentrations and osmolality in barley leaves studied at the single-cell level. Planta 192, 317–323.
26. Fromm J Eschrich W . 1988. Transport processes in stimulated and non-stimulated leaves of Mimosa pudica III. Displacement of ions during seismonastic leaf movements. Trees–Structure and Function 2, 65–72.
27. Geiger D Maierhofer T Al-Rasheid KAS etr al . 2011. Stomatal closure by fast abscisic acid signaling is mediated by the guard cell

anion channel SLAH3 and the receptor RCAR1. Science Signaling 4, ra32.

28. Gilliam JW . 1971. Rapid measurement of chlorine in plant materials. Soil Science Society of America Proceedings 35, 512–513.
29. Gitz DC Baker JT . 2009. Methods for creating stomatal impressions directly onto archivable slides. Agronomy Journal 101, 232–236.
30. Hänsch R Mendel RR . 2009. Physiological functions of mineral micronutrients (Cu, Zn, Mn, Fe, Ni, Mo, B, Cl). Current Opinion in Plant Biology 12, 259–266.
31. Hawkesford M Horst W Kichey T Lambers H Schjoerring J Skrumsager Moller I White P . 2012. Functions of macronutrients. In: Marschner P, ed. Marschner's mineral nutrition of higher plants, 3rd edn . San Diego: Academic Press, 135–190.
32. He J-M Ma X-G Zhang Y Sun T-F Xu F-F Chen Y-P Liu X Yue M . 2013. Role and interrelationship of G alpha protein, hydrogen peroxide, and nitric oxide in ultraviolet B-induced stomatal closure in Arabidopsis leaves. Plant Physiology 161, 1570–1583.
33. Heslop-Harrison JS Reger BJ . 1986. Chloride and potassium ions and turgidity in the grass stigma. Journal of Plant Physiology 124, 55–60.
34. Hogue E Wilcox GE Cantliff DJ . 1970. Effect of soil phosphorus levels on phosphate fractions in tomato leaves. Journal of the American Society for Horticultural Science 95, 174–176.
35. Iino M Long C Wang X . 2001. Auxin- and abscisic acid-dependent osmoregulation in protoplasts of Phaseolus vulgaris pulvini. Plant and Cell Physiology 42, 1219–1227.
36. Johnson CM Stout PR Broyer TC Carlton AB . 1957. Comparative chlorine requirements of different plant species. Plant and Soil 8, 337–353.
37. Kameli A Lösel DM . 1995. Contribution of carbohydrates and other solutes to osmotic adjustment in wheat leaves under water stress. Journal of Plant Physiology 145, 363–366.
38. Kirkby E Marschner P . 2012. Introduction, definition and classification of nutrients. In: Marschner P, ed. Marschner's mineral nutrition of higher plants, 3rd edn . San Diego: Academic Press, 3–5.

39. Kropman MF Bakker HJ . 2001. Dynamics of water molecules in aqueous solvation shells. Science 291, 2118–2120.

40. Langdale JA . 1994. In situ hybridization . In: Freeling MWalbot V, eds. The maize handbook . New York: Springer-Verlag, 165–179.

41. Leigh RA Wyn Jones RG . 1985. Cellular compartmentation in plant nutrition: the selective cytoplasm and the promiscuous vacuole. In: Tinker PBLaüchli A, eds. Advances in plant nutrition , Vol. 2. New York: Praeger Press, 249–279.

42. Leigh RA Tomos AD . 1993. Ion distribution in cereal leaves: pathways and mechanisms. Philosophical Transactions of the Royal Society B: Biological Sciences 341, 75–86.

43. Li AX Han YY Wang X Chen YH Zhao MR Zhou SM Wang W . 2014. Root-specific expression of wheat expansin gene TaEXPB23 enhances root growth and water stress tolerance in tobacco. Environmental and Experimental Botany 110, 73–84.

44. Liu F Stützel H . 2004. Biomass partitioning, specific leaf area, and water use efficiency of vegetable amaranth (Amaranthus spp.) in response to drought stress. Scientia Horticulturae 102, 15–27.

45. Marcelis LFM Heuvelink E Goudriaan J . 1998. Modelling biomass production and yield of horticultural crops: a review. Scientia Horticulturae 74, 83–111.

46. Maxwell K Johnson GN . 2000. Chlorophyll fluorescence—a practical guide. Journal of Experimental Botany 51, 659–668.

47. Meloni DA Oliva MA Ruiz HA Martinez CA . 2001. Contribution of proline and inorganic solutes to osmotic adjustment in cotton under salt stress. Journal of Plant Nutrition 24, 599–612.

48. Moya JL Gomez-Cadenas A Primo-Millo E Talon M . 2003. Chloride absorption in salt-sensitive Carrizo citrange and salt-tolerant Cleopatra mandarin citrus rootstocks is linked to water use. Journal of Experimental Botany 54, 825–833.

49. Moya JL Primo-Millo E Talon M . 1999. Morphological factors determining salt tolerance in citrus seedlings: the shoot to root ratio modulates passive root uptake of chloride ions and their accumulation in leaves. Plant, Cell and Environment 22, 1425–1433.

50. Neales TF Sharkey PJ . 1981. Effect of salinity on growth and on

mineral and organic-constituents of the halophyte Disphyma australe (Soland.) J.M. Black. Australian Journal of Plant Physiology 8, 165–179.

51. Negi J Matsuda O Nagasawa T Oba Y Takahashi H Kawai-Yamada M Uchimiya H Hashimoto M Iba K . 2008. CO2 regulator SLAC1 and its homologues are essential for anion homeostatsis in plant cells. Nature 452, 483–486.
52. Novozamsky I Van Eck R . 1977. Total sulfur determination in plant material. Fresenius Zeitschrift Fur Analytische Chemie 286, 367–368.
53. Perez-Perez JG Syvertsen JP Botia P Garcia-Sanchez F . 2007. Leaf water relations and net gas exchange responses of salinized Carrizo citrange seedlings during drought stress and recovery. Annals of Botany 100, 335–345.
54. Pitman M . 1982. Transport across plant roots. Quarterly Reviews of Biophysics 15, 481–554.
55. Radcliffe SA Miller AJ Ratcliffe RG . 2005. Microelectrode and 133Cs nuclear magnetic resonance evidence for variable cytosolic and cytoplasmic nitrate pools in maize root tips. Plant, Cell and Environment 28, 1379–1387.
56. Reinecke DM . 1999. 4-Chloroindole-3-acetic acid and plant growth. Plant Growth Regulation 27, 3–13.
57. Rosales MA Ocampo E Rodríguez-Valentín R Olvera-Carrillo Y Acosta-Gallegos J Covarrubias AA . 2012. Physiological analysis of common bean (Phaseolus vulgaris L.) cultivars uncovers characteristics related to terminal drought resistance. Plant Physiology and Biochemistry 56, 24–34.
58. Salisbury EJ . 1927. On the causes and ecological significance of stomatal frequency, with special reference to the woodland flora. Philosophical Transactions of the Royal Society B: Biological Sciences 216, 1–65.
59. Sardini A Bittar EE Michael P . 2006. Cell volume homeostasis: the role of volume-sensitive chloride channels. Advances in Molecular and Cell Biology 38, 199–214.
60. Scafaro AP Von Caemmerer S Evans JR Atwell BJ . 2011. Temperature response of mesophyll conductance in cultivated and

wild Oryza species with contrasting mesophyll cell wall thickness. Plant, Cell and Environment 34, 1999–2008.

61. Siddiq MY Glass ADM Ruth TJ . 1991 Studies of the uptake of nitrate in barley: III. Compartmentation of NO3. Journal of Experimental Botany 42, 1455–1463.
62. Smith JL Hare JD . 2004. Spectral properties, gas exchange, and water potential of leaves of glandular and non-glandular trichome types in Datura wrightii (Solanaceae). Functional Plant Biology 31, 267–273.
63. Steward FC . 1986. Solutes in cells: their responses during growth and development. In: Steward FC. Plant physiology: a treatise. Volume IX. Water and solutes in plants . San Diego: Academic Press, 551–594.
64. Storey R Walker RR . 1999. Citrus and salinity. Scientia Horticulturae 78, 39–81.
65. Tadeo F Cercos M Colmenero-Flores JM et al . 2008. Molecular physiology of development and quality of citrus. Advances in Botanical Research 47, 147–223.
66. Taguas JJ Rivero A . 1989. Hojas, ADN Sofware. Sevilla.
67. Terry N . 1977. Photosynthesis, growth, and role of chloride. Plant Physiology 60, 69–75.
68. Vahisalu T Kollist H Wang YF et al . 2008. SLAC1 is required for plant guard cell S-type anion channel function in stomatal signalling. Nature 452, 487–491.
69. Vos J Oyarzun PJ . 1987. Photosynthesis and stomatal conductance of potato leaves— effects of leaf age, irradiance, and leaf water potential. Photosynthesis Research 11, 253–264.
70. White PJ Broadley MR . 2001. Chloride in soils and its uptake and movement within the plant: a review. Annals of Botany 88, 967–988.
71. Whitehead DC . 1985. Chlorine deficiency in red-clover grown in solution culture. Journal of Plant Nutrition 8, 193–198.
72. Wright GC Rao RC Farquhar GD . 1994. Water-use efficiency and carbon isotope discrimination in peanut under water deficit conditions. Crop Science 34, 92–97.

73. Xu G Magen H Tarchitzky J Kafkafi U . 2000. Advances in chloride nutrition of plants. Advances in Agronomy 68, 96–150.

74. Yamagami M Haga K Napier RM Iino M . 2004. Two distinct signaling pathways participate in auxin-induced swelling of pea epidermal protoplasts. Plant Physiology 134, 735–747.

75. Yelenosky G Guy CL . 1989. Freezing tolerance of citrus, spinach, and petunia leaf tissue osmotic adjustment and sensitivity to freeze induced cellular dehydration. Plant Physiology 89, 444–451.

76. Yeo AR Flowers TJ . 1980. Salt tolerance in the halophyte Suaeda maritima L. Dum.: evaluation of the effect of salinity upon growth. Journal of Experimental Botany 31, 1171–1183.

77. Yeo AR Flowers TJ . 1986. Ion transport in Suaeda maritima: its relation to growth and implications for the pathway of radial transport of ions across the root. Journal of Experimental Botany 37, 143–159.

CHAPTER

# 11

# MEMBRANE TRANSPORTERS AND DROUGHT RESISTANCE – A COMPLEX ISSUE

Karolina M. Jarzyniak[1,2] and Michał Jasiński[1,2]

[1]Laboratory of Plant Molecular Physiology, Department of Natural Products Biochemistry, Institute of Bioorganic Chemistry Polish Academy of Sciences, Poznań, Poland

[2]Laboratory of Molecular Biology, Department of Biochemistry and Biotechnology, University of Life Sciences, Poznań, Poland

Land plants have evolved complex adaptation strategies to survive changes in water status in the environment. Understanding the molecular nature of such adaptive changes allows the development of rapid innovations to improve crop performance. Plant membrane transport systems play a significant role when adjusting to water scarcity. Here we put proteins participating in transmembrane allocations of various molecules in the context of stomatal, cuticular, and root responses, representing a part of the drought resistance strategy. Their role in the transport of signaling molecules, ions or osmolytes is summarized and the challenge of the forthcoming research, resulting from the recent discoveries, is highlighted.

**Keywords**: Drought avoidance, drought tolerance, transmembrane allocation, transport systems, abscisic acid

## INTRODUCTION

Drought is one of the most acute abiotic stressors, commonly encountered by plants; that adversely affects crop production worldwide. Today, the development of plants with an increased capacity to survive during water scarcity combined with a high yield potential is the principal objective of agrobiotechnology.

Over the course of evolution, plants have developed sophisticated mechanisms that allow adaptation and survival during periods of water deficit. These mechanisms serve as candidate pathways for the engineering of enhanced drought stress tolerance. In response to water deprivation, plants exhibit the following strategies: (i) escape, (ii) avoidance, and (iii) tolerance. Drought-escaping plants, such as desert ephemerals, annual crops and pasture plants, complete their life cycle before the onset of acute drought. This is achieved through high metabolic activity, rapid growth, and the acceleration of flowering. Avoidance responses occur in both annual and perennial plants. Such strategies rely on the maintenance of water uptake through the modulation of root architecture. Water loss is also limited by reducing stomatal and cuticular conductance and evaporative surfaces. Drought tolerance is associated with the ability of tissues to withstand low water potential; this is achieved via osmotic adjustment and the synthesis of low-molecular weight proteins that protect plants from damage caused by water deficiency (Bacelar et al., 2012).

Recent progress in plant functional genomics has enabled the identification and characterization of genes involved in many essential steps of the drought stress response. Increasing evidence indicates that various plant membrane transport systems play a significant role in adaptation to drought. Depending on energy needs, translocation through biological membranes occurs in a passive or active manner. Passive transport, also known as facilitated diffusion, occurs through channels and carriers. Active transport is performed by primary and secondary transporters that use ATP hydrolysis and ion gradients, respectively, to drive solutes across membranes (Schulz, 2011). The study of transporters related to signaling molecules, e.g., abscisic acid (ABA), as well as ions and osmolytes paves the way for the development of new drought-tolerant varieties. In this review, membrane transport systems will be presented according to their participation in drought avoidance and tolerance strategies. The role of various transport systems in ABA translocation, stomatal, cuticular, and root responses, as well as osmotic adjustment will be summarized.

## ABA IN DROUGHT AVOIDANCE AND TOLERANCE STRATEGIES

Abscisic acid is a ubiquitous plant hormone and signaling molecule; it controls plant growth and development and modulates the response to environmental stressors. As a phytohormone crucial in drought avoidance strategies, ABA elicits two distinct responses: rapid and gradual. The earliest plant reaction, regulated in an ABA-dependent manner, is the modulation of stomatal aperture; this minimizes the loss of water through transpiration. Exposure to ABA prompts guard cells to decrease their volume and close across the airway pore. This is achieved via changes in ion flux within the guard cell (Finkelstein, 2013). ABA gradually increases hydraulic conductivity and promotes cell elongation in the root, enabling the plant to recover after water shortage (Kudoyarova et al., 2011). By inducing the accumulation of osmotically active compounds, which protects cells from damage, ABA participates in the drought tolerance response (Finkelstein, 2013).

Biosynthetic pathways that result in free, active ABA have been previously reviewed (Finkelstein, 2013). ABA is synthesized from precursors, mainly carotenoids and xanthophylls found in plastids. The accumulation of synthesis enzymes has been observed in the veins of vegetative tissues, guard cells, maternal tissues, and embryos (Boursiac et al., 2013). Free ABA is also generated by the deconjugation of the ABA glucosyl ester (ABA-GE), a major ABA conjugate that functions as an ABA reservoir in the vacuole and endoplasmic reticulum (ER; Lee et al., 2006; Xu et al., 2012; Burla et al., 2013). According to the recent findings, the location of ABA biosynthesis depends on the water status in the environment. In the early stages of soil water deficiency, a root-derived $SO_4^{2-}$ signal is transported to the shoots. Consequently, ABA is synthesized in the veins of vegetative tissues and promotes stomatal closure (Ernst et al., 2010). ABA biosynthesis also occurs directly in the guard cells in response to low humidity; this ABA is assumed to be sufficient for stomata closure (Bauer et al., 2013). Continuing water shortage results in ABA biosynthesis in the roots and the redistribution of root-derived ABA through the xylem into the aerial parts of the plant (Goodger and Schachtman, 2010).

## ABA Transporters

Recently, the molecular basis of ABA transport has been defined. Two models of transmembrane ABA translocation, based on diffusion and the presence of primary and secondary transporters, are now widely accepted. ABA, as a weak acid, exists in an anionic ($ABA^-$), and in a protonated (ABAH) form. The latter, when uncharged, is able to diffuse through the plasma membrane (PM). The diffusion of weak acids has its limitations. An increase in pH (ABA pKa 4.7) due to drought stress can drastically decrease the pool of apoplastic-diffusible ABAH. This limiting step highlights the necessity of ABA transporters, which increase the amount of ABA accessible to intracellular receptors upon stress conditions (Boursiac et al., 2013).

### *Primary ABA Transporters*

In *Arabidopsis thaliana*, members of ATP-binding cassette (ABC) family have been shown to be involved in ABA transport. The ABC proteins have been grouped into eight major subfamilies (A–H); these subfamilies correspond to phylogenetic pathways and structural features. In regard to protein architecture, the largest known subfamily of ABC proteins is the G subfamily. The G subfamily (ABCG) has been divided into the following two distinct groups: (i) the half-size transporters, which were formerly known as the white brown complex (WBC), and (ii) the full-size transporters (also known as the pleiotropic drug resistance transporters – PDRs). ABCGs comprise a single or double set of two basic structural elements: a transmembrane domain (TMD) and a nucleotide-binding domain (NBD) (Verrier et al., 2008). The half-size AtABCG25/WBC26 transporter, which is expressed mainly in vascular tissues, is an ABA

Efflux carrier (see Figure 1). Overexpression of this transporter reduces water loss from leaves by facilitating the delivery of ABA to guard cells (Kuromori et al., 2010). By contrast, the full-size AtABCG40 transporter, which is primarily found in guard cells, imports ABA into stomatal cells (see Figure 1). This transporter is necessary for the proper plant response to ABA. Loss-of-function *abcg40* mutants have guard cells with reduced sensitivity to ABA and are more susceptible to drought stress (Kang et al., 2010). Kuromori et al. (2011) reported on a half-size ABCG transporter (AtABCG22/WBC23) that likely enhances ABA influx into guard cells (see Figure 1). A mutation in the gene encoding this protein results in an increase in water transpiration and drought susceptibility. However, the ABA transport activity of the corresponding protein remains to be examined.

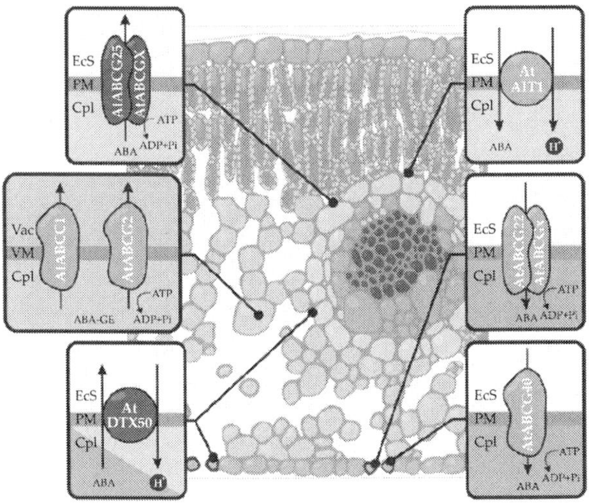


**Figure 1.** Schematic representation of abscisic acid (ABA) and ABA glucosyl ester (ABA-GE) translocation across the plasma membrane (PM) and the tonoplast (VM) in leaf tissue. The transporters that function as exporters (AtABCG25, AtDTX50) are shown in red, and importers (AtABCG40, AtABCG22, AtAIT1/NRT1.2, AtABCC1, AtABCC2) are presented in blue. To fulfill their role, the half-size ABC transporters form homo- or heterodimers. The composition of AtABCG25 and AtABCG22 dimers has not been yet determined. EcS, extracellular space; Cpl, cytoplasm; Vac, vacuole.

The role of ABC transporters in ABA transport remains to be functionally analyzed. These multifaceted proteins must be fit into a broader context of functional plasticity. It is tempting to question whether a given ABC transporter or a closely related group of paralogs, which are widely expressed in various organs (e.g., leaves and roots) and/or have defined substrate specificity (ABA), participates in a coordinated manner in response to drought.

## *Secondary ABA Transporters*

Using a modified yeast two-hybrid system, Kanno et al. (2012) identified a nitrate transporter (NRT1.2) belonging to the NRT1/PTR (Nitrate transporter1/Peptide transporter) family currently known as ABA-IMPORTING TRANSPORTER (AIT1; see Figure 1). The activity of the AIT1 promoter was observed in the vascular tissues of the inflorescence stems, leaves, and roots. After the characterization of loss-of-function mutants and transport assays, it was suggested that AIT1 is an ABA importer that is essential for the regulation of stomatal aperture in the inflorescence stems. Recently, Zhang et al. (2014) identified a new ABA transporter in *A. thaliana* known as DTX50 (see Figure 1). This protein belongs to the DTX/MATE (The Detoxification Efflux Carriers/ Multidrug and Toxic Compound Extrusion) family. Following ABA treatment, *AtDTX50*transcripts accumulated in both guard cells and the vascular bundles of leaves. Proteins encoded by *AtDTX50* localized to the PM. Transport experiments using *Escherichia coli* and *Xenopus* oocytes expressing *AtDTX50* revealed that it is an ABA exporter. AtDTX50 transport activity was the highest at an apoplastic pH of 7.0, which reflects drought conditions. Disruption of AtDTX50 activity resulted in slower wilting; this is a direct consequence of the inability to remove excess ABA and the hyperaccumulation of the hormone in leaves. In *A. thaliana* with dysfunctional DTX50, stomata closed more rapidly; this resulted in an increased tolerance to drought. Insertion mutants displayed growth retardation and ABA hypersensitivity. Overall, these results provided strong evidence that AtDTX50 plays an essential role in the control of ABA accumulation in various cell types; this protein is the second ABA Efflux transporter identified, after AtABCG25 (Zhang et al., 2014).

## ABA Conjugate Transporters

It has been proposed that ABA-GE conjugates are actively transported into the vacuoles of mesophyll cells by ABC transporters (see Figure 1) or proton gradient-driven transporters (Burla et al., 2013). Two members of the *A. thaliana* ABCC/MRP (multidrug resistance-associated protein) subfamily, AtABCC1/MRP1 and AtABCC2/MRP2, are localized in tonoplast and exhibit ABA-GE transport activity in a yeast heterologous expression system. According to microarray analysis, the expression of ABCC2 is induced upon drought. In addition to ABCCs, secondary transporters also participate in the vacuolar sequestration of ABA conjugates; these transporters are presumably members of the MATE family (Burla et al., 2013).

## Ion Movement and Stomatal Closure upon Drought

The reliable and fast adjustment of stomatal aperture is necessary for effective drought avoidance in plants. A rapid response reduces the amount of water lost, increases water use efficiency, and allows plants to survive (Farooq et al., 2009). There are two pathways leading to stomatal closure, passive, and active. In the passive pathway, turgor loss occurs without a reduction of the solute content of the guard cells (McAdam and Brodribb, 2014). Because the passive pathway cannot sufficiently protect the plant against drought, active solute loss is necessary. The movement of osmoregulatory ions ($K^+$, $Cl^-$, and malate$^{2-}$) across guard cell membranes and the gluconeogenic conversion of malate into starch, accompanied by water export through aquaporins (AQPs), results in a decrease in cell volume. Turgor is lost, and the stomata finally close. Mature guard cells have no plasmodesmata; the existence of channels, transporters, and pumps in the PM are required (Lawson and Blatt, 2014).

## $Ca^{2+}$ Permeable Channels

In the active solute loss scenario, ABA is imported into guard cells by ABCG40 in response to drought. This triggers the production of reactive oxygen species (ROS) such as hydrogen peroxide ($H_2O_2$) and nitric oxide (NO) and cytoplasmic $Ca^{2+}$ ($[Ca^{2+}]_{cyt}$) increases (Kim et al., 2010). The level of oxidative stress is controlled by a group of low molecular antioxidant compounds known as flavonols (Watkins et al.,

2014). The molecular nature of flavonol transporters in these particular cells remains to be elucidated. However, the action of such proteins indirectly influences the aperture of guard cells. Members of the MATE and ABC protein families are involved in the transport of secondary plant metabolites, including flavonoids (Zhao and Dixon, 2010). The accumulation of $[Ca^{2+}]_{cyt}$, is possible due to $Ca^{2+}$-permeable channels located in the tonoplast and the PM (see Figure 2). These proteins are encoded by genes belonging to the TPC1 (two-pore channel 1), CNGC (cyclic nucleotide-gated channel), and GLR (glutamate receptor-like) families. TPC1 has been classified as slow vacuolar (SV) non-selective cation channel that is regulated by luminal $Ca^{2+}$ (Islam et al., 2010). In *A. thaliana*, members of the CNGC and GLR families, namely AtCNGC5, AtCNGC6, and AtGLR3.1, are PM-located $Ca^{2+}$ channels (Cho et al., 2009; Wang et al., 2013). The distinguishing feature of CNGC proteins is cGMP-dependent activation (Wang et al., 2013), which is also required for an increase in $[Ca^{2+}]_{cyt}$ (Dubovskaya et al., 2011). The long-term $Ca^{2+}$-programmed stomatal response that prevents guard cell reopening is established by AtGLR3.1 (Cho et al., 2009).

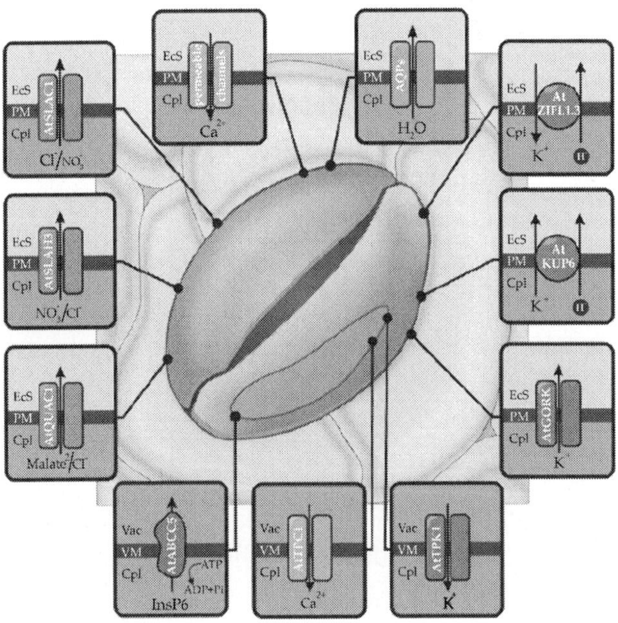


**Figure 2.** Schematic illustration of ion channels, aquaporins (AQPs) and transporters activated by drought or ABA and controlling stomatal closure. The

accumulation of cytoplasmic $Ca^{2+}$ ($[Ca^{2+}]_{cyt}$), is possible due to $Ca^{2+}$-permeable channels (yellow) localized in the PM (e.g., AtCNGC5, AtCNGC6, AtGLR3.1) and the tonoplast (AtTPC1). Increasing the amount of calcium activates S-type (AtSLAC1 and AtSLAH3) and R-type (AtQUAC1) channels, which are indicated in green. AtABCC5/MRP5 (orange), localized to the vacuolar membrane (VM), is a regulator of $Ca^{2+}$-permeable and S-type channels. The actions of the S- and R-type channels induce membrane depolarization and activate $K^+$ flow through AtGORK and AtKUP6 (pink) from guard cells. The release of $K^+$ from the vacuole is mediated by the AtTPK1 channel (pink). Additionally, stomatal closure is regulated by the action of the AtZIFL1.3 isoform, indicated in pink. AQPs are responsible for the outflow of water (e.g., VfPIP1), shown in blue. EcS, extracellular space; Cpl, cytoplasm; Vac, vacuole.

## Anion Channels (S-Type and R-Type)

The ABA-induced signal cascade results in the activation of S-type (slow-activating sustained) and R-type (rapid-transient) anion channels, which cause PM depolarization. Simultaneously, the inhibition of $H^+$-ATPase and inward $K^+$ channels (KAT1/KAT2) prevents the hyperpolarization of the PM (Osakabe et al., 2014). Recent studies have identified the genes encoding S and R-type anion channels. The *A. thaliana* SLAC1 (slow anion channel-associated 1) protein is a PM-localized (Negi et al., 2008; Vahisalu et al., 2008) S-type anion channel with a preference for $Cl^-$ and $NO_3^-$(see Figure 2; Geiger et al., 2009; Lee et al., 2009b). This protein operates mainly in response to ABA, $CO_2$, $Ca^{2+}$, NO, and $H_2O_2$ (Vahisalu et al., 2008). Mutations in SLAC1 result in the excessive accumulation of anions, such as $Cl^-$, in cell protoplasts and the disruption of stomatal closure (Negi et al., 2008; Vahisalu et al., 2008). Phosphorylation at distinct sites enables SLAC1 activation by both calcium-dependent and calcium-independent pathways downstream of ABA (Maierhofer et al., 2014). A second S-type anion channel, SLAH3 (SLAC homolog 3), is expressed in *A. thaliana* guard cells (Geiger et al., 2011). In contrast to its homolog, SLAH3 has higher permeability toward $NO_3^-$ than $Cl_2^-$ (see Figure 2; Geiger et al., 2009, 2011; Lee et al., 2009b). AtMRP5/ABCC5, a vacuolar inositol hexakisphosphate (InsP6) transporter from the ABC transporter family, may be a central regulator of $Ca^{2+}$-permeable and S-type anion channels in the PM of guard cells. This regulation is possible due to the complexing of divalent cations by InsP6 or the triggering a continuous Efflux of $Ca^{2+}$ into the cytosol by an InsP6-regulated channel (see Figure 2; Nagy et al., 2009). A member of the aluminum-activated malate transporter family in *A. thaliana*, QUAC1/ALMT12 (quick anion channel 1/aluminum-activated anion channel 12), has been identified and classified as malate-sensitive R-type anion channel (see Figure 2; Meyer et al., 2010) or a chloride and nitrate current facilitator (Sasaki et al., 2010). Loss of QUAC1 function results in impairment of stimulus-induced stomatal closure (owing to factors such as ABA, $CO_2$, $Ca^{2+}$; Meyer et al., 2010; Sasaki et al., 2010). Compared to our understanding of anion influx across the tonoplast, the molecular mechanisms of Efflux from the vacuole are poorly understood.

## Potassium Channels/Transporters

Plasma membrane depolarization activates the Efflux of potassium. This transport is facilitated by GORK (guard cell outward rectifying $K^+$) channels, which belong to the Shaker family of voltage-gated ion channels (see Figure 2). GORK transcripts have been widely observed in *A. thaliana*, especially in guard cells (Becker et al., 2003; Hosy et al., 2003), where they represent the only outward-rectifying $K^+$ channel (Eisenach et al., 2014). The inhibition of GORK activity in guard cells results in defects in $K^+$ Efflux and a lower rate of stomatal closure (Hosy et al., 2003). Osakabe et al. (2013) identified stress-responsive $K^+$ uptake permeases (KUPs), which share redundant functions with GORK channels (see Figure 2). The KUP6 and KUP8 proteins in *A. thaliana* function as $K^+/H^+$ symporters. KUP6, which is highly up-regulated in response to the ABA treatment and water deficit, is localized to the PMs of guard cells, root tip cells and vascular tissues. KUP6-overexpressing lines were more tolerant to drought. A *kup68g* double mutant line (with combined *kup6* and *kup8* mutations in addition to mutation in the *gork* potassium channel) had impaired stomatal closure and ABA sensitivity; this mutant also had greater stomatal conductance and water loss rates (Osakabe et al., 2013). An isoform of the ZIFL1 (zinc-induced facilitator-like1) transporter, which belongs to the major facilitator superfamily (MFS), proved to be a modulator of stomatal aperture and conferred drought stress tolerance. The PM-localized AtZIFL1.3 protein possesses $H^+$-coupled $K^+$transport activity, and its overexpression resulted in more efficient stomatal closure (see Figure 2; Remy et al., 2013). The main cellular depository of ions, including potassium, is the vacuole. The release of $K^+$ from the vacuole is mediated by channels and transporters localized to the tonoplast. TPK1, a member of the two pore $K^+$ channel (TPK) family, is responsible for vacuolar $K^+$ translocation in guard cells (see Figure 2). The removal of TPK1 was associated with a weak phenotype; most often the stomatal closure was slower, but the opening kinetics were normal (Gobert et al., 2007).

## Water Channels

The outflow of water results in a reduction in guard cell turgor and stomatal closure; this is mediated by AQPs, which are membrane channels that belong to the MIP (major intrinsic protein) family (see Figure 2; Heinen et al., 2009). AQPs may form homo- and heterotetramers, with each monomer defining a single pore. Based on amino acid similarity and membrane localization, AQPs are divided into five subgroups; for a review, see (Maurel et al., 2008; Chaumont and Tyerman, 2014). The expression of several genes encoding AQPs belonging to the PIP (PM intrinsic proteins) and TIP (tonoplast intrinsic proteins) subgroups was up-regulated in leaves, mainly in the guard cells, in drought stress conditions (Heinen et al., 2009). Heinen et al. (2014) have identified *Zea mays* PIPs that are localized to stomatal complexes (mainly ZmPIP1;1, ZmPIP1;3, ZmPIP2;2) and presumably participate in stomatal movement (Heinen et al., 2014). The expression of PIP1 from *Vicia faba* and *Brassica juncea*improved drought resistance in *A. thaliana* and *Nicotiana tabacum* plants, respectively, through the promotion of stomatal closure (Cui et al., 2008; Zhang et al., 2008).

## Channels/Transporters Influencing Stomatal Opening

Proteins that regulate stomatal movements by supporting swelling belong to a separate group of transporters. The disruption of $Cl^-$ (AtCLCc, AtALMT9), $K^+$ (AtNHX1, AtNHX2, AtCHX20), and $NO_3^-$ (AtNRT1.1, AtCLCa) channels/transporters, as well as the AtMRP4/AtABCC4 protein, results in a delay in stomatal opening (Guo et al., 2003; Klein et al., 2004; Padmanaban et al., 2007; Jossier et al., 2010; De Angeli et al., 2013; Andres et al., 2014; Wege et al., 2014). Phosphorylation-dependent changes in the activity of AtCLCa enable the protein to perform dual roles, in anion sequestration in the vacuole during stomatal opening and in anion release during stomatal closure, in response to ABA (Wege et al., 2014). Mutations in a malate importer protein (AtABCB14) result in accelerated stomatal closure (Lee et al., 2008). Mutations in AtALMT9, AtNRT1.1, and AtMRP4 transporters result in the impairment of stomatal responses and are associated with increased drought tolerance (Guo et al., 2003; Klein et al., 2004; De Angeli et al., 2013).

## CUTICLE DEPOSITION AS A FACET OF DROUGHT AVOIDANCE

The formation of the plant cuticle, which covers the surface of all aerial organs, is a crucial plant adaptation to terrestrial habitats. This specialized cell wall (CW) component, which is synthesized exclusively by aerial epidermal cells, consists of wax embedded in (intracuticular) and layered on (epicuticular) a lipid polyester matrix called cutin. The major function of the cuticle is to protect the plant from unfavorable environmental conditions, including drought, by limiting non-stomatal water transpiration (Yeats and Rose, 2013). In *A. thaliana*, the biosynthesis of cuticle components is induced by water deficiency (Kosma et al., 2009). Recent studies demonstrated that the increased accumulation of cuticular waxes is associated with drought resistance (Yang et al., 2011; Zhu et al., 2014) and ABA treatment (Seo et al., 2011).

During cuticle deposition, the trafficking of wax and cutin precursors occurs from the ER across the PM and CW to the plant surface (Yeats and Rose, 2013).

### Intracellular Translocation of Cuticular Components

To date, the intracellular mechanism of wax and cutin precursor trafficking remains unclear. Several hypothetical pathways to the PM have been suggested (see Figure 3). These precursors might reach the PM via the following routes: (i) direct relocation at ER-PM contact sites (Samuels and McFarlane, 2012); (ii) transport by cytosolic carrier proteins (e.g., Acyl-CoA binding proteins – ACBPs; Li-Beisson et al., 2010); (iii) vesicular translocation in oleosome bodies coated by oleosin-like proteins; (iv) translocation in uncoated vesicles that sequestrate the cuticular lipids into lipid rafts (Schulz and Frommer, 2004); or (v) Golgi mediated exocytosis (Samuels et al., 2008). Through mutant analysis combined with live cell imaging, electron microscopy and chemical phenotyping, McFarlane et al. (2014) demonstrated that the movement of intracellular wax requires vesicle traffic mediated by Golgi- and *trans*-Golgi networks. However, AtACBP1 has been shown to contribute to the formation of the stem cuticle; dysfunction in this protein resulted in a reduction in wax and cutin monomer composition. Moreover, AtACBP1 exhibits biding activity to wax precursors (Xue et al., 2014). The cytosol-localized protein LjLTP3, a member of the LTP (lipid transfer protein) family, is able to bind lipids and confers drought tolerance in *Lotus japonicus*.

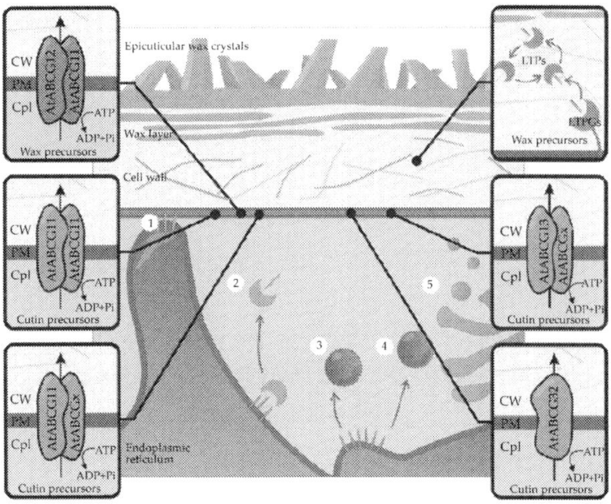


**Figure 3.** Schematic illustration of cuticular lipid translocation. Different scenarios describing the transport of wax precursors (orange disks) from the endoplasmic reticulum (ER) to the PM have been proposed. Cuticular lipids may be (1) relocated at ER-PM contact sites, (2) grasped by ACBPs (e.g., AtACBP1; green crescent-shaped bodies), transported by (3) coated oleophilic bodies, (4) uncoated vesicles, or (5) the Golgi-mediated secretory pathway. The translocation of wax and cutin precursors through the PM is a complex action involving several ABC transporters, which are indicated in orange and green, respectively. The latter are either half-size proteins (AtABCG11, AtABCG12, AtABCG13), which function as homo- or heterodimers, or full-size transporters (AtABCG32). Often composition of the homo- or heterodimers defines the profile of the transported substrate (see the text for details). The transport of wax precursors across the cell wall to the plant surface is mediated by LTPGs (e.g., AtLTPG1 and AtLTPG2) in association with yet unknown LTPs. CW, cell wall; Cpl, cytoplasm.

## Transport of Cutin and Wax Precursors across the Plasma Membrane

Members of the ABC transporter family mediate the flux of cuticular components through the PM. Two half-size ABC transporters, named AtABCG12/WBC12/CER5 (Pighin et al., 2004) and AtABCG11/WBC11/DSO/COF1 (Bird et al., 2007; Luo et al., 2007; Panikashvili et al., 2007; Ukitsu et al., 2007), are involved in the secretion of cuticle precursors into the apoplastic environment (see Figure 3). Mutation of both genes results in a significant reduction in the stem wax load (approx. 50%); a major depletion in C29 alkane, the major component of cuticular waxes in stems, is also observed (Pighin et al., 2004; Bird et al., 2007). The total amount of wax in the epidermis remains unchanged, indicating that the absence of these two proteins impedes wax transport rather than biosynthesis. AtABCG11 is also likely involved in the translocation of cutin monomers in vegetative and reproductive organs and the transport of suberin in roots (Panikashvili et al., 2007, 2010). The evidence that AtABCG11 forms homo- and heterodimers with AtABCG12 comes from bimolecular fluorescence complementation and protein trafficking assays (McFarlane et al., 2010). AtABCG11 also plays a dominant role in the created complex; ABCG12 could not traffic to the PM in the absence of ABCG11, whereas ABCG11 trafficking was independent of ABCG12 (McFarlane et al., 2010). The half-size ABCG transporter from *Solanum tuberosum* (StABCG1) was shown to contribute to the export of suberin components across the PM (Landgraf et al., 2014). Silencing of the *StABCG1* gene, which is expressed mainly in the roots and tuber periderm, results in a reduction in esterified suberin monomers; this is accompanied by an accumulation of the presumed suberin precursors. Root and tuber morphology was altered, and tubers were more susceptible to drought (Landgraf et al., 2014). The putative ortholog of AtABCG12 found in the moss species *Physcomitrella patens* is involved in the transport of a wax precursor. This finding suggests that the role of ABCG transporters, particularly half-size transporters, in the movement of cuticle components is evolutionarily conserved among land plants (Buda et al., 2013). Loss-of-function in *AtABCG13/WBC13* (another half-size ABCG transporter that belongs to the ABCG12 clade) results in a significant depletion in petal cutin monomers; this result indicates that the protein plays a role in cutin deposition in flowers (see Figure 3; Panikashvili et al., 2011). The full-size ABCG transporter (AtABCG32) has been shown to contribute to the formation of functional cuticles (see Figure 3; Bessire et al., 2011). Knockout mutations in AtABCG32 result in a significant increase in cuticle permeability in leaves and flowers, as well as in water loss from rosettes and herbicide sensitivity. The ultrastructural changes in the cuticle were associated with reduced amounts of minor aliphatic cutin monomers; these monomers are most likely exported from epidermal cells by AtABCG32. The AtABCG32 monocot orthologs OsABCG31 1 and HvABCG31 may also be involved in cuticle deposition. Loss-of-function mutations in these genes, both in *Oryza sativa* and in *Hordeum vulgare*, result in water retention deficiency phenotypes (Chen et al., 2011).

Members of the ABCG subfamily in *A. thaliana* are also required for proper pollen coat (extracellular matrix) and exine (outer wall) formation. These structures protect gametophytes in unfavorable environmental conditions. Dysfunction in AtABCG9/WBC9 and AtABCG31 proteins results in reduced steryl glycoside levels and altered pollen coat morphology. Correct morphology is essential for pollen coat maturation. A double mutant exhibited increased water loss upon exposure to dry air (Choi et al., 2014). The AtABCG26/WBC27 protein participates in exine deposition in the pollen, presumably by transporting sporopollenin precursors (Quilichini et al., 2010; Choi et al., 2011; Dou et al., 2011). An ortholog in *O. sativa* (OsABCG15) is crucial for anther cuticle and pollen exine formation (Wu et al., 2014).

## Transport of Cuticular Components through the Cell Wall to the Cuticle

Once cuticular components have been moved across the cell membrane, they must pass the hydrophilic CW to reach the cuticle. Movement through the extracellular matrix appears to be facilitated by LTPs. Two members of the LTP family (LTPG1 and LTPG2) are defined as GPI-anchored (glycosylphosphatidylinositol –anchored) proteins. These proteins are located at the exterior face of the PM and are possibly involved in wax export (see Figure 3; Debono et al., 2009; Lee et al., 2009a; Kim et al., 2012). Disruption of AtLTPG1 or its homolog AtLTPG2 modifies the structure of the cuticular layer and reduces the amount of C29 alkane; this is also seen in AtABCG12 and AtABCG11 mutants (Debono et al.,

2009; Lee et al., 2009a; Kim et al., 2012). The expression of LTPG1 can complement wax scarcities in *ltpg1* mutant lines (Debono et al., 2009). Compared to each single mutant, the double mutant exhibited a more severe phenotype (Kim et al., 2012). Taken together, it seems likely that LTPG1 and LTPG2 are functionally overlapping wax exporters. These proteins act either directly by associating with ABC transporters or indirectly by creating suitable environment for wax Efflux (Debono et al., 2009; Kim et al., 2012). The former model assumes that LTPGs are loaded with C29 alkane cargo and transfer it to other extracellular LTPs, which then carry the lipid load to the plant surface (Debono et al., 2009).

## ROOT RESPONSES UPON DROUGHT

In addition to mechanisms that reduce water loss in the aerial parts of plant, drought avoidance also requires strategies based on the regulation of root water uptake; this is achieved by modulating root growth and hydraulic conductivity (Aroca and Ruiz-Lozano, 2012).

### Transporters Participating in Root Growth and Development

To increase the volume of potentially accessible water, the integrated action of ABA and auxins promotes root growth and development (Xu et al., 2013b). Several transporters, which belong to distinct gene families, act as key modulators of auxin translocation and subsequent root morphology changes (Petrasek and Friml, 2009; Zazimalova et al., 2010). There are only few examples of extensive crosstalk between the mechanisms that regulate auxin transport, root development, and drought resistance. The expression of an *A. thaliana* vacuolar $H^+$-pyrophosphatase (AVP1) confers tolerance against drought in many plant species, including *Lycopersicon esculentum* (Park et al., 2005), *Saccharum officinarum* (Kumar et al., 2014b), *N. tabacum* (Arif et al., 2013), and *Gossypium hirsutum* (Pasapula et al., 2011). AVP1 overexpressing plants exhibited greater $PP_i$ -driven sequestration of ions and sugars into the vacuole, increased water retention and increased cell turgor. Auxin transport is also stimulated, enhancing root system development (Park et al., 2005). This adaptation to stress conditions is also possible due to the different subcellular and tissue distribution of isoforms of transporters with similar substrate profiles. The MFS carrier ZIFL1 protein (Remy et al., 2013) is one example. The two splice isoforms of ZIFL1, ZIFL1.1, and 1.3, both have $K^+$ transport activity that likely influences membrane proton gradients, resulting in distinct biological roles in roots and guard cells, respectively. The change of pH that occurs as a consequence of $K^+$ translocation results either in stomata closure or the modulation of auxin transport. In this way, plants can control stomata aperture and change root morphology if necessary.

The effects of ABA on root morphogenesis differ between various plant species, especially the effects on the formation of the lateral roots. In *A. thaliana*, ABA inhibits lateral root formation. In legumes such as *Medicago truncatula*, ABA promotes the growth of lateral roots and affects the nodulation process. ABA also acts as a negative regulator of the early stages of nodulation (affecting epidermal Nod factor signaling and cytokinin-activated cortex cell division) and suppresses symbiotic nitrogen fixation (Liang et al., 2007; Ding and Oldroyd, 2009). Nodule formation is an energetic cost that must be strictly controlled; nodulation is highly influenced by environmental conditions. It has been proposed that alternations in hormone levels, especially stress hormones such as ABA, serve as shoot derived signals directed at the roots (Ding and Oldroyd, 2009). This fluctuation in ABA levels is the bridge joining nodulation control and the environmental status of the plant. Despite their importance, the mechanisms behind the translocation/control of such signals are still largely unknown.

## Changing Root Hydraulic Conductivity – Aquaporins Take Action

Root hydraulic conductivity (L), the water flux across the root surface area, is governed by drought and ABA. In water deficit conditions, root hydraulic conductivity decreases to prevent water outflow from the roots due to progressively decreasing soil water potential (Aroca et al., 2012). Morphological changes in the root tissue and the existence of water channels constitute the main long-term and short-term determinants of root hydraulic conductivity, respectively. The formation of apoplastic barriers, such as Casparian bands and suberin lamellae, prevents cellular death by limiting water outflow into the soil (Lobet et al., 2014). The half-size ABCGs, StABCG1, and AtABCG11, affect the root suberin content. Therefore, these proteins might be responsible for the proper formation of suberin lamellae (Panikashvili et al., 2010; Landgraf et al., 2014). Rapid and reversible alterations in water permeability are done by controlling the expression and activity of AQPs. AQPs allow for the passive movement of water through a cell-to-cell pathway, which is the major water path during drought (Aroca and Ruiz-Lozano, 2012). The expression of the two aforementioned MIP subgroups, PIPs and TIPs, is greater in roots than in leaves. Moreover, the localization of PIPs has been associated with the presence of root hydrophobic barriers; this finding suggests that these proteins are involved in the regulation of water movement during drought (Chaumont and Tyerman, 2014). Upon water deficit, the expression of several PIPs is down-regulated to reduce water loss (Moshelion et al., 2014). According to Jang et al. (2004), nine out of the thirteen PIP genes expressed in the root (three from the PIP1 group and six from PIP2 group) showed significant reductions in mRNA accumulation in *A. thaliana* plants exposed to drought. Similarly, in *N. tabacum*, the expression of PIPs is significantly down-regulated in a manner corresponding to the level of drought stress (Mahdieh et al., 2008); this phenomenon has also been observed in *Fragaria vesca* (Surbanovski et al., 2013) and *Camellia sinensis* roots (Yue et al., 2014). Additionally, the ectopic overexpression of *A. thaliana* PIP1;4 and PIP2;5 in *N. tabacum* resulted in increased water loss under dehydration stress (Jang et al., 2007). By contrast, the heterologous expression of the *Triticum aestivum* aquaporin gene (TaAQP7), which belongs to PIP2 subgroup, conferred drought tolerance in transgenic *N. tabacum*; plants were able to maintain their water status, reduce the accumulation of ROS and prevent membrane damage via an ABA-dependent pathway (Zhou et al., 2012). Concerning TIP AQPs, plants may respond to drought through cell elongation, osmotic adjustment, and the mediation of water transport across the tonoplast (Wudick et al., 2009). The constitutive expression of TIP1 from *Panax ginseng*, TIP1;2 from the highly drought-resistant *Thellungiella salsuginea* and TIP2 from *Solanum lycopersicum* confer tolerance to water deprivation in *A. thaliana*; an increase in water absorption requires roots elongation and high water permeability in the tonoplast (Peng et al., 2007; Sade et al., 2009; Wang et al., 2014). Conversely, it has been demonstrated that TIPs from *Glycine soja* (TIP2;1) and *T. aestivum* (TIP2;2) have a negative impact on *A. thaliana* drought tolerance (Wang et al., 2011; Xu et al., 2013a).

## Arbuscular Mycorrhiza Improves Water Uptake Capacity

Increasing evidence indicates that arbuscular mycorrhiza (AM) play a crucial role in enhancing tolerance to drought (Abbaspour et al., 2012; Li et al., 2013, 2014; Barzana et al., 2014). Glomeromycota fungi are intimately associated with 80% of land plants; these organisms, supply water and nutrients to the roots of host plants in return for carbon compounds (Schmitz and Harrison, 2014). The improvement in water status is possible by harnessing the hyphae growing in the soil, enlarging the absorption surface, and exploiting the ability of the fungus to take up water despite the low soil water potential (Ruiz-Lozano et al., 2012). In addition, AM symbiosis affects the hydraulic properties of roots and maintains root water permeability by the regulating the expression of plant AQPs (PIPs and TIPs) and the activity of fungal AQPs (Li et al., 2013; Barzana et al., 2014). Considering the involvement of AM in plant response to drought stress, transport mechanisms play a crucial role in the establishment of plant–fungus interactions. The successful recognition of root-exuded plant metabolites, in particular flavonoids (F), 2-hydroxy fatty acids (2OH-FA), strigolactones (SL), and 16C cutin monomers, is a prerequisite for host-root colonization via hyphopodium development (Nadal and Paszkowski, 2013). Despite the importance of these transporters, the knowledge of this pre-symbiotic molecular crosstalk is limited. The isolation of the *Petunia hybrida* PDR1 transporter has shed light on SL transport, as it encodes for a putative root SL exporter (Kretzschmar et al., 2012). A corresponding protein has been found in the PM of hypodermal passage cells (HPC), the non-suberized cortical entry points for AM hyphae (Kretzschmar et al., 2012). The entrance into cortical cells, where hyphae develop a treelike structure, is called an arbuscule. Arbuscules form through the PM and tonoplast invagination. Formed arbuscules are enveloped by a plant-derived periarbuscular membrane (PAM), a key symbiosis interface that contains several plant transporters and mediates bi-directional nutrient flux (Recorbet et al., 2013).

# OSMOTIC ADJUSTMENT AT THE CELLULAR LEVEL AS A FACET OF DROUGHT TOLERANCE

Most plants have the capacity to tolerate low water content by increasing cellular osmolarity, thereby protecting cells from damage due to dehydration. This acclimating process involves the accumulation of an array of highly concentrated, osmotically active compounds in the cytosol, rather than in the vacuole, without the interference of primary metabolism. Osmotic regulators include inorganic ions and compatible solutes, such as sugars, sugar alcohols (polyols and cyclitols), amino acids [particularly proline (Pro)], and quaternary ammonium compounds (QACs), notably glycine betaine (GB; Sanders and Arndt, 2012). ABA participates in the accumulation of the aforementioned protectants. Therefore, ABA serves as a bridge between avoidance and tolerance responses. Strategies for achieving high levels of osmoregulatory solutes to cope with water deprivation are based either on altering metabolism or transport. Early studies indicated that the movement and differential distribution of these compounds, both within and between cells, is regulated by membrane-associated transporters; most of these transporters are secondary active transporters.

## Amino Acids and QACs Transporters

### *Proline Transporters*

Proline accumulation is the first response of plants subjected to a range of environmental stressors, including drought. The correlation between this widely occurring amino acid and drought tolerance has been demonstrated in several plant species (Javed et al., 2014; Kumar et al., 2014a; Ravikumar et al., 2014; Singh et al., 2014). In addition to its well-established role in osmotic adjustment, Pro contributes to stress tolerance by functioning as a molecular chaperon that stabilizes subcellular structures, as a free radical scavenger and as stress-signaling molecule. During periods of stress, Pro is synthesized in the chloroplast and cytosol and then transported. Pro is degraded within root mitochondria (Szabados and Savoure, 2010) to provide energy, which beneficial for the accumulation of root biomass (Kavi Kishor and Sreenivasulu, 2013).

Transporters that recognize Pro have been classified into the following three subfamilies: (i) the amino acid permease (AAP) subfamily; (ii) the lysine–histidine transporter (LHT) subfamily; and (iii) the Pro transporter (ProT) subfamily. Members belonging to the AAP and LHT subfamilies are low selectivity proline transporters and participate in the translocation of neutral amino acids (Rentsch et al., 2007; Lehmann et al., 2010). Three ProTs have been identified in *A. thaliana* (Grallath et al., 2005). The expression profiles of the identified genes revealed that the ProT1 and ProT2 mRNA is accumulated when Pro levels are elevated. During progressive drought, only the ProT2 gene was strongly up-regulated. Grallath et al. (2005) observed that GUS expression driven by the ProT2 promoter occurred in the root epidermis and cortex. The corresponding protein has been found in the PM. AtProT2 exhibits Pro transport activity (Lehmann et al., 2011), and may facilitate Pro import across the PM under stress conditions (see Figure 4).

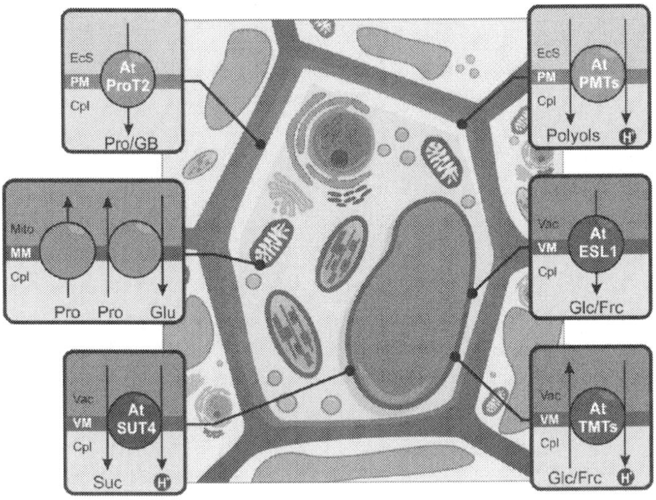


**Figure 4.** Schematic illustration of the sub-cellular distribution of osmotically active compounds upon drought stress. Proline (Pro) and glycine betaine (GB) transport (pink) across the PM is mediated by Pro transporters (e.g., AtProT2). Pro translocation through the mitochondrial membrane (MM) is possible due to Pro carriers that function as uniporters and Proline/Glutamate (Pro/Glu) antiporters. Glucose (Glc)/fructose (Frc) influx and Efflux across the tonoplast (VM) is driven by Tonoplast Monosaccharide Transporters (e.g., AtTMT1 and AtTMT2) and Early Responsive to Dehydration six-like 1 carrier (AtESL1), indicated in brown. Sucrose (Suc) transporters (e.g., AtSUT4, red) participate

in Suc export from the vacuole. The import of polyols into the cell is possible due to polyol/monosaccharide transporters (e.g., AtPMT1, AtPMT2, AtPMT5), shown in blue. EcS, extracellular space; Cpl, cytoplasm; Mito, mitochondria; Vac, vacuole.

Transport mechanisms across chloroplast membranes, where Pro is synthesized, remain elusive. Two transport systems for Pro uptake into the mitochondria have been demonstrated (Di Martino et al., 2006). A study using *Triticum durum* seedlings confirmed that Pro enters the mitochondria through the uniporter (see Figure 4); in the mitochondria, Pro is further catabolized to glutamate (Glu). Glu is then exported from the mitochondria in exchange for Pro via the Pro/Glu antiporters (see Figure 4).

### Glycine Betaine Transporters

Mass synthesis of GB occurs in response to water limitations and ABA treatment (Zhang et al., 2012). Comprehensive studies using transgenic plants such as *Spinacia oleracea, S. lycopersicum,* and *O. sativa* have demonstrated that GB accumulation results in improved plant tolerance to drought stress (Shirasawa et al., 2006; Zhang et al., 2011). The transport activity of GB has been demonstrated using ProTs from *A. thaliana*, *Beta vulgaris*, *H. vulgare*, *S. lycopersicum*, and *Avicennia marina* (Lehmann et al., 2011). The uptake of GB and Pro has been shown to be regulated by pH (Takabe, 2012).

## Sugar and Sugar Alcohol Transporters

An increased synthesis of sugar and polyol compounds is triggered by various abiotic stresses. However, types of accumulated sugars may differ among plant species and nature of abiotic stressors (Zhuo et al., 2013; Folgado et al., 2014; Moyankova et al., 2014). Within the cell, carbohydrates are stored in chloroplasts and in the vacuole in the forms of starch and soluble sugars, respectively. In unfavorable conditions, the cytosolic concentrations of sugars and their derivatives are regulated by membrane-embedded transporters. There are three major families of mono- and disaccharide transporters, as follows: monosaccharide/polyol transporters (MST), sucrose transporters/carriers (SUT/SUC) and the SWEET sugar family of transporters (Lalonde and Frommer, 2012). The vast majority of sugar transport is an active process, requiring a proton motive force. However, recent findings indicate that sugar-specific facilitated diffusion transporters also exist in plant cells.

## *Monosaccharide Transporters*

Two tonoplast-localized, secondary transporters from *A. thaliana* are candidate vacuolar hexose loading proteins (see Figure 4). These transporters are TMT1 and TMT2 (Tonoplast Monosaccharide Transporters), which belong to the MST family. The expression of these proteins was up-regulated upon drought treatment. Mutation in the TMTs resulted in the decreased accumulation of glucose and fructose. The direct involvement of AtTMTs in energy-dependent vacuolar monosaccharide transport has been demonstrated in a glucose uptake assay using leaf mesophyll vacuoles isolated from wild-type and knockout lines (Wormit et al., 2006).

While the influx of hexoses into the vacuole is thermodynamically active, Efflux is assumed to be a passive process. Efflux is mediated by members of the early responsive to dehydration 6-like (ERD6) protein group, one of the clades within the MST family. Three members have been characterized thus far; only two have been described in the context of drought stress (Yamada et al., 2010; Poschet et al., 2011). The first identified gene, *ERD6*, was isolated from a cDNA library obtained from *A. thaliana* plants subjected to water deprivation (Kiyosue et al., 1998). Further phylogenetic analysis of the ERD6-like family revealed a close homologue of ERD6, the tonoplast-localized ESL1 (ERD six-like 1; see Figure 4; Yamada et al., 2010). The accumulation of *AtESL1* mRNA was observed after drought and the exogenous application of ABA. GUS staining was detected mainly in pericycle and xylem parenchyma cells of roots. To obtain direct evidence that AtESL1 is a functional monosaccharide transporter, a heterologous expression system in BY-2 cells was used. Translocation ability was not altered by a reduction in the proton gradient, indicating that ESL1 acts as a facilitated diffusion carrier (Yamada et al., 2010).

## *Disaccharide Transporters*

Our understanding of the involvement of SUT in drought tolerance is limited. Among several described SUTs, only one transporter from *P.*

*tremula* (PtaSUT4), one from *O. sativa* (OsSUT2) and two from *A. thaliana* (AtSUC2 and AtSUC4), may be relevant in a drought tolerance context; these proteins are tonoplast symporters. *PtaSUT4* is predominantly present in the tonoplast of mesophyll cells in stems and source leaves (Payyavula et al., 2011). RNAi-mediated silencing of *PtaSUT4* resulted in the increased vacuolar sequestration of SUC; this promoted leaf growth and reduced of water uptake and movement in the xylem. Simultaneously, concentration of raffinose family oligosaccharides (RFOs) has also been reduced, implying the role of PtaSUT4 in regulation of osmotic gradients between cellular compartments in response to water stress (Frost et al., 2012). RFOs serve as compatible solutes for protection against abiotic stresses especially drought (Taji et al., 2002), and are synthesized in the cytosol mainly from sucrose (den Ende, 2013). Therefore, suppression of *PtaSUT4* caused a decline in cytosolic sucrose pool necessary for RFOs biosynthesis upon stress conditions. The SUT protein from *O. sativa* (OsSUT2) is a tonoplast-localized $H^+$-Suc symporter (Eom et al., 2011). GUS activity was detectable in all of the examined *O. sativa* tissues, the highest *OsSUT2* expression was observed in leaf mesophyll cells. In an analysis of primary carbon metabolites, an increased accumulation of Suc in the leaves of *ossut2* mutant plants was observed; this might be the result of decreased sugar export activity in the vacuole. Changes in *OsSUT2* expression upon drought treatment have not been observed (Eom et al., 2011). In a parallel study, Ibraheem et al. (2011) demonstrated the strong up-regulation of *OsSUT2* during water stress. Two *A. thaliana* SUT proteins involved in the drought stress response have been identified (Gong et al., 2014). The loss-of-function mutant lines for AtSUC2 and AtSUC4 displayed hypersensitivity to drought and ABA treatment; this was due to changes in sucrose distribution in the shoots and roots and to sucrose accumulation in source organs. In response to drought and ABA treatment, *AtSUT4* and *AtSUT2* expression was induced in the minor veins of leaves where high-capacity sucrose transport is needed for phloem loading (Gong et al., 2014). The heterologous expression of AtSUC4 in yeast confirmed Suc transport ability (see Figure 4; Weise et al., 2000). Overall, the results suggest a putative role for AtSUC2 and AtSUC4 in Suc flux under drought stress conditions.

### *Sugar Alcohol Transporters*

Six genes belonging to the polyol/monosaccharide transporter (PMT) subfamily have been identified in *A. thaliana*; these belong to the MST family. However, only AtPMT5, AtPMT1, and AtPMT2 have been partially characterized (see Figure 4; Klepek et al., 2005, 2010; Reinders et al., 2005). Simultaneous analyses of AtPMT5 described this protein as a broad-spectrum $H^+$ symporter for polyols, cyclitols and numerous monosaccharides that resides in the PM (Klepek et al., 2005; Reinders et al., 2005). Considering its wide substrate specificity, a ubiquitous expression pattern and an unchanged mutant phenotype under testing conditions (e.g., drought), the physiological functions of AtPMT5 remain unclear and require further investigation (Klepek et al., 2005). AtPMT1 and AtPMT2 catalyze the energy-dependent PM transport of xylitol and fructose in pollen grains and young xylem; these proteins may also play a role in plant CW modifications (Klepek et al., 2010).

## CONCLUSION

It is predicted that water deficits will continue to be a major abiotic factor affecting global crop yields. Genetic engineering for drought resistance using reported candidate genes in major food crops is in progress.

Recent discoveries in plant transporters, e.g., the coordinated action of ABA exporters and importers that control the stomatal pore, points directly toward candidate genes that may have innovative applications. However, the underlying message from basic research is that many fundamental mechanisms influencing drought adaptation remain to be uncovered. The existence of homologs from multigenic families (e.g., ABC proteins) and spliced isoforms (e.g., ZIFL1) illustrate that molecular transport can have various local tissue/organ effects that result in coordinated plant adaptation to drought stress. Understanding the complexity of such responses will be the main challenge in upcoming years.

The knowledge acquired must be adapted to particular evolutionary backgrounds. For legumes (the second most important family of crop plants after *Poaceae*), targeting drought tolerance via the genetic manipulation of a particular organ (e.g., root) during especially susceptible stages such as nodulation could be important. However, the legume-specific reactions of roots to ABA, combined with nodulation control and environmental status, force researchers to propose new legume-specific approaches. It is expected that the research into the fundamental mechanisms of plant membrane transport processes will continue. New crop varieties will be produced, and new avenues toward more sustainable and productive agriculture will open up, in spite of the impending challenges.

## CONFLICT OF INTEREST STATEMENT

The authors declare that the research was conducted in the absence of any commercial or financial relationships that could be construed as a potential conflict of interest.

## ACKNOWLEDGMENT

National Science Centre Grants supported this work: 2013/10/M/NZ3/00260.

## REFERENCES

1. Abbaspour, H., Saeidi-Sar, S., Afshari, H., and Abdel-Wahhab, M. A. (2012). Tolerance of mycorrhiza infected pistachio (*Pistacia vera* L.) seedling to drought stress under glasshouse conditions. *J. Plant Physiol.* 169, 704–709. doi: 10.1016/j.jplph.2012.01.014
2. Andres, Z., Perez-Hormaeche, J., Leidi, E. O., Schlucking, K., Steinhorst, L., Mclachlan, D. H.,et al. (2014). Control of vacuolar dynamics and regulation of stomatal aperture by tonoplast potassium uptake. *Proc. Natl. Acad. Sci. U.S.A.* 111, E1806–E1814. doi: 10.1073/pnas.1320421111
3. Arif, A., Zafar, Y., Arif, M., and Blumwald, E. (2013). Improved growth, drought tolerance, and ultrastructural evidence of increased turgidity in tobacco plants overexpressing *Arabidopsis*vacuolar pyrophosphatase (AVP1). *Mol. Biotechnol.* 54, 379–392. doi: 10.1007/s12033-012-9577–9579
4. Aroca, R., Porcel, R., and Ruiz-Lozano, J. M. (2012). Regulation of root water uptake under abiotic stress conditions. *J. Exp. Bot.* 63, 43–57. doi: 10.1093/jxb/err266
5. Aroca, R., and Ruiz-Lozano, J. (2012). "Regulation of root water uptake under drought stress conditions," in *Plant Responses to*

*Drought Stress*, ed. R. Aroca (Berlin: Springer), 113–127. doi: 10.1007/978-3-642-32653-0_4

6. Bacelar, E. V. A., Moutinho-Pereira, J., Gonçalves, B. C., Brito, C. Q., Gomes-Laranjo, J., Ferreira, H. F.,et al. (2012). "Water use strategies of plants under drought conditions," in *Plant Responses to Drought Stress*, ed. R. Aroca (Berlin: Springer), 145–170. doi: 10.1007/978-3-642-32653-0_6
7. Barzana, G., Aroca, R., Bienert, G. P., Chaumont, F., and Ruiz-Lozano, J. M. (2014). New insights into the regulation of aquaporins by the arbuscular mycorrhizal symbiosis in maize plants under drought stress and possible implications for plant performance. *Mol. Plant Microbe Interact.* 27, 349–363. doi: 10.1094/MPMI-09-13-0268-R
8. Bauer, H., Ache, P., Lautner, S., Fromm, J., Hartung, W., Al-Rasheid, K. A.,et al. (2013). The stomatal response to reduced relative humidity requires guard cell-autonomous ABA synthesis. *Curr. Biol.* 23, 53–57. doi: 10.1016/j.cub.2012.11.022
9. Becker, D., Hoth, S., Ache, P., Wenkel, S., Roelfsema, M. R., Meyerhoff, O.,et al. (2003). Regulation of the ABA-sensitive *Arabidopsis* potassium channel gene GORK in response to water stress. *FEBS Lett.* 554, 119–126. doi: 10.1016/S0014-5793(03)01118-9
10. Bessire, M., Borel, S., Fabre, G., Carraca, L., Efremova, N., Yephremov, A.,et al. (2011). A member of the PLEIOTROPIC DRUG RESISTANCE family of ATP binding cassette transporters is required for the formation of a functional cuticle in *Arabidopsis*. *Plant Cell* 23, 1958–1970. doi: 10.1105/tpc.111.083121
11. Bird, D., Beisson, F., Brigham, A., Shin, J., Greer, S., Jetter, R.,et al. (2007). Characterization of *Arabidopsis* ABCG11/WBC11, an ATP binding cassette (ABC) transporter that is required for cuticular lipid secretion. *Plant J.* 52, 485–498. doi: 10.1111/j.1365-313X.2007.03252.x
12. Boursiac, Y., Leran, S., Corratge-Faillie, C., Gojon, A., Krouk, G., and Lacombe, B. (2013). ABA transport and transporters. *Trends Plant Sci.* 18, 325–333. doi: 10.1016/j.tplants.2013.01.007

13. Buda, G. J., Barnes, W. J., Fich, E. A., Park, S., Yeats, T. H., Zhao, L.,et al. (2013). An ATP binding cassette transporter is required for cuticular wax deposition and desiccation tolerance in the moss physcomitrella patens. *Plant Cell.* 25, 4000–4013. doi: 10.1105/tpc.113.117648

14. Burla, B., Pfrunder, S., Nagy, R., Francisco, R. M., Lee, Y., and Martinoia, E. (2013). Vacuolar transport of abscisic acid glucosyl ester is mediated by ATP-binding cassette and proton-antiport mechanisms in *Arabidopsis*. *Plant Physiol.* 163, 1446–1458. doi: 10.1104/pp.113.222547

15. Chaumont, F., and Tyerman, S. D. (2014). Aquaporins: highly regulated channels controlling plant water relations. *Plant Physiol.* 164, 1600–1618. doi: 10.1104/pp.113.233791

16. Chen, G., Komatsuda, T., Ma, J. F., Nawrath, C., Pourkheirandish, M., Tagiri, A.,et al. (2011). An ATP-binding cassette subfamily G full transporter is essential for the retention of leaf water in both wild barley and rice. *Proc. Natl. Acad. Sci. U.S.A.* 108, 12354–12359. doi: 10.1073/pnas.1108444108

17. Cho, D., Kim, S. A., Murata, Y., Lee, S., Jae, S. K., Nam, H. G.,et al. (2009). De-regulated expression of the plant glutamate receptor homolog AtGLR3.1 impairs long-term $Ca^{2+}$-programmed stomatal closure. *Plant J.* 58, 437–449. doi: 10.1111/j.1365-313X.2009.03789.x

18. Choi, H., Jin, J. Y., Choi, S., Hwang, J. U., Kim, Y. Y., Suh, M. C.,et al. (2011). An ABCG/WBC-type ABC transporter is essential for transport of sporopollenin precursors for exine formation in developing pollen. *Plant J.* 65, 181–193. doi: 10.1111/j.1365-313X.2010.04412.x

19. Choi, H., Ohyama, K., Kim, Y. Y., Jin, J. Y., Lee, S. B., Yamaoka, Y.,et al. (2014). The role of *Arabidopsis* ABCG9 and ABCG31 ATP binding cassette transporters in pollen fitness and the deposition of steryl glycosides on the pollen coat. *Plant Cell* 26, 310–324. doi: 10.1105/tpc.113.118935

20. Cui, X. H., Hao, F. S., Chen, H., Chen, J., and Wang, X. C. (2008). Expression of the *Vicia faba* VfPIP1 gene in *Arabidopsis*

*thaliana* plants improves their drought resistance. *J. Plant Res.* 121, 207–214. doi: 10.1007/s10265-007-0130-z

21. De Angeli, A., Zhang, J., Meyer, S., and Martinoia, E. (2013). AtALMT9 is a malate-activated vacuolar chloride channel required for stomatal opening in *Arabidopsis*. *Nat. Commun.* 4, 1804. doi: 10.1038/ncomms2815
22. Debono, A., Yeats, T. H., Rose, J. K., Bird, D., Jetter, R., Kunst, L.,et al. (2009). *Arabidopsis* LTPG is a glycosylphosphatidylinositol-anchored lipid transfer protein required for export of lipids to the plant surface. *Plant Cell* 21, 1230–1238. doi: 10.1105/tpc.108.064451
23. den Ende, W. V. (2013). Multifunctional fructans and raffinose family oligosaccharides. *Front. Plant Sci.* 4:247. doi: 10.3389/fpls.2013.00247
24. Di Martino, C., Pizzuto, R., Pallotta, M. L., De Santis, A., and Passarella, S. (2006). Mitochondrial transport in proline catabolism in plants: the existence of two separate translocators in mitochondria isolated from durum wheat seedlings. *Planta* 223, 1123–1133. doi: 10.1007/s00425-005-0166-z
25. Ding, Y., and Oldroyd, G. E. (2009). Positioning the nodule, the hormone dictum. *Plant Signal. Behav.* 4, 89–93. doi: 10.4161/psb.4.2.7693
26. Dou, X.-Y., Yang, K.-Z., Zhang, Y., Wang, W., Liu, X.-L., Chen, L.-Q.,et al. (2011). WBC27, an adenosine Tri-phosphate-binding cassette protein, controls pollen wall formation and patterning in *Arabidopsis*. *J. Integr. Plant Biol.* 53, 74–88. doi: 10.1111/j.1744-7909.2010.01010.x
27. Dubovskaya, L. V., Bakakina, Y. S., Kolesneva, E. V., Sodel, D. L., Mcainsh, M. R., Hetherington, A. M.,et al. (2011). cGMP-dependent ABA-induced stomatal closure in the ABA-insensitive *Arabidopsis* mutant abi1-1. *New Phytol.* 191, 57–69. doi: 10.1111/j.1469-8137.2011.03661.x
28. Eisenach, C., Papanatsiou, M., Hillert, E.-K., and Blatt, M. R. (2014). Clustering of the $K^+$ channel GORK of *Arabidopsis* parallels its gating by extracellular $K^+$. *Plant J.* 78, 203–214. doi: 10.1111/tpj.12471
29. Eom, J. S., Cho, J. I., Reinders, A., Lee, S. W., Yoo, Y., Tuan, P. Q.,et

al. (2011). Impaired function of the tonoplast-localized sucrose transporter in rice, OsSUT2, limits the transport of vacuolar reserve sucrose and affects plant growth. *Plant Physiol.* 157, 109–119. doi: 10.1104/pp.111.176982

30. Ernst, L., Goodger, J. Q., Alvarez, S., Marsh, E. L., Berla, B., Lockhart, E.,et al. (2010). Sulphate as a xylem-borne chemical signal precedes the expression of ABA biosynthetic genes in maize roots. *J. Exp. Bot.* 61, 3395–3405. doi: 10.1093/jxb/erq160
31. Farooq, M., Wahid, A., Kobayashi, N., Fujita, D., and Basra, S. M. A. (2009). Plant drought stress: effects, mechanisms and management. *Agron. Sustain. Dev.* 29, 185–212. doi: 10.1051/agro:2008021
32. Finkelstein, R. (2013). Abscisic acid synthesis and response. *Arabidopsis Book* 11, e0166. doi: 10.1199/tab.0166
33. Folgado, R., Sergeant, K., Renaut, J., Swennen, R., Hausman, J. F., and Panis, B. (2014). Changes in sugar content and proteome of potato in response to cold and dehydration stress and their implications for cryopreservation. *J. Proteom.* 98, 99–111. doi: 10.1016/j.jprot.2013.11.027
34. Frost, C. J., Nyamdari, B., Tsai, C. J., and Harding, S. A. (2012). The tonoplast-localized sucrose transporter in Populus (PtaSUT4) regulates whole-plant water relations, responses to water stress, and photosynthesis. *PLoS ONE* 7:e44467. doi: 10.1371/journal.pone.0044467
35. Geiger, D., Maierhofer, T., Al-Rasheid, K. A., Scherzer, S., Mumm, P., Liese, A.,et al. (2011). Stomatal closure by fast abscisic acid signaling is mediated by the guard cell anion channel SLAH3 and the receptor RCAR1. *Sci. Signal.* 4, ra32. doi: 10.1126/scisignal.2001346
36. Geiger, D., Scherzer, S., Mumm, P., Stange, A., Marten, I., Bauer, H.,et al. (2009). Activity of guard cell anion channel SLAC1 is controlled by drought-stress signaling kinase-phosphatase pair. *Proc. Natl. Acad. Sci. U.S.A.* 106, 21425–21430. doi: 10.1073/pnas.0912021106
37. Gobert, A., Isayenkov, S., Voelker, C., Czempinski, K., and Maathuis, F. J. (2007). The two-pore channel TPK1 gene encodes the vacuolar

K$^+$ conductance and plays a role in K$^+$homeostasis. *Proc. Natl. Acad. Sci. U.S.A.* 104, 10726–10731. doi: 10.1073/pnas.0702595104

38. Gong, X., Liu, M., Zhang, L., Ruan, Y., Ding, R., Ji, Y.,et al. (2014). *Arabidopsis AtSUC2* and *AtSUC4*, encoding sucrose transporters, are required for abiotic stress tolerance in an ABA-dependent pathway. *Physiol. Plant.* doi: 10.1111/ppl.12225 [Epub ahead of print].
39. Goodger, J. Q., and Schachtman, D. P. (2010). Re-examining the role of ABA as the primary long-distance signal produced by water-stressed roots. *Plant Signal. Behav.* 5, 1298–1301. doi: 10.4161/psb.5.10.13101
40. Grallath, S., Weimar, T., Meyer, A., Gumy, C., Suter-Grotemeyer, M., Neuhaus, J. M.,et al. (2005). The AtProT family. Compatible solute transporters with similar substrate specificity but differential expression patterns. *Plant Physiol.* 137, 117–126. doi: 10.1104/pp.104.055079
41. Guo, F. Q., Young, J., and Crawford, N. M. (2003). The nitrate transporter AtNRT1.1 (*CHL*1) functions in stomatal opening and contributes to drought susceptibility in *Arabidopsis*. *Plant Cell* 15, 107–117. doi: 10.1105/tpc.006312
42. Heinen, R. B., Bienert, G. P., Cohen, D., Chevalier, A. S., Uehlein, N., Hachez, C.,et al. (2014). Expression and characterization of plasma membrane aquaporins in stomatal complexes of *Zea mays*. *Plant Mol. Biol.* 86, 335–350. doi: 10.1007/s11103-014-0232-7
43. Heinen, R. B., Ye, Q., and Chaumont, F. (2009). Role of aquaporins in leaf physiology. *J. Exp. Bot.* 60, 2971–2985. doi: 10.1093/jxb/erp171
44. Hosy, E., Vavasseur, A., Mouline, K., Dreyer, I., Gaymard, F., Poree, F.,et al. (2003). The *Arabidopsis* outward K$^+$ channel GORK is involved in regulation of stomatal movements and plant transpiration. *Proc. Natl. Acad. Sci. U.S.A.* 100, 5549–5554. doi: 10.1073/pnas.0733970100
45. Ibraheem, O., Dealtry, G., Roux, S., and Bradley, G. (2011). The effect of drought and salinity on the expressional levels of sucrose transporters in rice (*'Oryza sativa'* Nipponbare) cultivar plants. *Plant Omics* 4, 68–74.

46. Islam, M. M., Munemasa, S., Hossain, M. A., Nakamura, Y., Mori, I. C., and Murata, Y. (2010). Roles of AtTPC1, vacuolar two pore channel 1, in *Arabidopsis* stomatal closure. *Plant Cell Physiol.* 51, 302–311. doi: 10.1093/pcp/pcq001
47. Jang, J. Y., Kim, D. G., Kim, Y. O., Kim, J. S., and Kang, H. (2004). An expression analysis of a gene family encoding plasma membrane aquaporins in response to abiotic stresses in *Arabidopsis thaliana*. *Plant Mol. Biol.* 54, 713–725. doi: 10.1023/B:PLAN.0000040900.61345.a6
48. Jang, J. Y., Lee, S. H., Rhee, J. Y., Chung, G. C., Ahn, S. J., and Kang, H. (2007). Transgenic *Arabidopsis* and tobacco plants overexpressing an aquaporin respond differently to various abiotic stresses. *Plant. Mol. Biol.* 64, 621–632. doi: 10.1007/s11103-007-9181-8
49. Javed, S., Ashraf, M. Y., Meraj, M., Bukhari, S. A., and Zovia, I. (2014). Salinity and drought induced antioxidant responses in different cultivars of sa□ower (*Carthamus tinctorius* L.). *Curr. Pharm. Biotechnol.* 14, 814–819. doi: 10.2174/1389201014666131227114228
50. Jossier, M., Kroniewicz, L., Dalmas, F., Le Thiec, D., Ephritikhine, G., Thomine, S.,et al. (2010). The *Arabidopsis* vacuolar anion transporter, AtCLCc, is involved in the regulation of stomatal movements and contributes to salt tolerance. *Plant J.* 64, 563–576. doi: 10.1111/j.1365-313X.2010.04352.x
51. Kang, J., Hwang, J., Lee, M., Kim, Y., Assmann, S., Martinoia, E.,et al. (2010). PDR-type ABC transporter mediates cellular uptake of the phytohormone abscisic acid. *Proc. Natl. Acad. Sci. U.S.A.* 107, 2355–2360. doi: 10.1073/pnas.0909222107
52. Kanno, Y., Hanada, A., Chiba, Y., Ichikawa, T., Nakazawa, M., Matsui, M.,et al. (2012). Identification of an abscisic acid transporter by functional screening using the receptor complex as a sensor. *Proc. Natl. Acad. Sci. U.S.A.* 109, 9653–9658. doi: 10.1073/pnas.1203567109
53. Kavi Kishor, P. B., and Sreenivasulu, N. (2013). Is proline accumulation per se correlated with stress tolerance or is proline homeostasis a more critical issue? *Plant Cell Environ.* 37, 300–311.

doi: 10.1111/pce.12157

54. Kim, H., Lee, S. B., Kim, H. J., Min, M. K., Hwang, I., and Suh, M. C. (2012). Characterization of glycosylphosphatidylinositol-anchored lipid transfer protein 2 (LTPG2) and overlapping function between LTPG/LTPG1 and LTPG2 in cuticular wax export or accumulation in *Arabidopsis thaliana*. *Plant Cell Physiol.* 53, 1391–1403. doi: 10.1093/pcp/pcs083
55. Kim, T. H., Bohmer, M., Hu, H., Nishimura, N., and Schroeder, J. I. (2010). Guard cell signal transduction network: advances in understanding abscisic acid, $CO_2$, and $Ca^{2+}$ signaling. *Annu. Rev. Plant Biol.* 61, 561–591. doi: 10.1146/annurev-arplant-042809-112226
56. Kiyosue, T., Abe, H., Yamaguchi-Shinozaki, K., and Shinozaki, K. (1998). ERD6, a cDNA clone for an early dehydration-induced gene of *Arabidopsis,* encodes a putative sugar transporter. *Biochim. Biophys. Acta* 1370, 187–191. doi: 10.1016/S0005-2736(98)00007-8
57. Klein, M., Geisler, M., Suh, S. J., Kolukisaoglu, H. U., Azevedo, L., Plaza, S.,et al. (2004). Disruption of AtMRP4, a guard cell plasma membrane ABCC-type ABC transporter, leads to deregulation of stomatal opening and increased drought susceptibility. *Plant J.* 39, 219–236. doi: 10.1111/j.1365-313X.2004.02125.x
58. Klepek, Y. S., Geiger, D., Stadler, R., Klebl, F., Landouar-Arsivaud, L., Lemoine, R.,et al. (2005). *Arabidopsis* POLYOL TRANSPORTER5, a new member of the monosaccharide transporter-like superfamily, mediates $H^+$-Symport of numerous substrates, including myo-inositol, glycerol, and ribose. *Plant Cell* 17, 204–218. doi: 10.1105/tpc.104.026641
59. Klepek, Y. S., Volke, M., Konrad, K. R., Wippel, K., Hoth, S., Hedrich, R.,et al. (2010). *Arabidopsis thaliana* POLYOL/MONOSACCHARIDE TRANSPORTERS 1 and 2: fructose and xylitol/H+ symporters in pollen and young xylem cells. *J. Exp. Bot.* 61, 537–550. doi: 10.1093/jxb/erp322
60. Kosma, D. K., Bourdenx, B., Bernard, A., Parsons, E. P., Lu, S., Joubes, J.,et al. (2009). The impact of water deficiency on leaf cuticle lipids of *Arabidopsis*. *Plant Physiol.* 151, 1918–1929. doi: 10.1104/pp.109.141911

61. Kretzschmar, T., Kohlen, W., Sasse, J., Borghi, L., Schlegel, M., Bachelier, J. B.,et al. (2012). A petunia ABC protein controls strigolactone-dependent symbiotic signalling and branching. *Nature* 483, 341–344. doi: 10.1038/nature10873

62. Kudoyarova, G., Veselova, S., Hartung, W., Farhutdinov, R., Veselov, D., and Sharipova, G. (2011). Involvement of root ABA and hydraulic conductivity in the control of water relations in wheat plants exposed to increased evaporative demand. *Planta* 233, 87–94. doi: 10.1007/s00425-010-1286–1287

63. Kumar, D., Datta, R., Sinha, R., Ghosh, A., and Chattopadhyay, S. (2014a). Proteomic profiling of gamma-ECS overexpressed transgenic *Nicotiana* in response to drought stress. *Plant Signal. Behav.* 9. [Epub ahead of print].

64. Kumar, T., Uzma, Khan, M. R., Abbas, Z., and Ali, G. M. (2014b). Genetic improvement of sugarcane for drought and salinity stress tolerance using arabidopsis vacuolar pyrophosphatase (AVP1) gene. *Mol. Biotechnol.* 56, 199–209. doi: 10.1007/s12033-013-9695-z

65. Kuromori, T., Miyaji, T., Yabuuchi, H., Shimizu, H., Sugimoto, E., and Kamiya, A. (2010). ABC transporter AtABCG25 is involved in abscisic acid transport and responses. *Proc. Natl. Acad. Sci. U.S.A.* 107, 2361–2366. doi: 10.1073/pnas.0912516107

66. Kuromori, T., Sugimoto, E., and Shinozaki, K. (2011). *Arabidopsis* mutants of AtABCG22, an ABC transporter gene, increase water transpiration and drought susceptibility. *Plant J.* 67, 885–894. doi: 10.1111/j.1365-313X.2011.04641.x

67. Lalonde, S., and Frommer, W. B. (2012). SUT sucrose and MST monosaccharide transporter inventory of the *Selaginella* genome. *Front. Plant Sci.* 3:24. doi: 10.3389/fpls.2012.00024

68. Landgraf, R., Smolka, U., Altmann, S., Eschen-Lippold, L., Senning, M., Sonnewald, S.,et al. (2014). The ABC transporter ABCG1 Is required for suberin formation in potato tuber periderm. *Plant Cell* 26, 3403–3415. doi: 10.1105/tpc.114.124776

69. Lawson, T., and Blatt, M. R. (2014). Stomatal size, speed, and responsiveness impact on photosynthesis and water use

efficiency. *Plant Physiol.* 164, 1556–1570. doi: 10.1104/pp.114.237107

70. Lee, K. H., Piao, H. L., Kim, H.-Y., Choi, S. M., Jiang, F., Hartung, W.,et al. (2006). Activation of glucosidase via stress-induced polymerization rapidly increases active pools of abscisic acid. *Cell* 126, 1109–1120. doi: 10.1016/j.cell.2006.07.034
71. Lee, M., Choi, Y., Burla, B., Kim, Y. Y., Jeon, B., Maeshima, M.,et al. (2008). The ABC transporter AtABCB14 is a malate importer and modulates stomatal response to $CO_2$. *Nat. Cell Biol.*10, 1217–1223. doi: 10.1038/ncb1782
72. Lee, S. B., Go, Y. S., Bae, H. J., Park, J. H., Cho, S. H., Cho, H. J.,et al. (2009a). Disruption of glycosylphosphatidylinositol-anchored lipid transfer protein gene altered cuticular lipid composition, increased plastoglobules, and enhanced susceptibility to infection by the fungal pathogen *Alternaria brassicicola. Plant Physiol.* 150, 42–54. doi: 10.1104/pp.109.137745
73. Lee, S. C., Lan, W., Buchanan, B. B., and Luan, S. (2009b). A protein kinase-phosphatase pair interacts with an ion channel to regulate ABA signaling in plant guard cells. *Proc. Natl. Acad. Sci. U.S.A.* 106, 21419–21424. doi: 10.1073/pnas.0910601106
74. Lehmann, S., Funck, D., Szabados, L., and Rentsch, D. (2010). Proline metabolism and transport in plant development. *Amino Acids* 39, 949–962. doi: 10.1007/s00726-010-0525–523
75. Lehmann, S., Gumy, C., Blatter, E., Boeffel, S., Fricke, W., and Rentsch, D. (2011). In planta function of compatible solute transporters of the AtProT family. *J. Exp. Bot.* 62, 787–796. doi: 10.1093/jxb/erq320
76. Li-Beisson, Y., Shorrosh, B., Beisson, F., Andersson, M. X., Arondel, V., Bates, P. D.,et al. (2010). Acyl-lipid metabolism. *Arabidopsis Book* 8, e0133. doi: 10.1199/tab.0133
77. Li, T., Hu, Y. J., Hao, Z. P., Li, H., and Chen, B. D. (2013). Aquaporin genes GintAQPF1 and GintAQPF2 from *Glomus intraradices* contribute to plant drought tolerance. *Plant Signal. Behav.* 8, e24030. doi: 10.4161/psb.24030
78. Li, T., Lin, G., Zhang, X., Chen, Y., Zhang, S., and Chen, B. (2014). Relative importance of an arbuscular mycorrhizal fungus

(*Rhizophagus intraradices*) and root hairs in plant drought tolerance. *Mycorrhiza* 24, 595–602. doi: 10.1007/s00572-014-0578-3

79. Liang, Y., Mitchell, D. M., and Harris, J. M. (2007). Abscisic acid rescues the root meristem defects of the *Medicago truncatula* latd mutant. *Dev. Biol.* 304, 297–307. doi: 10.1016/j.ydbio.2006.12.037

80. Lobet, G., Couvreur, V., Meunier, F., Javaux, M., and Draye, X. (2014). Plant water uptake in drying soils. *Plant Physiol.* 164, 1619–1627. doi: 10.1104/pp.113.233486

81. Luo, B., Xue, X. Y., Hu, W. L., Wang, L. J., and Chen, X. Y. (2007). An ABC transporter gene of *Arabidopsis thaliana*, AtWBC11, is involved in cuticle development and prevention of organ fusion. *Plant Cell Physiol.* 48, 1790–1802. doi: 10.1093/pcp/pcm152

82. Mahdieh, M., Mostajeran, A., Horie, T., and Katsuhara, M. (2008). Drought stress alters water relations and expression of PIP-type aquaporin genes in *Nicotiana tabacum* plants. *Plant Cell Physiol.* 49, 801–813. doi: 10.1093/pcp/pcn054

83. Maierhofer, T., Diekmann, M., Offenborn, J. N., Lind, C., Bauer, H., Hashimoto, K.,et al. (2014). Site- and kinase-specific phosphorylation-mediated activation of SLAC1, a guard cell anion channel stimulated by abscisic acid. *Sci. Signal.* 7, ra86. doi: 10.1126/scisignal.2005703

84. Maurel, C., Verdoucq, L., Luu, D. T., and Santoni, V. (2008). Plant aquaporins: membrane channels with multiple integrated functions. *Annu. Rev. Plant Biol.* 59, 595–624. doi: 10.1146/annurev.arplant.59.032607.092734

85. McAdam, S. A., and Brodribb, T. J. (2014). Separating active and passive influences on stomatal control of transpiration. *Plant Physiol.* 164, 1578–1586. doi: 10.1104/pp.113.231944

86. McFarlane, H. E., Shin, J. J., Bird, D. A., and Samuels, A. L. (2010). *Arabidopsis* ABCG transporters, which are required for export of diverse cuticular lipids, dimerize in different combinations. *Plant Cell* 22, 3066–3075. doi: 10.1105/tpc.110.077974

87. McFarlane, H. E., Watanabe, Y., Yang, W., Huang, Y., Ohlrogge, J., and Samuels, A. L. (2014). Golgi- and trans-Golgi network-mediated vesicle trafficking is required for wax secretion from epidermal

cells. *Plant Physiol.* 164, 1250–1260. doi: 10.1104/pp.113.234583

88. Meyer, S., Mumm, P., Imes, D., Endler, A., Weder, B., Al-Rasheid, K. A.,et al. (2010). AtALMT12 represents an R-type anion channel required for stomatal movement in *Arabidopsis* guard cells. *Plant J.* 63, 1054–1062. doi: 10.1111/j.1365-313X.2010.04302.x
89. Moshelion, M., Halperin, O., Wallach, R., Oren, R., and Way, D. A. (2014). Role of aquaporins in determining transpiration and photosynthesis in water-stressed plants: crop water-use efficiency, growth and yield. *Plant Cell Environ.* doi: 10.1111/pce.12410 [Epub ahead of print].
90. Moyankova, D., Mladenov, P., Berkov, S., Peshev, D., Georgieva, D., and Djilianov, D. (2014). Metabolic profiling of the resurrection plant *Haberlea rhodopensis* during desiccation and recovery. *Physiol. Plant.* 152, 675–687. doi: 10.1111/ppl.12212
91. Nadal, M., and Paszkowski, U. (2013). Polyphony in the rhizosphere: presymbiotic communication in arbuscular mycorrhizal symbiosis. *Curr. Opin. Plant Biol.* 16, 473–479. doi: 10.1016/j.pbi.2013.06.005
92. Nagy, R., Grob, H., Weder, B., Green, P., Klein, M., Frelet-Barrand, A.,et al. (2009). The *Arabidopsis* ATP-binding cassette protein AtMRP5/AtABCC5 is a high affinity inositol hexakisphosphate transporter involved in guard cell signaling and phytate storage. *J. Biol. Chem.* 284, 33614–33622. doi: 10.1074/jbc.M109.030247
93. Negi, J., Matsuda, O., Nagasawa, T., Oba, Y., Takahashi, H., Kawai-Yamada, M.,et al. (2008). $CO_2$ regulator SLAC1 and its homologues are essential for anion homeostasis in plant cells. *Nature* 452, 483–486. doi: 10.1038/nature06720
94. Osakabe, Y., Arinaga, N., Umezawa, T., Katsura, S., Nagamachi, K., Tanaka, H.,et al. (2013). Osmotic stress responses and plant growth controlled by potassium transporters in *Arabidopsis*. *Plant Cell* 25, 609–624. doi: 10.1105/tpc.112.105700
95. Osakabe, Y., Yamaguchi-Shinozaki, K., Shinozaki, K., and Tran, L. S. (2014). ABA control of plant macroelement membrane transport systems in response to water deficit and high salinity. *New Phytol.* 202, 35–49. doi: 10.1111/nph.12613
96. Padmanaban, S., Chanroj, S., Kwak, J. M., Li, X., Ward, J. M., and

Sze, H. (2007). Participation of endomembrane cation/$H^+$ exchanger AtCHX20 in osmoregulation of guard cells. *Plant Physiol.* 144, 82–93. doi: 10.1104/pp.106.092155

97. Panikashvili, D., Savaldi-Goldstein, S., Mandel, T., Yifhar, T., Franke, R. B., Hofer, R.,et al. (2007). The *Arabidopsis* DESPERADO/AtWBC11 transporter is required for cutin and wax secretion. *Plant Physiol.* 145, 1345–1360. doi: 10.1104/pp.107.105676
98. Panikashvili, D., Shi, J. X., Bocobza, S., Franke, R. B., Schreiber, L., and Aharoni, A. (2010). The *Arabidopsis* DSO/ABCG11 transporter affects cutin metabolism in reproductive organs and suberin in roots. *Mol. Plant* 3, 563–575. doi: 10.1093/mp/ssp103
99. Panikashvili, D., Shi, J. X., Schreiber, L., and Aharoni, A. (2011). The *Arabidopsis* ABCG13 transporter is required for flower cuticle secretion and patterning of the petal epidermis. *New Phytol.* 190, 113–124. doi: 10.1111/j.1469-8137.2010.03608.x
100. Park, S., Li, J., Pittman, J. K., Berkowitz, G. A., Yang, H., Undurraga, S.,et al. (2005). Up-regulation of a $H^+$-pyrophosphatase ($H^+$-PPase) as a strategy to engineer drought-resistant crop plants. *Proc. Natl. Acad. Sci. U.S.A.* 102, 18830–18835. doi: 10.1073/pnas.0509512102
101. Pasapula, V., Shen, G., Kuppu, S., Paez-Valencia, J., Mendoza, M., Hou, P.,et al. (2011). Expression of an *Arabidopsis* vacuolar $H^+$-pyrophosphatase gene (AVP1) in cotton improves drought- and salt tolerance and increases fibre yield in the field conditions. *Plant Biotechnol. J.* 9, 88–99. doi: 10.1111/j.1467-7652.2010.00535.x
102. Payyavula, R. S., Tay, K. H., Tsai, C. J., and Harding, S. A. (2011). The sucrose transporter family in Populus: the importance of a tonoplast PtaSUT4 to biomass and carbon partitioning. *Plant J.* 65, 757–770. doi: 10.1111/j.1365-313X.2010.04463.x
103. Peng, Y., Lin, W., Cai, W., and Arora, R. (2007). Overexpression of a *Panax ginseng* tonoplast aquaporin alters salt tolerance, drought tolerance and cold acclimation ability in transgenic *Arabidopsis* plants. *Planta* 226, 729–740. doi: 10.1007/s00425-007-0520-4
104. Petrasek, J., and Friml, J. (2009). Auxin transport routes in plant development. *Development* 136, 2675–2688. doi: 10.1242/dev.030353

105. Pighin, J. A., Zheng, H., Balakshin, L. J., Goodman, I. P., Western, T. L., Jetter, R.,et al. (2004). Plant cuticular lipid export requires an ABC transporter. *Science* 306, 702–704. doi: 10.1126/science.1102331

106. Poschet, G., Hannich, B., Raab, S., Jungkunz, I., Klemens, P. A., Krueger, S.,et al. (2011). A novel *Arabidopsis* vacuolar glucose exporter is involved in cellular sugar homeostasis and affects the composition of seed storage compounds. *Plant Physiol.* 157, 1664–1676. doi: 10.1104/pp.111.186825

107. Quilichini, T. D., Friedmann, M. C., Samuels, A. L., and Douglas, C. J. (2010). ATP-binding cassette transporter G26 is required for male fertility and pollen exine formation in *Arabidopsis*. *Plant Physiol.* 154, 678–690. doi: 10.1104/pp.110.161968

108. Ravikumar, G., Manimaran, P., Voleti, S. R., Subrahmanyam, D., Sundaram, R. M., Bansal, K. C.,et al. (2014). Stress-inducible expression of AtDREB1A transcription factor greatly improves drought stress tolerance in transgenic indica rice. *Transgenic Res.* 23, 421–439. doi: 10.1007/s11248-013-9776-6

109. Recorbet, G., Abdallah, C., Renaut, J., Wipf, D., and Dumas-Gaudot, E. (2013). Protein actors sustaining arbuscular mycorrhizal symbiosis: underground artists break the silence. *New Phytol.* 199, 26–40. doi: 10.1111/nph.12287

110. Reinders, A., Panshyshyn, J. A., and Ward, J. M. (2005). Analysis of transport activity of *Arabidopsis* sugar alcohol permease homolog AtPLT5. *J. Biol. Chem.* 280, 1594–1602. doi: 10.1074/jbc.M410831200

111. Remy, E., Cabrito, T. R., Baster, P., Batista, R. A., Teixeira, M. C., Friml, J.,et al. (2013). A major facilitator superfamily transporter plays a dual role in polar auxin transport and drought stress tolerance in *Arabidopsis*. *Plant Cell* 25, 901–926. doi: 10.1105/tpc.113.110353

112. Rentsch, D., Schmidt, S., and Tegeder, M. (2007). Transporters for uptake and allocation of organic nitrogen compounds in plants. *FEBS Lett.* 581, 2281–2289. doi: 10.1016/j.febslet.2007.04.013

113. Ruiz-Lozano, J. M., Porcel, R., Bárzana, G., Azcón, R., and Aroca, R. (2012). "Contribution of arbuscular mycorrhizal symbiosis to plant drought tolerance: state of the art," in *Plant Responses to*

*Drought Stress*, ed. R. Aroca (Berlin: Springer), 335–362.

114. Sade, N., Vinocur, B. J., Diber, A., Shatil, A., Ronen, G., Nissan, H.,et al. (2009). Improving plant stress tolerance and yield production: is the tonoplast aquaporin SlTIP2; 2 a key to isohydric to anisohydric conversion? *New Phytol.* 181, 651–661. doi: 10.1111/j.1469-8137.2008.02689.x
115. Samuels, L., Kunst, L., and Jetter, R. (2008). Sealing plant surfaces: cuticular wax formation by epidermal cells. *Annu. Rev. Plant Biol.* 59, 683–707. doi: 10.1146/annurev.arplant.59.103006.093219
116. Samuels, L., and McFarlane, H. E. (2012). Plant cell wall secretion and lipid traffic at membrane contact sites of the cell cortex. *Protoplasma* 249(Suppl. 1), S19–S23. doi: 10.1007/s00709-011-0345-7
117. Sanders, G., and Arndt, S. (2012). "Osmotic Adjustment Under Drought Conditions," in *Plant Responses to Drought Stress*, ed. R. Aroca (Berlin: Springer), 199–229. doi: 10.1007/978-3-642-32653-0_8
118. Sasaki, T., Mori, I. C., Furuichi, T., Munemasa, S., Toyooka, K., Matsuoka, K.,et al. (2010). Closing plant stomata requires a homolog of an aluminum-activated malate transporter. *Plant Cell Physiol.* 51, 354–365. doi: 10.1093/pcp/pcq016
119. Schmitz, A. M., and Harrison, M. J. (2014). Signaling events during initiation of arbuscular mycorrhizal symbiosis. *J. Integr. Plant Biol.* 56, 250–261. doi: 10.1111/jipb.12155
120. Schulz, B. (2011). "Functional classification of plant plasma membrane transporters," in *The Plant Plasma Membrane*, eds A. S. Murphy, B. Schulz, and W. Peer (Berlin: Springer), 131–176.
121. Schulz, B., and Frommer, W. B. (2004). Plant biology. A plant ABC transporter takes the lotus seat. *Science* 306, 622–625. doi: 10.1126/science.1105227
122. Seo, P. J., Lee, S. B., Suh, M. C., Park, M. J., Go, Y. S., and Park, C. M. (2011). The MYB96 transcription factor regulates cuticular wax biosynthesis under drought conditions in *Arabidopsis*. *Plant Cell* 23, 1138–1152. doi: 10.1105/tpc.111.083485
123. Shirasawa, K., Takabe, T., Takabe, T., and Kishitani, S. (2006). Accumulation of glycinebetaine in rice plants that overexpress

choline monooxygenase from spinach and evaluation of their tolerance to abiotic stress. *Ann. Bot.* 98, 565–571. doi: 10.1093/aob/mcl126

124. Singh, R., Pandey, N., Naskar, J., and Shirke, P. A. (2014). Physiological performance and differential expression profiling of genes associated with drought tolerance in contrasting varieties of two *Gossypium* species. *Protoplasma* doi: 10.1007/s00709-014-0686-0 [Epub ahead of print].
125. Surbanovski, N., Sargent, D. J., Else, M. A., Simpson, D. W., Zhang, H., and Grant, O. M. (2013). Expression of *Fragaria vesca* PIP aquaporins in response to drought stress: PIP down-regulation correlates with the decline in substrate moisture content. *PLoS ONE* 8:e74945. doi: 10.1371/journal.pone.0074945
126. Szabados, L., and Savoure, A. (2010). Proline: a multifunctional amino acid. *Trends Plant Sci.* 15, 89–97. doi: 10.1016/j.tplants.2009.11.009
127. Taji, T., Ohsumi, C., Iuchi, S., Seki, M., Kasuga, M., Kobayashi, M.,et al. (2002). Important roles of drought- and cold-inducible genes for galactinol synthase in stress tolerance in *Arabidopsis thaliana*. *Plant J.* 29, 417–426. doi: 10.1046/j.0960-7412.2001.01227.x
128. Takabe, T. (2012). Engineering of betaine biosynthesis and transport for abiotic stress tolerance in plants. *J. Plant Biochem. Biotechnol.* 21, 58–62. doi: 10.1007/s13562-012-0143-0
129. Ukitsu, H., Kuromori, T., Toyooka, K., Goto, Y., Matsuoka, K., Sakuradani, E.,et al. (2007). Cytological and biochemical analysis of COF1, an *Arabidopsis* mutant of an ABC transporter gene. *Plant Cell Physiol.* 48, 1524–1533. doi: 10.1093/pcp/pcm139
130. Vahisalu, T., Kollist, H., Wang, Y. F., Nishimura, N., Chan, W. Y., Valerio, G.,et al. (2008). SLAC1 is required for plant guard cell S-type anion channel function in stomatal signalling. *Nature* 452, 487–491. doi: 10.1038/nature06608
131. Verrier, P., Bird, D., Burla, B., Dassa, E., Forestier, C., and Geisler, M. (2008). Plant ABC proteins: a unified nomenclature and updated inventory. *Trends Plant Sci.* 13, 151–159. doi: 10.1016/j.tplants.2008.02.001
132. Wang, L. L., Chen, A. P., Zhong, N. Q., Liu, N., Wu, X. M.,

Wang, F.,et al. (2014). The *Thellungiella salsuginea* tonoplast aquaporin TsTIP1;2 functions in protection against multiple abiotic stresses. *Plant Cell Physiol.* 55, 148–161. doi: 10.1093/pcp/pct166

133. Wang, X., Li, Y., Ji, W., Bai, X., Cai, H., Zhu, D.,et al. (2011). A novel *Glycine soja* tonoplast intrinsic protein gene responds to abiotic stress and depresses salt and dehydration tolerance in transgenic *Arabidopsis thaliana. J. Plant Physiol.* 168, 1241–1248. doi: 10.1016/j.jplph.2011.01.016

134. Wang, Y. F., Munemasa, S., Nishimura, N., Ren, H. M., Robert, N., Han, M.,et al. (2013). Identification of cyclic GMP-activated nonselective Ca2+-permeable cation channels and associated CNGC5 and CNGC6 genes in *Arabidopsis* guard cells. *Plant Physiol.* 163, 578–590. doi: 10.1104/pp.113.225045

135. Watkins, J. M., Hechler, P. J., and Muday, G. K. (2014). Ethylene-induced flavonol accumulation in guard cells suppresses reactive oxygen species and moderates stomatal aperture. *Plant Physiol.* 164, 1707–1717. doi: 10.1104/pp.113.233528

136. Wege, S., De Angeli, A., Droillard, M. J., Kroniewicz, L., Merlot, S., Cornu, D.,et al. (2014). Phosphorylation of the vacuolar anion exchanger AtCLCa is required for the stomatal response to abscisic acid. *Sci. Signal.* 7, ra65. doi: 10.1126/scisignal.2005140

137. Weise, A., Barker, L., Kuhn, C., Lalonde, S., Buschmann, H., Frommer, W. B.,et al. (2000). A new subfamily of sucrose transporters, SUT4, with low affinity/high capacity localized in enucleate sieve elements of plants. *Plant Cell* 12, 1345–1355. doi: 10.1105/tpc.12.8.1345

138. Wormit, A., Trentmann, O., Feifer, I., Lohr, C., Tjaden, J., Meyer, S.,et al. (2006). Molecular identification and physiological characterization of a novel monosaccharide transporter from *Arabidopsis* involved in vacuolar sugar transport. *Plant Cell* 18, 3476–3490. doi: 10.1105/tpc.106.047290

139. Wu, L., Guan, Y., Wu, Z., Yang, K., Lv, J., Converse, R.,et al. (2014). OsABCG15 encodes a membrane protein that plays an important role in anther cuticle and pollen exine formation in rice. *Plant Cell Rep.* 33, 1881–1899. doi: 10.1007/s00299-014-1666-8

140. Wudick, M. M., Luu, D. T., and Maurel, C. (2009). A look

inside: localization patterns and functions of intracellular plant aquaporins. *New Phytol.* 184, 289–302. doi: 10.1111/j.1469-8137.2009.02985.x

141. Xu, C., Wang, M., Zhou, L., Quan, T., and Xia, G. (2013a). Heterologous expression of the wheat aquaporin gene TaTIP2; 2 compromises the abiotic stress tolerance of *Arabidopsis thaliana*. *PLoS ONE* 8:e79618. doi: 10.1371/journal.pone.0079618
142. Xu, W., Jia, L., Shi, W., Liang, J., Zhou, F., Li, Q.,et al. (2013b). Abscisic acid accumulation modulates auxin transport in the root tip to enhance proton secretion for maintaining root growth under moderate water stress. *New Phytol.* 197, 139–150. doi: 10.1111/nph.12004
143. Xu, Z. Y., Lee, K. H., Dong, T., Jeong, J. C., Jin, J. B., Kanno, Y.,et al. (2012). A vacuolar beta-glucosidase homolog that possesses glucose-conjugated abscisic acid hydrolyzing activity plays an important role in osmotic stress responses in *Arabidopsis*. *Plant Cell* 24, 2184–2199. doi: 10.1105/tpc.112.095935
144. Xue, Y., Xiao, S., Kim, J., Lung, S. C., Chen, L., Tanner, J. A.,et al. (2014). *Arabidopsis* membrane-associated acyl-CoA-binding protein ACBP1 is involved in stem cuticle formation. *J. Exp. Bot.* 65, 5473–5483. doi: 10.1093/jxb/eru304
145. Yamada, K., Osakabe, Y., Mizoi, J., Nakashima, K., Fujita, Y., Shinozaki, K.,et al. (2010). Functional analysis of an *Arabidopsis thaliana* abiotic stress-inducible facilitated diffusion transporter for monosaccharides. *J. Biol. Chem.* 285, 1138–1146. doi: 10.1074/jbc.M109.054288
146. Yang, J., Isabel Ordiz, M., Jaworski, J. G., and Beachy, R. N. (2011). Induced accumulation of cuticular waxes enhances drought tolerance in *Arabidopsis* by changes in development of stomata. *Plant Physiol. Biochem.* 49, 1448–1455. doi: 10.1016/j.plaphy.2011.09.006
147. Yeats, T. H., and Rose, J. K. (2013). The formation and function of plant cuticles. *Plant Physiol.* 163, 5–20. doi: 10.1104/pp.113.222737
148. Yue, C., Cao, H., Wang, L., Zhou, Y., Hao, X., Zeng, J.,et al. (2014). Molecular cloning and expression analysis of tea plant aquaporin (AQP) gene family. *Plant Physiol. Biochem.* 83, 65–76. doi: 10.1016/j.plaphy.2014.07.011

149. Zazimalova, E., Murphy, A. S., Yang, H., Hoyerova, K., and Hosek, P. (2010). Auxin transporters–why so many? *Cold Spring Harb. Perspect. Biol.* 2, a001552. doi: 10.1101/cshperspect.a001552
150. Zhang, H., Zhu, H., Pan, Y., Yu, Y., Luan, S., and Li, L. (2014). ADTX/MATE-type transporter facilitates abscisic acid Efflux and modulates ABA sensitivity and drought tolerance in *Arabidopsis*. *Mol. Plant* 7, 1522–1532. doi: 10.1093/mp/ssu063.
151. Zhang, L., Gao, M., Hu, J., Zhang, X., Wang, K., and Ashraf, M. (2012). Modulation role of abscisic acid (ABA) on growth, water relations and glycinebetaine metabolism in two maize (*Zea mays* L.) cultivars under drought stress. *Int. J. Mol. Sci.* 13, 3189–3202. doi: 10.3390/ijms13033189
152. Zhang, N., Si, H.-J., Wen, G., Du, H.-H., Liu, B.-L., and Wang, D. (2011). Enhanced drought and salinity tolerance in transgenic potato plants with a BADH gene from spinach. *Plant Biotechnol. Rep.* 5, 71–77. doi: 10.1007/s11816-010-0160–161
153. Zhang, Y., Wang, Z., Chai, T., Wen, Z., and Zhang, H. (2008). Indian mustard aquaporin improves drought and heavy-metal resistance in tobacco. *Mol. Biotechnol.* 40, 280–292. doi: 10.1007/s12033-008-9084–9081
154. Zhao, J., and Dixon, R. A. (2010). The 'ins' and 'outs' of flavonoid transport. *Trends Plant Sci.* 15, 72–80. doi: 10.1016/j.tplants.2009.11.006
155. Zhou, S., Hu, W., Deng, X., Ma, Z., Chen, L., Huang, C.,et al. (2012). Overexpression of the wheat aquaporin gene, TaAQP7, enhances drought tolerance in transgenic tobacco. *PLoS ONE*7:e52439. doi: 10.1371/journal.pone.0052439
156. Zhu, L., Guo, J., Zhu, J., and Zhou, C. (2014). Enhanced expression of EsWAX1 improves drought tolerance with increased accumulation of cuticular wax and ascorbic acid in transgenic *Arabidopsis*. *Plant Physiol. Biochem.* 75, 24–35. doi: 10.1016/j.plaphy.2013.11.028
157. Zhuo, C., Wang, T., Lu, S., Zhao, Y., Li, X., and Guo, Z. (2013). A cold responsive galactinol synthase gene from *Medicago falcata* (MfGolS1) is induced by myo-inositol and confers multiple tolerances to abiotic stresses. *Physiol. Plant* 149, 67–78. doi: 10.1111/ppl.12019

CHAPTER

# 12

# BALANCING WATER UPTAKE AND LOSS THROUGH THE COORDINATED REGULATION OF STOMATAL AND ROOT DEVELOPMENT

Christopher Hepworth[1], Carla Turner[2], Marcela Guimaraes Landim[2,3], Duncan Cameron[1], Julie E. Gray[2]

[1] Department of Animal and Plant Sciences, University of Sheffield, Sheffield, United Kingdom

[2] Department of Molecular Biology and Biotechnology, University of Sheffield, Sheffield, United Kingdom

[3] University of Brasília, Brasília, Federal District, Biological institute, Brazil

## ABSTRACT

Root development is influenced by nutrient and water availabilities. Plants are able to adjust many attributes of their root in response to environmental signals including the size and shape of the primary root, lateral roots and root hairs. Here we investigated the response of roots to changes in the levels of leaf transpiration associated with altered stomatal frequency. We found that plants with high stomatal density and conductance produce a larger rooting area and as a result have enhanced phosphate uptake capacity whereas plants with low stomatal conductance produce a smaller root. Manipulating the growth environment of plants indicated that enhanced root growth is most likely a result of an increased demand for water rather than phosphate. Plants manipulated to have an increase or reduction in root hair growth show a reduction or increase respectively, in stomatal conductance and density. Our results demonstrate that plants can balance their water uptake and loss through coordinated regulation of both stomatal and root development.

## INTRODUCTION

Recent advances in molecular genetics provide the resources to dissect the effect of water loss on plant growth and performance. Work in the genetic model species *Arabidopsis thaliana*, to unravel the molecular basis of stomatal development, has resulted in a series of mutants which range in stomatal density (*D*), stomatal conductance, transpiration (*E*), water use efficiency and carbon assimilation in the same genetic background [1–4]. Arabidopsis plant lines with either low or high *D* have been produced by manipulating the level of expression of epidermal patterning factors (EPF). In the current study we focused on genotypes with manipulated levels of EPF1 and EPF2; peptides which normally act to suppress inappropriate stomatal development. Plants manipulated to ectopically overexpress *EPF2* have substantially reduced *D* on leaves, and in our previous study, had approximately 50% of the *E* of wild-type mature leaves [5]. In contrast, *epf1epf2* double mutant plants, lacking both *EPF1* and *EPF2*expression, have a high *D* and approximately 170% of the *E* of wild-type leaves.

Here we investigated direct uptake of phosphate (*P*) by the roots of plants with manipulated stomatal densities. *P* is an essential plant nutrient which is actively taken up by transporters predominately located on the surface of root hairs. A relationship between *D* and nutrient accumulation via mass flow has been previously reported [5]. We sought to address whether this relationship extended to direct nutrient uptake via the roots by examining the accumulation of a radioactive phosphate isotope, supplied to the roots as $^{33}$P-orthophosphate. Furthermore, It is generally accepted that the presence of root hairs is important for both water and nutrient uptake by increasing the extent of the rhizosphere and maximising the root to soil interface. Root hairs may also act as environmental sensors in particular of water stress [6]. The molecular genetic basis of root development has been well studied [7] and in particular a family of transcription factors including ROOT HAIR DEFECTIVE 6-LIKE 4 (RSL4), ROOT HAIR DEFECTIVE 6 (RHD6) and RHD SIX LIKE1 (RSL1) have been shown to regulate root hair patterning/development [8, 9]. Plants lacking the expression of these genes display shorter root hairs whilst those overexpressing *RSL4* develop longer root hairs.

To our knowledge no study has previously compared the root structure or phosphate uptake between plants of the same species with significantly differing levels of *D* or *E* grown under the same conditions. Likewise it is currently unknown what effect altering root hair development may have on *D* or *E*. Via manipulation of both the stomatal and root hair development pathways, we sought to address these questions.

## RESULTS AND DISCUSSION

Five hours after transfer of *Arabidopsis thaliana* seedling roots to an aqueous nutrient solution, plants with high *D* (*epf1epf2*) had taken up over twice as much radioactive phosphate as wild-type. Low *D* plants, (the *EPF2*OE genotype) showed only a slight reduction in phosphate (*P*) accumulation, and this was not significantly different to the level of *P* taken up by wild-type plants under the same conditions (Figure 1). This suggested that, whilst an increase in *D* and *E* may promote *P* accumulation the opposite did not hold true; a large decrease in *E* had no significant effect on *P* uptake.

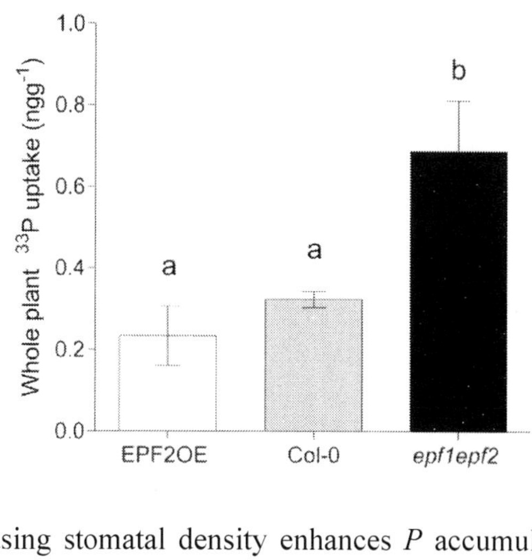


**Figure 1.** Increasing stomatal density enhances *P* accumulation. Mean total

[$^{33}$P] concentrations (ng g$^{-1}$) taken up directly by whole plants over 5 hours (n = 5). Bars with no letters in common are significantly different, P<0.05, (Tukeys test after one-way ANOVA). Error bars represent SE.

We then investigated whether the increase in *P* accumulation seen in plants with increased *D*could be a result of an increased root surface area for nutrient uptake. First we sought to confirm that *EPF1* and *EPF2*, which have previously been reported to be expressed in immature stomata and stomatal precursor cells [10–12] do not have a direct effect on root anatomy and physiology. Histological staining of plants expressing β-glucuronidase (GUS) under the control of either the *EPF1* or *EPF2* gene promoter regions (Hunt & Gray, 2009) indicated that neither of these genes are expressed in seedling roots (Figure 2a) which is in line with transcriptomic studies [13]. Thus, it appears unlikely that the absence of *EPF1* and *EPF2*expression in the *epf1epf2* mutant could have a direct effect on the roots. It is of course possible that, these secreted peptides could travel to the roots, or that the constitutive over-expression of *EPF2* in the *EPF2*OE mutant could have a direct effect on root nutrient uptake. Indeed EPFL9, a positive regulator of stomatal development, has been previously shown to have a role in root cortex development [14]. However, no significant differences in *P* accumulation were observed in the EPF2 over-expressing genotype (Figure 1).

a

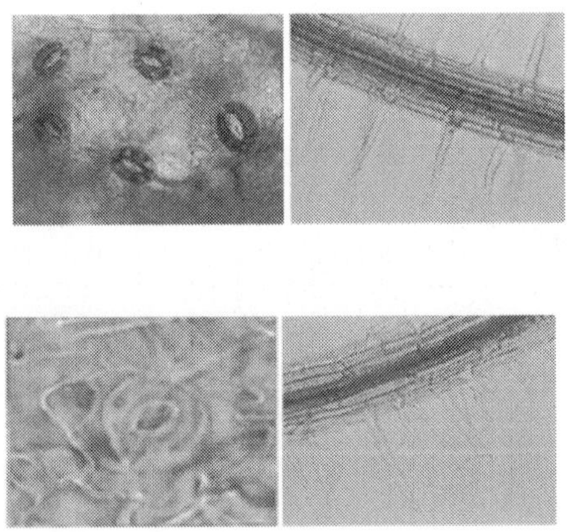

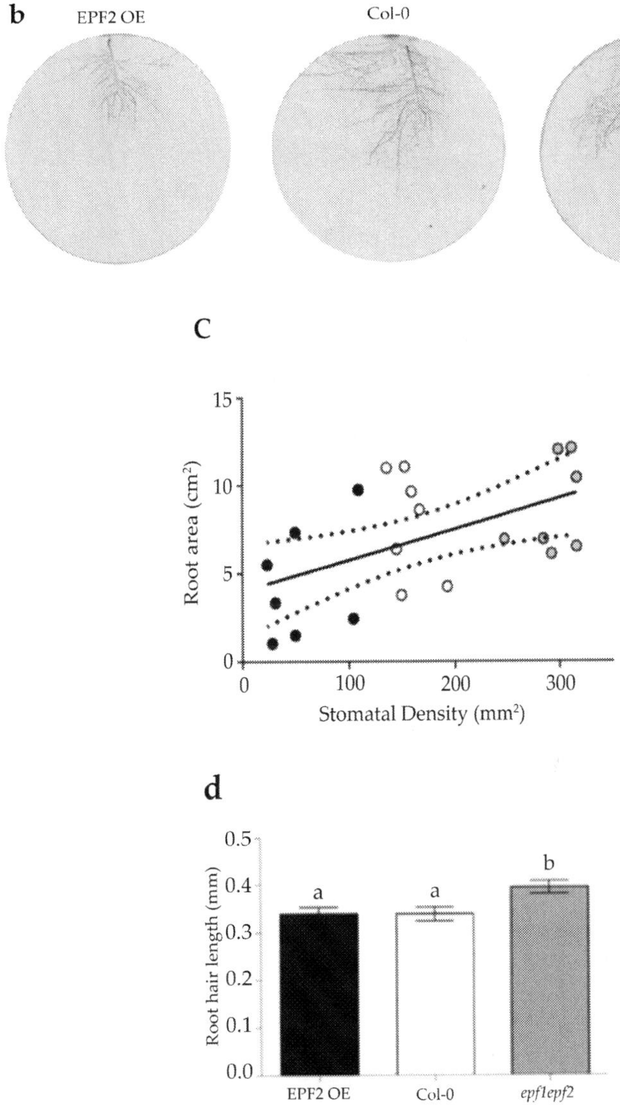
b
EPF2 OE
Col-0
epf1epf2
C
Root area (cm²)
15
10
5
0
0
100
200
300
Stomatal Density (mm²)
d
Root hair length (mm)
0.5
0.4
0.3
0.2
0.1
0.0
a
a
b
EPF2 OE
Col-0
epf1epf2

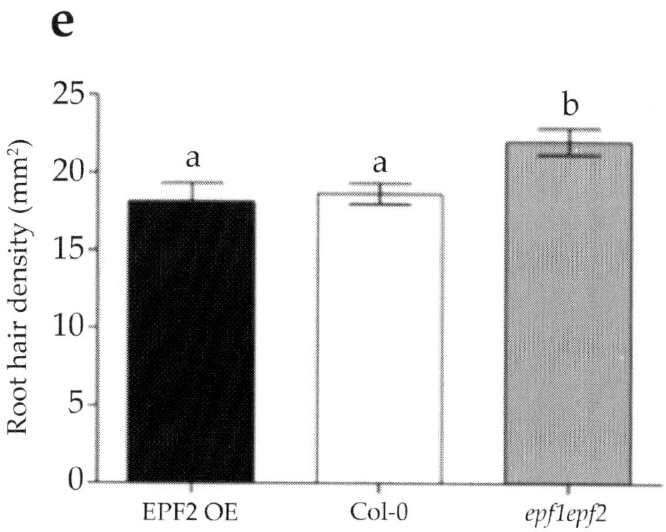


**Figure 2.** Stomatal density correlates with root area. (*A*) Histochemical staining of seedlings expressing *pEPF1::GUS* (top) indicates *EPF1*expression in guards cells but not in roots, and *pEPF2::GUS* (bottom) indicates *EPF2*expression in early stomatal lineage cells and not in roots. (*B*) Representative scans to illustrate root size of plants grown in rhizotrons. (*C*) Linear regression showing a significant relationship between stomatal density and the rooting area of plants grown in rhizotrons (P = 0.0083). (*D*) Increased root hair length in plants with increased stomatal density. (*E*) Increased root hair density in plants with increased stomatal density. (*F*) Representative micrographs of plants grown in agar to assess root hair length and density. Bars with no letters in common are significantly different, P<0.05, (Tukeys test after one-way ANOVA). Error bars represent SE.

We examined the roots of our plants in several different ways. Each of our measurements suggested that plants with increased *D* had an increased root area. First we grew plants in rhizotrons with their roots supported on glass fibre disks. After 7 weeks when the leaf rosette was fully expanded we scanned the glass fibre disks to assess root area, and took leaf impressions to record *D*. A linear relationship was found suggesting a positive correlation between root area and *D* (Figure 2b and 2c). Next, we examined the root hairs of young plants grown on nutrient agar plates. We found that both the length and density of root hairs were significantly enhanced in plants with increased *D* (*epf1epf2*) in comparison to wild-type whilst low *D* plants (EPF2OE) shown no root hair phenotype (Figure 2d, 2e and 2f). Together these results indicated that plants with high *D* and thus an increased capacity for water-loss also displayed increased root area and root hair growth. Since the active transport of *P* occurs primarily at the root hairs this result, could at least in part, explain the enhanced *P* accumulation that we observed in the *epf1epf2* mutant (Figure 1).

To further investigate the links between *D*, *E*, root architecture, and *P* accumulation, we carried out the converse experiment to examine whether plants with altered root hair growth exhibited any alteration in their stomatal characteristics. Remarkably, we found that plants specifically manipulated to have an increased density of root hairs (*RSL4*OE) [8] had significantly lower stomatal conductance values than a mutant manipulated to have low root hair density (*rsl4rhd6rsl1*) (Figure 3a). These plants were then assayed for difference in *D* and, as with stomatal conductance, *D* was inversely related to root hair growth (Figure 3b). It is unlikely that the alteration in stomatal development observed in the root hair mutants occurs as a direct result of gene mis-expression, as the *RSL1/RSL4/RHD6* genes do not appear to be expressed in the shoots of *Arabidopsis thaliana* and the overexpression of the *RSL4* gene does not alter shoot morphology [8, 15].

a

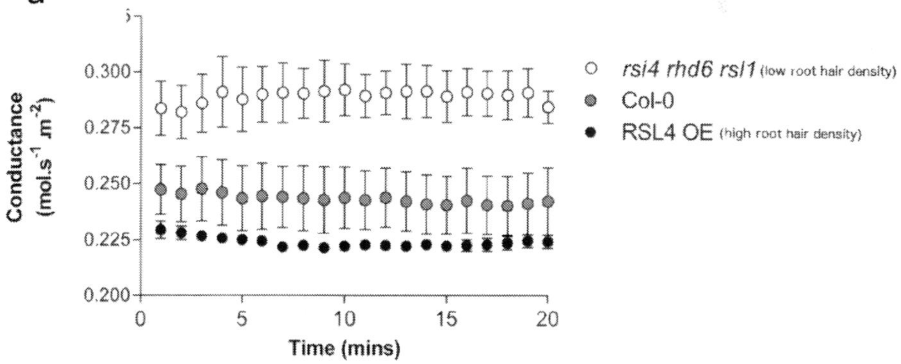

b

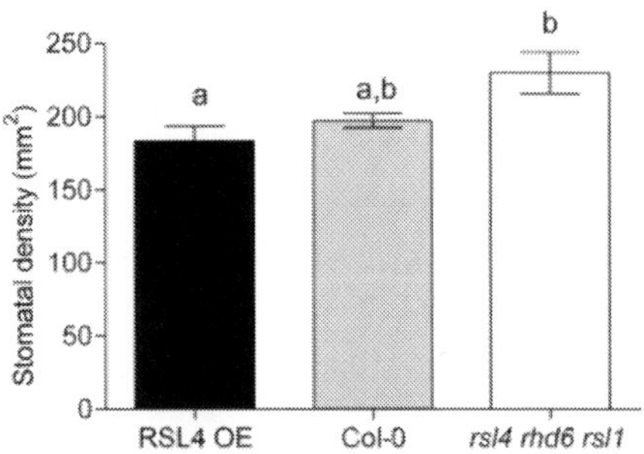


**Figure 3.** Plants with altered root hair density and length display correspondingly opposite alterations in stomatal conductance and stomatal density. (*A*) Mean steady state stomatal conductance rates for genotypes with decreased (*rsl4rhd6rsl1*; white symbols), or increased root hair length and density (*RSL4*OE; black symbols) in comparison to controls (Col-0; grey symbols) at 1000 μmol.s$^{-1}$m$^{-2}$ light, 400 ppm $CO_2$ over 20 minutes. (*B*) Mature leaf abaxial stomatal densities of RSL4 overexpression or *rsl4rhd6rs1* genotypes. Bars with no letters in common are significantly different, $P<0.05$, (Tukeys test after one-way ANOVA). Error bars represent SE.

Our results reveal a previously undescribed interplay between *D* and rooting area that appears to have a functional effect on *P* accumulation. The exact mechanism and stimuli for these observations is currently unclear. It appears that the high rate of water loss found in *epf1epf2*plants may stimulate root growth in an attempt to maximise water uptake and as a result enhances *P* uptake capability. This may allow *epf1epf2* plants to be intrinsically more capable of accumulating nutrients both through increased rates of mass flow as in Hepworth et al.,2015 and through direct root interception.

The enhanced root growth phenotype we observe in *epf1epf2* plants is reminiscent of roots grown under *P* starvation [16]. Thus an alternative explanation might be that the *epf1epf2* plant has an enhanced requirement for *P* and therefore produces more root hairs. The additional guard cells produced in *epf1epf2* may require additional energy in the form of ATP to power their turgor-driven aperture changes [17]. It is therefore not clear whether the enhanced *P*uptake capability that we observed in *epf1epf2* plants is due to their enhanced *E* levels necessitating a higher water uptake capacity and/or a higher requirement for *P* in these plants.

To investigate these two possibilities (Figure 4a), we first grew plants under conditions that promoted different levels of *E* by altering the distance of plants from a moving air source (a rotating fan) thus manipulating airflow at the height of the leaf boundary layer. We found that plants grown under growth conditions which promoted *E* (those closest to the fan) produced the longest roots and those sheltered from the moving air produced the shortest roots (Figure 4b). The trend towards larger roots under fast moving air was evident across all three genotypes; plants with low, normal or high *D* increased their root length by 40%, 33%, and 30% under fast moving air in comparison to the same genotypes grown normally (Figure 4b *left*).

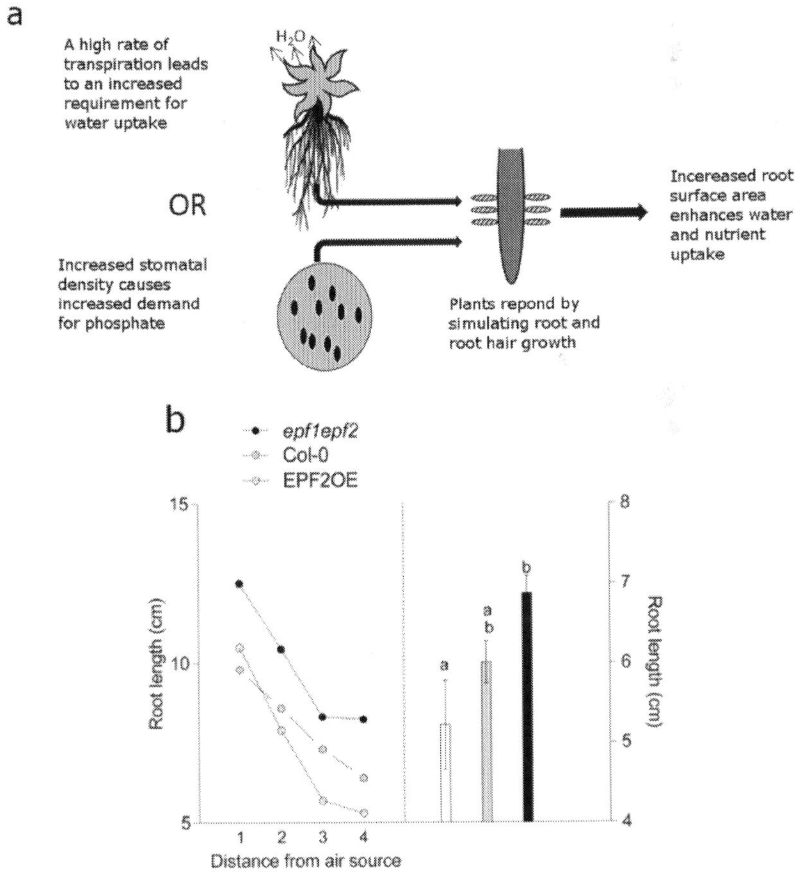

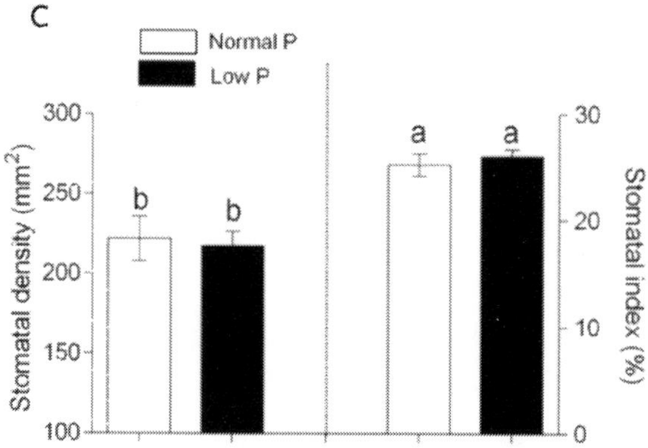


**Figure 4.** The promotion of root hair development is more likely to be due to an increased capacity for water loss rather than an enhanced demand for phosphate. (*A*) Two proposed models to explain the enhanced root development and phosphate accumulation observed in the *epf1epf2* mutant. (*Top*) Increased water loss promotes root hair development. (*Bottom*) increased frequency of stomata incurs a phosphate cost which leads to phosphate stress and the subsequent promotion of root hairs. (*B*, *Left)* A relationship exists between conditions which reduce leaf boundary layer resistance and rooting length. The X axis denotes distance from a moving air source with 1 being the closest. (*B*, *Right*) Plants grown under identical conditions without the presence of a moving air source. A relationship between transpiration/stomatal density and rooting length exists. (*C*) Plants grown under phosphate limiting conditions display normal levels of stomatal density. Bars with no letters in common are significantly different, $P<0.05$, (Tukeys test after one-way ANOVA). Error bars represent SE.

In a separate experiment, plants were grown under two different *P* availabilities, the lower of which significantly limited growth (S1 and S2 Figs). The *D* of plants at growth limiting levels of *P* availability was unaffected indicating that limiting *P* availability does not impair stomatal development in *Arabidopsis thaliana* (Figure 4c). These results suggest that the driving force for increased root growth in high *D* plants is more likely to be to accommodate for high levels of water loss, than to satisfy a demand for high levels of *P* needed for stomatal development and function as previously suggested [2, 18]. The nature of the shoot-to-root, and root-to-shoot, signals which we propose act to balance *E* and water uptake capacities are currently unknown. However plant hormones, including abscisic acid which is well known to modulate root architecture [19–21] stomatal conductance [22] and more recently stomatal development [23,

24] are obvious candidates.

This study has demonstrated that plants manipulated to have high maximal *E* respond by promoting root and root hair growth which enhances *P* accumulation, whereas those manipulated to have a larger rooting area, (and potentially increased nutrient and water uptake capacity), adapt their development to reduce stomatal conductance and *D*. We also show that in the case of *EPF2*OE, it is possible to reduce stomatal development to an extent that water use efficiency is substantially increased [1], nutrient accumulation is not impaired [5] and root hair development is not significantly altered (Figure 2). Moreover, our results indicate that attempts to enhance yield, carbon assimilation or plant cooling through increasing water loss, may indirectly promote root growth, and therefore increase water and nutrient uptake capacities.

# METHODS

## Plant Growth

All plant genotypes were in Col-0 background and have been previously described; stomatal development mutants by Hunt & Gray 2009 and root hair density mutants by Yi et al 2010. Seeds were stratified in the dark for 72 hours at 4°. Plants were grown in environmentally controlled growth chambers (Conviron BDR16) at 22°C/16°C, 9 hours light, 15 hours dark. To examine root hairs, plants were grown on 0.5x MS nutrient agar under sterile conditions. Plants used in Figure 1 were germinated on 0.5x MS agar before being transplanted into autoclaved sand and watered every two days with 20% Long Aston's solution [25] until 21 days old. For growth under low or normal phosphate availabilities (Figure 4c); 20% Long Aston's solution was prepared containing phosphate concentrations of 0.34mM or 0.034mM for normal and low phosphate conditions respectively. Plants were subjected to differing levels of area flow (Figure 4b) by positioning a fan (LLOYTRON F 039 12") 15cm from the first row of plants. Each row contained 1 plant of each genotype at an exact distance of 7cm between each of 4 plant rows. An identical collection of plants were grown in an adjacent position but were shielded from the movement of air. All plants were rotated every day within their rows (but not between rows) and kept at a fixed distance away from the moving air source.

## Stomatal Density

Dental resin (Coltene Whaledent, Switzerland) was applied to the abaxial surface of a fully expanded leaf and left to dry for a minimum of 5mins. Dry dental resin prints were removed from the leaf and nail vanish applied to the resin where the leaf is at its maximum width. Stomatal counts were determined by light microscopy (Olympus BX51) on the nail vanish peels and images recorded with a digital microscope eyepiece (HiROCAM MA88-300A 3.0 Mega Pixels). A minimum of 3 areas per leaf, 3 leaves per plant and 3 plants per genotype were used for analysis.

## Phosphate Accumulation

Holes were made in the lids of 55mm diameter petri dishes; one larger hole so that the roots of the plants could be placed inside the dish with the rosette remaining outside, and a second smaller hole for introducing radioactive orthophosphate into the solution in the base of the dish surrounding the roots. A total of 0.5MBq $H_3PO_4$ (specific activity 148 TBq $mmol^{-1}$) in 1ml $dH_20$ was injected into the smaller hole along with 3ml of 20% long Ashton's solution and both holes were then sealed. At this point the rosette was outside of the petri dish and not in contact with the orthophosphate solution. 5 plants of each genotype and 3 blank controls (dishes containing radioactivity but no plant) were assayed using identical Petri dishes, radioactive solution and sealed in the same manner. After 5 hours, plants were carefully removed from dishes, excess solution was fully blotted onto absorbent paper before material was dried, harvested and digested. Radioactivity and *P* content was quantified by liquid scintillation counting as in Field et al 2012. Readings were corrected for decay and subtracted from blanks before absolute phosphorus was calculated.

## Histochemical Staining

*EPF1::GUS* and *EPF2::GUS* promoter reporter expressing plants (Hunt and Gray 2009) were grown vertically on agar for ten days, removed, placed in X-Gluc reagent (50mM potassium phosphate, 1mM potassium ferrocyanide, 1mM potassium ferricyanide, 0.2% triton x-100, 2mM 5-bromo-4-chloro-3-indolyl-b-d-glucuronic acid and 10mM EDTA) and vacuum infiltrated for 15 minutes. Samples were incubated 37°C in the dark for 3 hours, cleared with 70% ethanol and images captured with an Olympus BX51 microscope connected to a DP51 digital camera.

## Rhizotron Analysis

10 day old seedlings were transferred from 0.5x MS agar plates to rhizotrons consisting of a 150mm diameter petri dish filled with vermiculite, and a rock wool plug placed over the 1cm$^2$holes at the top and bottom of the dish. Glass microfibre paper (Whatman) was placed between the vermiculite and the plastic of the petri dish. Seedlings roots were placed between glass fibre paper and the plastic base of the petri dish with the shoot placed through the hole at the top of the petri dish. The dish was wrapped in tin foil so that the roots were in darkness and positioned vertically. Seedlings were watered biweekly with 0.25x MS solution. Rhizotrons were opened after 5 weeks and the glass fibre disk with roots attached scanned and analyzed using GIMP and ImageJ software.

## Root Hair Measurements

Four Arabidopsis Col-0 plants were sown alongside 4 mutant plants, stratified, and grown vertically on 0.5x MS agar in 120mm x 120mm square Petri dishes. Root hair length and density were determined from a region 1cm above the first root hair from the root tip by light microscopy (Olympus BX51) and images taken with digital microscope eyepiece (HiROCAM MA88-300A 3.0 Mega Pixels) while still attached to the agar plates. At least 3 plates per genotype were measured.

## Infra-red Gas analysis

Measurements were taken with a LI-6400 portable photosynthesis system (Lincoln, NE) on mature, fully expanded leaves of 4 plants per genotype. Relative humidity was kept at 65%-75% using self-indicating desiccant, flow rate was 500μmol s$^{-1}$ and block temperature 20°C. Reference $CO_2$ was set to 450ppm, light intensity to 1000 μmol.m$^{-2}$.s$^{-1}$ and the plant left to equilibrate within the cuvette for at least 45 minutes. Measurements were taken every 30 seconds and the IRGA matched every ten minutes.

## ACKNOWLEDGMENTS

The authors wish to thank Professor Liam Dolan from the University of Oxford for providing the root hair genotypes described in this study.

## AUTHOR CONTRIBUTIONS

Conceived and designed the experiments: CH CT DC JEG. Performed the experiments: CH CT ML. Analyzed the data: CH CT JEG. Contributed reagents/materials/analysis tools: JEG DC. Wrote the paper: CH JEG.

## SUPPORTING INFORMATION

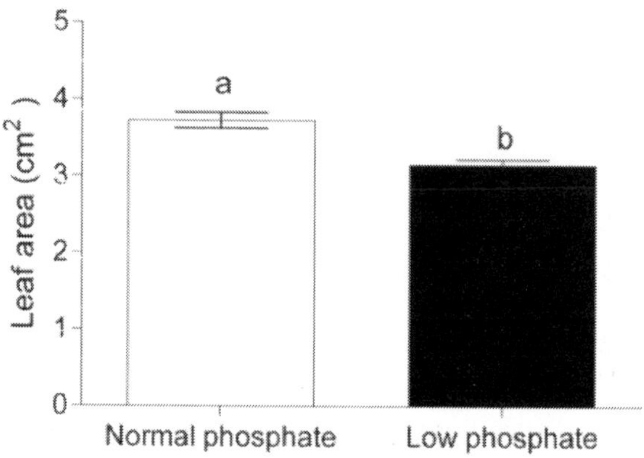


**S1 Figure.** The leaf area of plants grown under non-limiting or limiting concentrations of phosphate.

**S2 Figure.** Digital pictures of plants grown under non-limiting (a) or limiting (b) concentrations of phosphate.

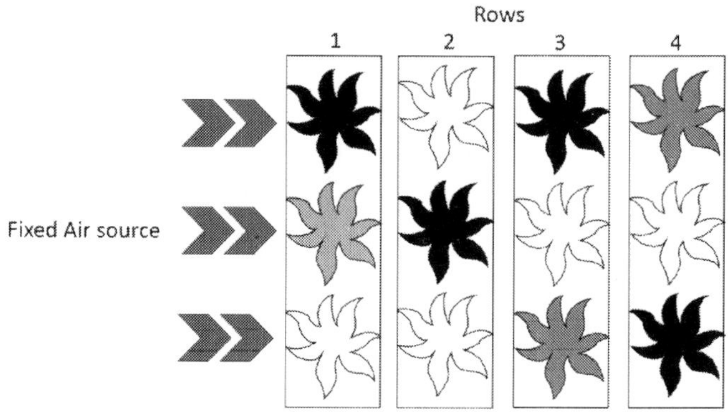


**S3 Figure.** Stylized diagram of the experiment setup reported in Fig 4b.

**S1 Table.** Raw data used to generate the means and errors for the graphs presented in this study

| | Leaf area ($cm^2$) | |
|---|---|---|
| a) | Normal phosphate | Low phosphate |
| | 3.6 | 3 |
| | 4 | 3.1 |
| | 3.9 | 3.2 |
| | 4.1 | 3.4 |
| | 3.8 | 3.1 |
| | 3.7 | 3 |
| | 3.5 | 2.9 |
| | 3.2 | 3.5 |

## REFERENCES

1. Franks PJ, W. Doheny-Adams T, Britton-Harper ZJ, Gray JE. Increasing water-use efficiency directly through genetic manipulation of stomatal density. New Phytologist. 2015;207(1):188–95. pmid:25754246
2. Doheny-Adams T, Hunt L, Franks PJ, Beerling DJ, Gray JE. Genetic manipulation of stomatal density influences stomatal size, plant growth and tolerance to restricted water supply across a growth carbon dioxide gradient. Philosophical Transactions of the Royal

Society of London B: Biological Sciences. 2012;367(1588):547–55. pmid:22232766

3. Dow GJ, Berry JA, Bergmann DC. The physiological importance of developmental mechanisms that enforce proper stomatal spacing in Arabidopsis thaliana. New Phytologist. 2014;201(4):1205–17. pmid:24206523
4. Tanaka Y, Sugano SS, Shimada T, Hara-Nishimura I. Enhancement of leaf photosynthetic capacity through increased stomatal density in Arabidopsis. New Phytologist. 2013;198(3):757–64. pmid:23432385
5. Hepworth C, Doheny-Adams T, Hunt L, Cameron DD, Gray JE. Manipulating stomatal density enhances drought tolerance without deleterious effect on nutrient uptake. New Phytologist. 2015;208(2):336–41. pmid:26268722
6. Kwasniewski M, Daszkowska-Golec A, Janiak A, Chwialkowska K, Nowakowska U, Sablok G, et al. Transcriptome analysis reveals the role of the root hairs as environmental sensors to maintain plant functions under water-deficiency conditions. Journal of Experimental Botany. 2015.
7. Datta S, Kim C, Pernas M, Pires N, Proust H, Tam T, et al. Root hairs: development, growth and evolution at the plant-soil interface. Plant and Soil. 2011;346(1–2):1–14.
8. Yi K, Menand B, Bell E, Dolan L. A basic helix-loop-helix transcription factor controls cell growth and size in root hairs. Nat Genet. 2010;42(3):264–7. http://www.nature.com/ng/journal/v42/n3/suppinfo/ng.529_S1.html. pmid:20139979
9. Menand B, Yi K, Jouannic S, Hoffmann L, Ryan E, Linstead P, et al. An Ancient Mechanism Controls the Development of Cells with a Rooting Function in Land Plants. Science (New York, NY). 2007;316(5830):1477–80.
10. Hara K, Kajita R, Torii KU, Bergmann DC, Kakimoto T. The secretory peptide gene EPF1 enforces the stomatal one-cell-spacing rule. Genes & Development. 2007;21(14):1720–5. PMC1920166.
11. Hara K, Yokoo T, Kajita R, Onishi T, Yahata S, Peterson KM, et al. Epidermal cell density is autoregulated via a secretory peptide, EPIDERMAL PATTERNING FACTOR 2 in Arabidopsis leaves.

Plant & cell physiology. 2009;50(6):1019–31. Epub 2009/05/14. pmid:19435754.

12. Hunt L, Gray JE. The signaling peptide EPF2 controls asymmetric cell divisions during stomatal development. Current biology: CB. 2009;19(10):864–9. Epub 2009/04/29. pmid:19398336.
13. Winter D, Vinegar B, Nahal H, Ammar R, Wilson GV, Provart NJ. An "Electronic Fluorescent Pictograph" Browser for Exploring and Analyzing Large-Scale Biological Data Sets. PloS one. 2007;2(8):e718. pmid:17684564
14. Cui H, Kong D, Wei P, Hao Y, Torii KU, Lee JS, et al. SPINDLY, ERECTA, and Its Ligand STOMAGEN Have a Role in Redox-Mediated Cortex Proliferation in the Arabidopsis Root. Molecular Plant. 2014;7(12):1727–39. PMC4261839. pmid:25267734
15. Czechowski T, Bari RP, Stitt M, Scheible W-R, Udvardi MK. Real-time RT-PCR profiling of over 1400 Arabidopsis transcription factors: unprecedented sensitivity reveals novel root- and shoot-specific genes. The Plant Journal. 2004;38(2):366–79. pmid:15078338
16. Peret B, Clement M, Nussaume L, Desnos T. Root developmental adaptation to phosphate starvation: better safe than sorry. Trends Plant Sci. 2011;16(8):442–50. Epub 2011/06/21. pmid:21684794.
17. Franks PJ, Beerling DJ. Maximum leaf conductance driven by CO2 effects on stomatal size and density over geologic time. Proceedings of the National Academy of Sciences. 2009;106(25):10343–7.
18. Franks PJ, Drake PL, Beerling DJ. Plasticity in maximum stomatal conductance constrained by negative correlation between stomatal size and density: an analysis using Eucalyptus globulus. Plant, cell & environment. 2009;32(12):1737–48.
19. Muraro D, Byrne H, King J, Bennett M. The role of auxin and cytokinin signalling in specifying the root architecture of Arabidopsis thaliana. Journal of Theoretical Biology. 2013;317:71–86. pmid:23026765
20. Harris J. Abscisic Acid: Hidden Architect of Root System Structure. Plants. 2015;4(3):548. pmid:27135341
21. McAdam SA, Brodribb TJ, Ross JJ. Shoot derived abscisic acid promotes root growth. Plant, cell & environment. 2015. Epub 2015/10/31. pmid:26514625.

22. Dodd IC. Abscisic acid and stomatal closure: a hydraulic conductance conundrum? New Phytologist. 2013;197(1):6–8. pmid:23181678
23. Zhang JY, He SB, Li L, Yang HQ. Auxin inhibits stomatal development through MONOPTEROS repression of a mobile peptide gene STOMAGEN in mesophyll. Proc Natl Acad Sci U S A. 2014;111(29):E3015–23. Epub 2014/07/09. pmid:25002510; PubMed Central PMCID: PMCPmc4115497.
24. Chater C, Peng K, Movahedi M, Dunn Jessica A, Walker Heather J, Liang Y-K, et al. Elevated CO-Induced Responses in Stomata Require ABA and ABA Signaling. Current Biology. 2015;25(20):2709–16. pmid:26455301
25. Hewitt. Sand and water culture methods used in the study of plant nutrition. 2nd edn. ed. 2nd edn. London: Commonwealth Agricultural Bureau1966.

CHAPTER

# 13

# LEAF HYDRAULIC CONDUCTANCE DECLINES IN COORDINATION WITH PHOTOSYNTHESIS, TRANSPIRATION AND LEAF WATER STATUS AS SOYBEAN LEAVES AGE REGARDLESS OF SOIL MOISTURE

Anna M. Locke[1] and Donald R. Ort[1,2,3]

[1] Department of Plant Biology, University of Illinois, Urbana, IL 61801, USA

[2] Institute for Genomic Biology, University of Illinois, Urbana, IL 61801, USA

[3] Global Change and Photosynthesis Research Unit, Agricultural Research Service, United States Department of Agriculture, Urbana, IL 61801, USA

## ABSTRACT

Photosynthesis requires sufficient water transport through leaves for stomata to remain open as water transpires from the leaf, allowing $CO_2$ to diffuse into the leaf. The leaf water needs of soybean change over time because of large microenvironment changes over their lifespan, as leaves mature in full sun at the top of the canopy and then become progressively shaded by younger leaves developing above. Leaf hydraulic conductance ($K_{leaf}$), a measure of the leaf's water transport capacity, can often be linked to changes in microenvironment and transpiration demand. In this study, we tested the hypothesis that $K_{leaf}$ would decline in coordination with transpiration demand as soybean leaves matured and aged. Photosynthesis ($A$), stomatal conductance ($g_s$) and leaf water potential ($\Psi_{leaf}$) were also measured at various leaf ages with both field- and chamber-grown soybeans to assess transpiration demand. $K_{leaf}$ was found to decrease as soybean leaves aged from maturity to shading to senescence, and this decrease was strongly correlated with midday $A$. Decreases in $K_{leaf}$ were further correlated with decreases in $g_s$, although the relationship was not as strong as that with $A$. Separate experiments investigating the response of $K_{leaf}$ to drought demonstrated no acclimation of $K_{leaf}$ to drought conditions to protect against cavitation or loss of $g_s$ during drought and confirmed the effect of leaf age in $K_{leaf}$ observed in the field. These results suggest that the decline of leaf hydraulic conductance as leaves age keeps hydraulic supply in balance with demand without $K_{leaf}$ becoming limiting to transpiration water flux.

**Keywords:** Development, drought, leaf age, leaf hydraulic conductance, leaf water potential, photosynthesis, senescence, stomatal conductance.

## INTRODUCTION

Greater than 99% of the water absorbed by a plant's roots is lost to the atmosphere through transpiration. This is an unavoidable consequence of allowing $CO_2$ diffusion into leaves for photosynthesis, but it is also necessary for leaf cooling and plant nutrient uptake. Water requirements change over the plant's and leaf's lifespan, as fluctuating microenvironments around leaves alter transpiration demand on daily and seasonal timescales (Hinckley and Ritchie, 1970; Barrett *et al.*, 1996). Evapotranspiration and stomatal conductance are known to decrease as leaves age in many species (Constable and Rawson, 1980; Sobrado, 1994; Kositsup *et al.*, 2010). A plant's carbon needs and photosynthetic capacity also change throughout development. Photosynthesis ($A$) often declines over the growing season after leaves have reached full expansion. In some cases this decline in $A$ is coordinated with stomatal conductance ($g_s$) and may limit $A$ (Kriedmann *et al.*, 1970; Aslam *et al.*, 1977; Constable and Rawson, 1980; Field and Mooney, 1983; Vos and Oyarzun, 1987; Kositsup *et al.*, 2010). Leaf hydraulic conductance ($K_{leaf}$) is a measure of the efficiency of water transport through the leaf, calculated as water flux through the leaf divided by the water potential driving force. $K_{leaf}$ is dynamic, depending on variable aquaporin expression and activation (Maurel *et al.*, 2008) as well as cavitation and embolism refilling in the xylem (Canny, 1997; McCully *et al.*, 1998; Holbrook *et al.*, 2001). The relative impacts of these $K_{leaf}$ regulators, particularly embolism and refilling in the xylem, are not fully understood and are under intense study (Wheeler *et al.*, 2013; Scoffoni and Sack, 2014). Because liquid water transport through the leaf is critical to maintain open stomata for $CO_2$ acquisition, $K_{leaf}$ is strongly linked with $A$, and $K_{leaf}$ and maximum photosynthetic capacity are correlated across many species (Brodribb *et al.*, 2005; Sack and Holbrook, 2006).

Given its link with $A$ and transpiration, hydraulic conductance is expected to change as leaves age. $K_{leaf}$ decreased as leaves age in several evergreen and deciduous tree species (Salleo *et al.*, 2002; Lo Gullo *et al.*, 2005). This decline is, in some cases, linked with photosynthetic parameters (Brodribb and Holbrook, 2003b). Xylem blockage by tyloses that progressively decrease $K_{leaf}$ may also be a component of senescence (Cochard and Tyree, 1990; Salleo *et al.*, 2002). However, the dynamics of $K_{leaf}$ with leaf age over a growing season have yet to be examined in any herbaceous or annual species, such as *Glycine max* (soybean). Soybean leaves experience much more dramatic microenvironment changes over their lifespan than most tree leaves, as soybean leaves mature in full sun and quickly become shaded and shielded from wind and precipitation by newer leaves above in a dense canopy, and both $A$ and $g_s$ have been shown to decrease as soybean leaves age (Woodward and Rawson, 1976; Reich *et al.*, 1985; Burkey and Wells, 1991). Thus, if a season-long $K_{leaf}$ decline exists, it may be even more drastic than in these tree species, and more likely to become limiting to transpiration and thereby photosynthesis.

We hypothesized that $K_{leaf}$ and transpiration demand in soybean would remain in balance as leaves aged. This was tested with field-grown and chamber-grown soybean. Because photosynthesis is dependent on water transport through the leaf, a decrease in $K_{leaf}$ could indicate a hydraulic component to loss of photosynthetic capacity and leaf senescence. Thus, this study tests whether photosynthesis in older leaves could be hydraulically limited, which would lead to lower total canopy photosynthesis. Although similar studies have been carried out in tree species, the potential limitation of canopy photosynthesis by $K_{leaf}$

in older leaves has to our knowledge not yet been examined in any key crop species.

Drought is the main yield-reducing environmental stress facing crops (Boyer, 1982), and rising greenhouse gas concentrations exacerbate this stress by altering global climate patterns, which is expected to increase the frequency of extreme weather events, including drought (Burke *et al.*, 2006; Meehl *et al.*, 2007). During severe drought, tracheary element cavitation is likely to occur at high xylem tensions, reducing $K_{leaf}$ (Machado and Tyree, 1994; Meinzer, 2002), although genotypic differences in hydraulic properties among cultivars can affect drought tolerance within a species (Silva *et al.*, 2004; Sadok and Sinclair, 2010a). In some species stomata show a direct decrease in $g_s$ in response to leaf water potential ($\Psi_{leaf}$), while $K_{leaf}$ does not decrease until a threshold $\Psi_{leaf}$ is reached, thereby delaying extensive vessel cavitation (Nardini and Salleo, 2000; Cochard, 2002; Brodribb and Holbrook, 2003a). However, $K_{leaf}$ decreases with soil drying in many woody and herbaceous species (Linton and Nobel, 2001; Brodribb and Holbrook, 2003a; Lo Gullo *et al.*, 2003; Blackman *et al.*, 2009; Ferrio *et al.*, 2012). $K_{leaf}$ also declined with $\Psi_{leaf}$ across a range of deciduous and evergreen trees and shrubs (Nardini *et al.*2001; Brodribb and Holbrook 2006; Johnson et al. 2011, 2012; Guyot *et al.* 2012; Bucci *et al.* 2012). Similarly, vapour pressure deficit-induced xylem cavitation resulted in stomatal closure for *Laurus nobilis* L. plants grown in constantly wet soil (Salleo *et al.*, 2000).

As elevated [$CO_2$] generally decreases stomatal conductance, it could protect the plant from drought by conserving soil moisture as well as slowing the decrease of $\Psi_{leaf}$ under conditions of limited water or high vapour pressure deficit (Allen *et al.*, 1998, 2003; Leakey *et al.*, 2006). Elevated [$CO_2$] has been observed to decrease hydraulic conductance of either whole plants or leaves in several species (Bunce, 1996; Bunce and Ziska, 1998; Domec *et al.*, 2009). However, we have previously observed a lack of $K_{leaf}$ acclimation to growth at elevated [$CO_2$] for field-grown soybean (Locke *et al.*, 2013) while stomatal conductance consistently decreased (Leakey *et al.*, 2006). Thus, restricted transpiration during growth at elevated [$CO_2$] could protect against $K_{leaf}$ decline during drought.

In addition to investigating leaf age effects on $K_{leaf}$ we tested the hypotheses that soybean $K_{leaf}$ will decline during drought and that growth at elevated [$CO_2$] will protect leaves from experiencing this decline. These experiments were essential for interpreting $K_{leaf}$ data in leaf-age-targeted field experiments during drought years. Drought experiments were conducted both in the field under open air conditions as well as in environmentally controlled growth chambers. Because of the link between leaf hydraulics and gas exchange, measuring the responses of $K_{leaf}$ to declines in soil moisture could help predict hydraulic limitation to photosynthesis during drought.

# MATERIALS AND METHODS

## Leaf Age Experiments

For the leaf age field experiment, soybean cultivar 93B15 (maturity group III) with indeterminate growth (Pioneer Hi-Bred, Johnston, IA, USA) was planted at the Soybean Free Air Concentration Enrichment (SoyFACE) facility in Savoy, IL, USA, on 8 June 2011. The experiment was conducted in six 6 m × 6 m blocks. Soybean was grown in yearly rotation with *Zea mays* (corn) according to standard agricultural practice in central Illinois, USA. Rows were spaced 0.76 m apart, and rows were thinned to one plant per 25cm (52400 plants/ha) when seedlings reached developmental stage VC. Leaves at the third and tenth nodes from the ground were marked with flagging tape tied around their petioles as they emerged.

This field experiment was repeated in 2013, when soybean cultivar 93Y40 (maturity group III) was planted on 6 June. A single plot, measuring 10 m × 15 m, was again thinned to a density of one plant per 25cm (52400 plants/ha). Leaves at the third and tenth nodes were tagged for sampling throughout the growing season.

These field experiments were not irrigated, as is standard practice for soybean farming in central Illinois. This region experienced a progressively worsening drought during the 2011 growing season, and less severe drought but nonetheless progressively drying soils again in 2013. In addition to conducting separate drought experiments, the leaf age experiment was repeated in growth chambers with consistently well-watered plants to confirm that observed results were attributable to leaf age and not to drought.

For the leaf age growth chamber experiment, soybean cultivar PI 154197 (maturity group 00) with determinate growth (Pioneer Hi-Bred) was planted in 14 l pots on 11 January 2013. This cultivar was selected to ensure that the plants did not outgrow the growth chambers. Twelve plants were grown in each of eight growth chambers. Chamber conditions from the time of seed planting were 25 °C, 60% relative humidity and approximately 1000 μmol $m^{-2}s^{-1}$ daytime photosynthetically active radiation (PAR). Plants were fertilized every other day with 50% Long-Ashton solution, amended with 10mM $NH_4NO_3$ (Hewitt, 1966). Leaves at the fifth and eighth nodes from the soil were tagged as they emerged. Plants were rotated within the growth chambers every 2 days and among the growth chambers every 4 days to minimize differential chamber effects.

Vapour pressure deficit was 0.6 kPa lower on average in the growth chambers than at midday above the canopy in the field on all but the final day of field measurements (data not shown). However, older, lower-canopy leaves in the field likely experienced greater uncoupling from atmospheric conditions than lower-canopy leaves in the growth chamber, because growth chamber fans mixed air throughout the entire chamber, keeping lower leaves at a similar temperature and relative humidity as upper canopy leaves even when the leaves were shaded. Thus, direct comparisons between $K_{leaf}$ values in the chamber and in the field are not valid, but differences among the growth chamber measurements should represent endogenous patterns in the plant, which are also reflected in field $K_{leaf}$ measurements.

$K_{leaf}$, $\Psi_{leaf}$ and leaf gas exchange were measured at three developmental stages for leaves at each tagged node: when the leaf was the youngest mature leaf on the plant, at the top of the canopy (stage A); when the leaf was older and shaded but still fully green (stage B); and when the leaf had visibly begun to senesce, considered to be at least 50% yellowed (stage C). Two sets of stage B measurements were taken in the 2013 field experiment, termed B1 and B2. In the chamber experiment measurements were only taken for stages A and B, because a fungal infection in the leaves prevented measurements during senescence. Leaves were assumed to be 10 days old when they reached maturity (stage A); stage B measurements were taken at 19–43 days old in the field experiments and at 23–24 days old in the chamber experiment; and stage C measurements were taken at

54–76 days old in the field experiments.

## Gas Exchange

$A$ and $g_s$ were measured with a LI-6400 photosynthesis system equipped with a leaf chamber fluorometer (LI-COR Inc., Lincoln, NE, USA). Measurements were taken between 12:00 and 14:00, as this time typically corresponds to peak $A$. [$CO_2$] was 400 ppm and relative humidity was maintained between 50 and 70% in the cuvette for all measurements. Light and block temperature were set to the ambient temperatures experienced by the leaf (Supplementary Table 1). These leaves were tagged after measurement so that the exact same leaves could be sampled for $K_{leaf}$ measurement the next morning. Intrinsic water-use efficiency (WUE) was calculated as $A/g_s$.

To determine if declining $A$ was merely the result of measurement at decreasing ambient light intensity rather than a down-regulation of photosynthetic capacity, light curves were measured on field-grown leaves at all growth stages in 2013. Leaves were excised before sunrise and their petioles were re-cut under water. $A$ was measured over a range of PAR levels from 2000 to 0 μmol $m^{-2}$ $s^{-1}$ for each leaf.

## Leaf Water Potential

Tissue was harvested for measurement of midday $\Psi_{leaf}$ with thermocouple psychrometers (C30, Wescor Inc., Logan, UT, USA) at the same time as gas exchange measurements were taken. In the field experiment, four leaves were sampled per block; in the chamber experiment, three leaves were sampled per chamber. For each leaf, three 1.2cm discs were removed and sealed into a steel chamber with the thermocouple psychrometer within 15 s of sampling. These chambers were allowed to equilibrate to 25 °C for 2.5–3h before leaf water potential was recorded by a datalogger (CR-7, Campbell Scientific, Logan, UT, USA). Leaf water potentials were then calculated based on a sucrose calibration performed with the psychrometers prior to the experiment.

In the growth chamber experiment, leaf osmotic potential ($\Psi_{\Pi}$) was measured subsequent to the water potential determinations. Following the $\Psi_{leaf}$ measurements, the steel psychrometer chambers were held in liquid nitrogen for 60 s to lyse the cells and eliminate cell wall turgor pressure. The chambers were then thawed overnight to re-equilibrate

$$\Psi_P = \Psi_{leaf} - \Psi_\Pi$$

## Leaf Hydraulic Conductance

$K_{leaf}$ was measured using the evaporative flux method (Sack *et al.*, 2002; Locke *et al.*, 2013). In this method, water flux through the leaf is measured while the leaf is placed in an environment favorable to transpiration. Leaves were harvested pre-sunrise in the field and before morning growth lights turned on in chamber experiments to ensure that as much embolism refilling as possible had occurred overnight. Leaves were cut with a razor blade at the base of the petiole and immediately placed in distilled water. Petioles were re-cut 2cm shorter under water upon return to the laboratory to remove major cavitation introduced during sampling; 2cm is sufficient to remove introduced embolism, as average vessel length in soybean petioles is less than 1mm (Ghorashy *et al.*, 1969). Leaves which were not sufficiently re-cut typically wilted quickly upon connection to the evaporative flux apparatus and were not included in the analysis. For water flux measurements, petioles were connected to tubing (Tygon R-3693, Saint-Groban Performance Plastics Corporation, Aurora, OH, USA) that led to a reservoir of water on a high-precision balance (± 0.01mg; XS 250, Mettler Toledo, Columbus, OH, USA). Crevices in the petioles were filled with petroleum jelly, and petioles were wrapped with Parafilm (Pechiney Plastic Packaging Company, Chicago, IL, USA) to ensure a tight seal with the tubing. Leaves were illuminated with approximately 700 μmol $m^{-2}$ $s^{-1}$ PAR from a 750W halogen lamp, with a clear water dish directly below the lamp to dissipate heat and a fan blowing on the leaf to reduce the leaf boundary layer. While 700 μmol $m^{-2}$ $s^{-1}$ PAR is usually not photosynthetically saturating for a soybean leaf, it is high enough to stimulate transpiration, and the consistent light level across all measurements ensured that comparisons among leaves are valid. The change in water mass was logged every 30 s by a datalogger (CR1000, Campbell Scientific) simultaneously for four balances, and flow rates were monitored on a single computer. Flow rate typically stabilized in 30–60min, at which point the leaf temperature was recorded (572 Handheld Infrared Thermometer, Fluke Corporation, Everett, WA, USA). Transpiration was sufficient to keep the leaf temperature 1–4 °C lower than ambient temperature (data not shown).

$\Psi_{leaf}$ was measured with thermocouple psychrometers as described above. Four psychrometers were used per trifoliate leaf, with three leaf discs per psychrometer chamber. Leaf margins were left intact so that leaves could be photographed, and leaf area was calculated using the freeware ImageJ (http://imagej.nih.gov/ij/).

To calculate $K_{leaf}$, flow rate was divided by $\Psi_{leaf}$ of the leaf during the $K_{leaf}$ measurement and leaf area. This value was temperature-normalized to account for the viscosity of water, which decreases approximately 2% per 1 °C increase (Yang and Tyree, 1993).

## Drought Experiments

A field experiment was conducted in Drought by Rain Interception-FACE (DRI-FACE) plots at the SoyFACE facility in 2010, which independently tested whether drought affects $K_{leaf}$. Soybean cultivar 93B15 (Pioneer Hi-Bred) was planted on 27 May 2010 in 38cm row spacing, and $CO_2$ fumigation began on 9 June 2010 and continued through senescence. $CO_2$ was fumigated with a target of 585 ppm in elevated-[$CO_2$] plots as described by Leakey *et al.* (2004). The DRI treatment was implemented with retractable 4.6 m × 9.2 m rain-interception awnings placed within the ambient- and elevated-[$CO_2$] plots. The awnings were controlled by a computer and deployed automatically when precipitation was detected by rain sensors and PAR was below 50 μmol $m^{-2}s^{-1}$, as described in detail by Gray *et al.* (2013). This low light threshold ensured that at most 0.05% of growing-season PAR was intercepted by the awnings (Gray *et al.*, 2013). Intercepted rain was diverted 20 m away from the reduced precipitation (RP) plots by gutters. This rain interception treatment resulted in a persistent and progressively increasing disparity between control precipitation (CP) and RP plots over the course of the growing season. Ambient- and elevated-[$CO_2$] treatments were applied in a randomized complete block design with four blocks, while precipitation treatments were applied as a split plot within the ambient- and elevated-[$CO_2$] plots. $K_{leaf}$ was measured on uppermost, fully mature leaves on 3 days over the course of the growing season.

For a drought experiment in growth chambers, soybeans were planted in 14 l pots. Twelve plants were grown in each of four growth chambers (GC-15, Environmental Growth Chambers, Chagrin Falls, OH, USA), with six pots assigned to drought treatment and six treated as controls.

Chamber conditions from the time of seed planting were 25 °C, 60% relative humidity and approximately 1000 μmol $m^{-2}$ $s^{-1}$ daytime PAR. Plants were fertilized every other day with 50% Long-Ashton solution, amended with 10mM $NH_4NO_3$(Hewitt, 1966). Plants were rotated within the growth chambers every 2 days and among the growth chambers every 4 days to minimize differential chamber effects.

Drought treatment was created by withholding water for a period of 4–5 days, until drought plants began to visibly lose turgor. Control pots were watered as normal during dry-down periods. $K_{leaf}$, $g_s$ and $\Psi_{leaf}$ measurements were taken at the end of each dry-down period, and then all plants were re-watered. Dry-down periods were spaced 6 days apart to allow sufficient time for plants to re-hydrate and for a new leaf to mature in well-watered conditions. $K_{leaf}$ and midday $\Psi_{leaf}$ were measured as described above.

For the drought growth chamber experiment, $g_s$ was measured with a steady-state diffusion porometer (model SC-1, Decagon Devices, Pullman, WA, USA). The instrument was allowed to equilibrate to growth chamber conditions for at least 30min before measurement, and measurements were taken on plants while inside the growth chamber. Abaxial conductance was measured on the uppermost fully expanded leaf for 12 plants per treatment.

### Statistical Analyses

Differences among growth stages were analyzed by repeated measures with the SAS MIXED procedure (SAS Inc., Cary, NC, USA). Node and growth stage were treated as fixed effects, and plots and chambers were considered random blocking effects. Correlations between $K_{leaf}$ and gas exchange parameters were tested using the REG procedure.

## RESULTS

### $K_{leaf}$ Decreases as Soybean Leaves Age

In field-grown (Figure 1A, B) and in chamber-grown (Figure 1C) soybean, $K_{leaf}$ changed as leaves aged ($P < 0.0001$). Leaf stages at measurement are denoted as stage A (youngest, fully mature leaf at the top of the canopy), stage B (older, fully green) and stage C (visibly senescent).

The decrease in $K_{leaf}$ appeared to be consistent for field-grown plants in 2011 (Figure 1A) and chamber-grown plants (Figure 1C), while for field-grown plants in 2013 a decrease was either not apparent until the end of the season, as for node 3, or was interrupted by peak $K_{leaf}$ during stage B, as for node 10 (Figure 1B). For field-grown soybean, $K_{leaf}$ decreased by 56–76% from stage A to stage C. $K_{leaf}$ decreased more rapidly over time for the determinate cultivar in the growth chamber experiment than the indeterminate cultivars grown in the field (Figure 2), consistent with the shorter maturity group (00) grown in the chamber experiments versus the field experiments (maturity group III). $K_{leaf}$ at leaf stage A was lower at upper canopy nodes in the 2011 field experiment and in growth chambers, but this was not observed in the 2013 field experiment. While $K_{leaf}$ values at stages B and C were comparable across experiments all three experiments, despite different genotypes and growing conditions, $K_{leaf}$ at stage A was substantially lower in the 2013 field experiment, causing the less consistent season-long decrease in $K_{leaf}$.

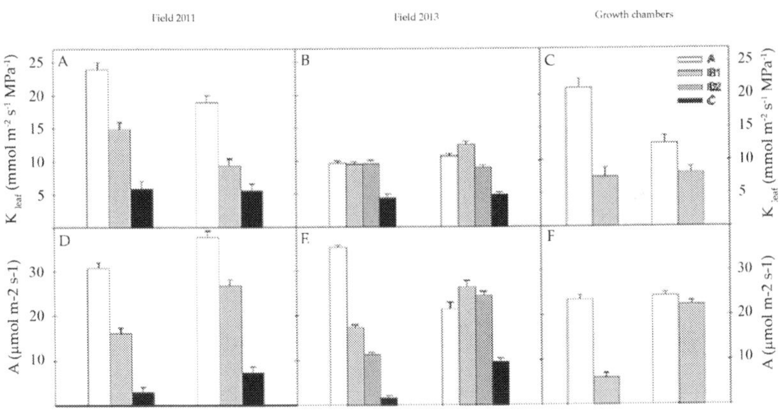

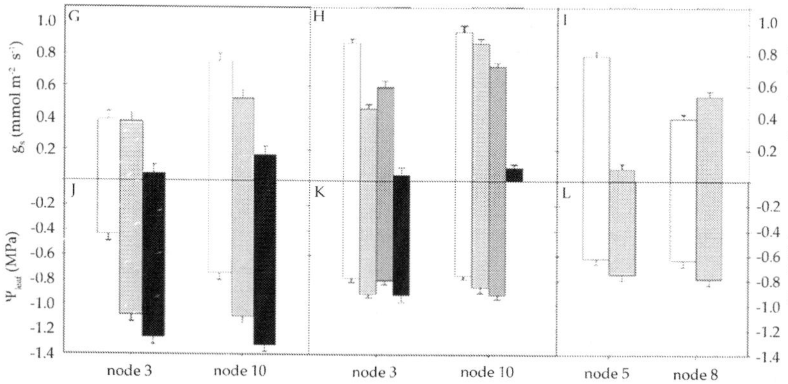


**Figure 1.** $K_{leaf}$, $A$, $g_s$ and $\Psi_{leaf}$ for each measured leaf stage in all experiments (2011 field experiment, A, D, G, J; 2013 field experiment, B, E, H, K; growth chamber experiment, C, F, I, L). $K_{leaf}$ (A–C) was measured with the evaporative flux method for leaves sampled before sunrise, $A$ (D–F) and $g_s$ (G–I) were measured with a LI-COR 6400 Portable Photosynthesis System and midday $\Psi_{leaf}$ (J–L) was measured with thermocouple psychrometers. When two stage B measurements were taken, the earlier stage B measurement is represented by light grey and the later measurement is represented by dark grey. Error bars represent standard error.

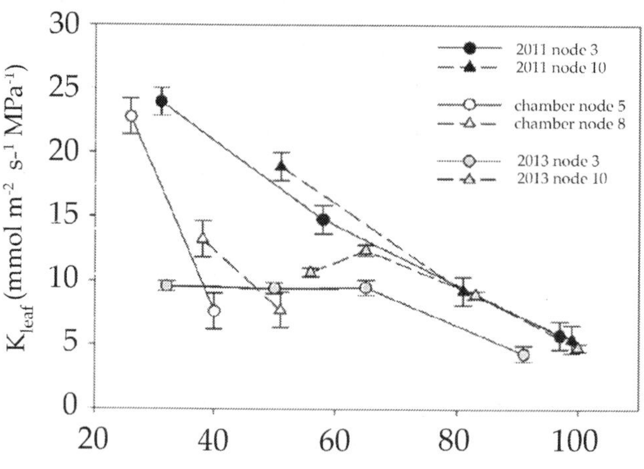


**Figure 2.** $K_{leaf}$ represented by days after planting. 2011 field experiment, closed symbols; growth chamber experiment, open symbols; 2013 field experiment,

grey symbols. Circles and triangles denote lower and upper nodes measured. Error bars represent standard error.

## Photosynthetic Capacity Decreased Coordinately with $K_{leaf}$ as Leaves Aged

*A* was measured at midday, around the time of peak diurnal photosynthesis. This evaluation of maximum photosynthesis along with maximum (pre-sunrise) $K_{leaf}$ allows for the examination of season-long trends in leaf hydraulic and photosynthetic capacities. *A* also decreased consistently as leaves aged in all three experiments ($P < 0.0001$), except for at node 10 in the 2013 field experiment (Figure 1D–F). The pairwise decrease was small ($P = 0.20$) for leaves at node 8 in the growth chamber experiment, which did not become shaded by younger leaves as they aged as a result of determinate growth. $K_{leaf}$ and *A* were significantly correlated across leaf ages for all nodes measured in all experiments, although strength of these relationships varied widely among nodes, with $R^2$ varying from 0.13 to 0.62 (Table 1, Figure 3A). In 2013, the light response of photosynthesis was measured at each leaf stage to determine if lower photosynthetic rates in older leaves is simply the result of low light beneath the canopy not maximizing photosynthetic capacity, or if photosynthetic capacity was actually down-regulated in older leaves. Light saturated photosynthesis was found to decrease consistently as the leaves aged for both node 3 (Figure 4A) and node 10 (Figure 4B). Leaves at node 3, stage A, reached a maximum photosynthetic rate of 29.9 μmol $m^{-2}$ $s^{-1}$ at 2000 μmol $m^{-2}$ $s^{-1}$ PAR, while maximum photosynthesis for the same leaves at stage C was only 5.2 μmol $m^{-2}$ $s^{-1}$ at 500 μmol $m^{-2}$ $s^{-1}$ PAR.

**Table 1.** Correlation coefficients for the relationships between $K_{leaf}$ and A or $g_s$ across the entire growing season for each node in every experiment

| | Node | $R^2$ | |
|---|---|---|---|
| | | $K_{leaf}$ vs $A$ | $K_{leaf}$ vs $g_s$ |
| Field 2011 | 3 | 0.61*** | 0.30*** |
| | 10 | 0.62*** | 0.53*** |
| Field 2013 | 3 | 0.23*** | 0.34*** |
| | 10 | 0.59*** | 0.71*** |

| Growth chamber | 5 | 0.51*** | 0.59*** |
|---|---|---|---|
| | 8 | 0.13* | n.s. |

Leaf gas exchange was measured at midday, and these exact same leaves were sampled for $K_{leaf}$ measurement before sunrise the next morning. Asterisks indicate the significance of the correlation (*$P < 0.05$, ***$P < 0.0001$; n.s., not significant).

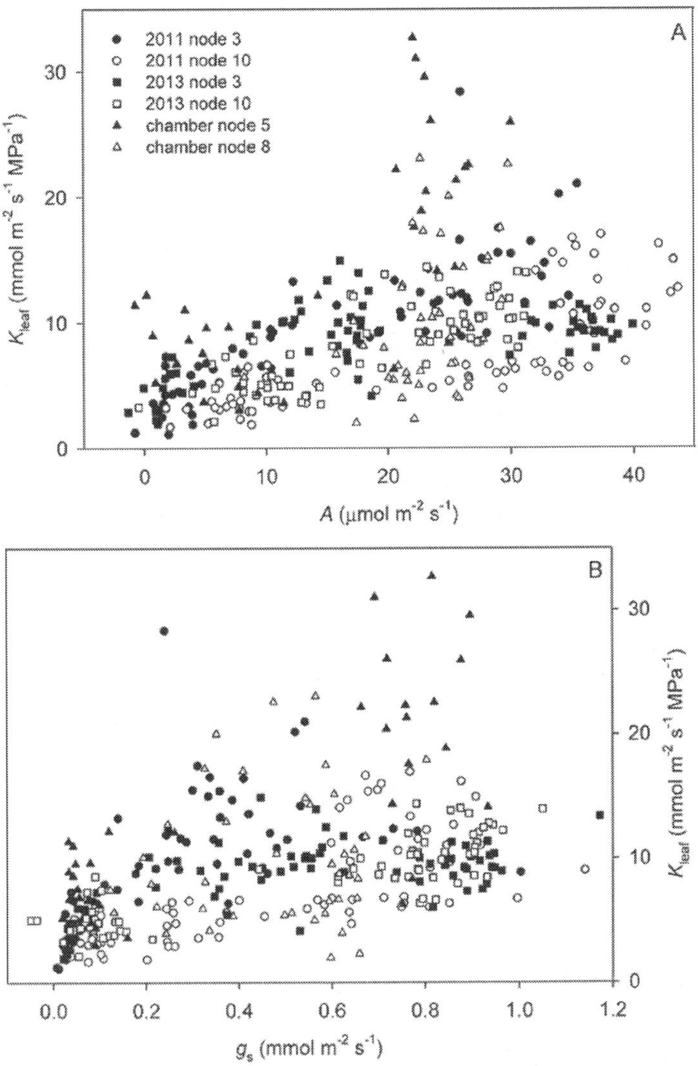


**Figure 3.** $K_{leaf}$ and $A$ (A) and $K_{leaf}$ and $g_s$ (B) for all leaves measured across the growing season. $A$ and $g_s$ were measured at midday, and maximum $K_{leaf}$ was

measured for the exact same leaves sampled before sunrise the next morning. Measurements at all leaf ages are included for the experiment/node combinations shown. Regressions were calculated separately for each experiment/node, and the correlation statistics for $K_{leaf}/A$ and $K_{leaf}/g_s$ relationships at each node are listed in Table 1.

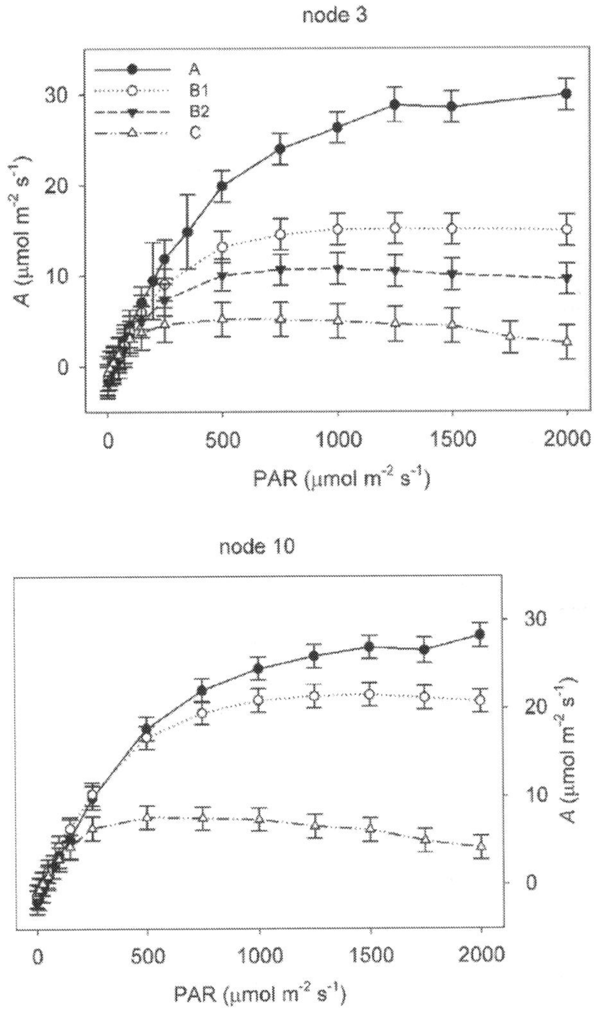


**Figure 4.** Changes in leaf photosynthetic capacity as leaves age. $A$ was measured at a range of PAR from 2000 to 0 µmol m$^{-2}$ s$^{-1}$ for leaves at each stage. Leaves were excised before sunrise and light-acclimated before measurement. Measurements were taken at stage A (closed circles), stage B/B1 (open circles), stage B2 (closed triangles), and stage C (open triangles). Error bars represent standard error.

Stomatal conductance decreased overall as leaves aged for field-grown soybean in 2011 ($P < 0.0001$, Figure 1G). For chamber-grown soybean (Figure 1I), however, $g_s$ decreased 90% from stage A to stage B for node 5 (pairwise $P < 0.0001$), while $g_s$ increased 36% from stage A to stage B for node 8 (pairwise $P < 0.01$). The increase or decrease in $g_s$ between specific stages did not as closely follow the patterns of $K_{leaf}$ as those of $A$ did, and the correlations between $K_{leaf}$ and $g_s$ were thus sometimes weaker than correlations between $K_{leaf}$ and $A$ (Table 1). In field-grown soybean, $K_{leaf}$ correlated with $g_s$ for both node 3 ($R^2 = 0.30$) and node 10 ($R^2 = 0.53$), but these correlations were not as strong as those between $K_{leaf}$ and $A$ (Table 1, Figure 3B). $K_{leaf}$ and $g_s$ were not correlated for chamber-grown soybean (Table 1, Figure 3B).

WUE was calculated from gas exchange data and did not change across experiments as leaves aged in either field- or chamber-grown soybean. Contrasts within nodes showed that WUE differs among stages ($P < 0.0001$ for all nodes), but the direction of these changes was not consistent and the significance does not hold across either experiment. $K_{leaf}$ and WUE were not correlated for field- or chamber-grown soybean (data not shown).

## $\Psi_{leaf}$ Declines as Soybean Leaves Age and Is Driven By Decreasing Osmotic Potential

$\Psi_{leaf}$ decreased as leaves aged for field-grown soybean in 2011 ($P < 0.0001$) and chamber-grown plants ($P < 0.0001$), but it did not change significantly as leaves aged in the 2013 field experiment (Table 2, Figure 1J–L). From stage A to stage B in the field experiment, $\Psi_{leaf}$ decreased 0.65MPa at node 3 and 0.35MPa at node 10. From stage A to stage C, $\Psi_{leaf}$ decreased 0.83MPa at node 3 and 0.58MPa at node 10. For chamber-grown soybeans, $\Psi_{leaf}$ decreased 0.13MPa at node 5 and 0.15MPa at node 8. In the 2013 field experiment and the growth chamber experiment, $\Psi_{\Pi}$ and $\Psi_{P}$ were also measured to determine what was driving changes in $\Psi_{leaf}$. In growth chambers, $\Psi_{\Pi}$ decreased as leaves aged as did $\Psi_{leaf}$, while $\Psi_{P}$ remained steady across the growing season. $\Psi_{\Pi}$ decreased by 0.11MPa at node 5 and by 0.19MPa at node 8 ($P < 0.001$). In the 2013 field experiment, osmotic potential also changed significantly as leaves aged, but the decrease was only steady at node 10, while $\Psi_{\Pi}$ actually increased from stages B1 to B2 and B2 to C at node 3 (Table 2).

**Table 2.** Mean values for leaf water potential, osmotic potential and turgor pressure ± standard error for the 2013 field experiment and the growth chamber experiment

| Experiment | Node | Stage | Water potential | Osmotic potential | Turgor pressure |
|---|---|---|---|---|---|
| Field 2013 | 3 *$\Psi_{\Pi}$ | A | −0.78±0.03 | −1.00±0.03 | 0.22±0.03 |
| | | B1 | −0.91±0.03 | −1.08±0.04 | 0.22±0.03 |
| | | B2 | −0.80±0.04 | −0.97±0.04 | 0.17±0.04 |
| | | C | −0.92±0.06 | −0.93±0.11 | 0.09±0.09 |
| | 10 *$\Psi_{\Pi}$ | A | −0.76±0.03 | −0.96±0.03 | 0.20±0.02 |
| | | B1 | −0.85±0.05 | −0.99±0.05 | 0.14±0.03 |
| | | B2 | −0.92±0.03 | −1.08±0.05 | 0.16±0.04 |
| Growth chambers | 5 | A | −0.62±0.04 | −0.79±0.05 | 0.17±0.03 |
| | | B | −0.75±0.05 | −0.90±0.05 | 0.15±0.04 |
| | 8 | A | −0.63±0.05 | −0.93±0.05 | 0.29±0.04 |
| | | B | −0.79±0.05 | −1.12±0.05 | 0.35±0.04 |

Asterisks indicate a significant change ($P < 0.05$) in a parameter for a particular node.

## Soybean $K_{leaf}$ Does Not Acclimate to Drought

In the field, DRI awnings intercepted 41% of growing-season precipitation, resulting in soil moisture decreases of up to 50%, as reported in detail by Gray *et al.* (2013). $K_{leaf}$ for field-grown soybean decreased significantly over the course of the growing season when measured on youngest fully expanded leaves ($P < 0.0001$) (Figure 5). Pre-dawn $K_{leaf}$ decreased from an average across treatments of 15.4 mmol m$^{-2}$ s$^{-1}$ MPa$^{-1}$ at 60 days after planting to 10.8 and 5.9 mmol m$^{-2}$ s$^{-1}$ MPa$^{-1}$ at 76 and 97 days after planting (Figure 5A, B). Therefore, treatment effects were analyzed separately for each measurement day. $K_{leaf}$ was not affected by persistent drought on any measurement day, and this was the case at both ambient (Figure 5A) and elevated [$CO_2$] (Figure 5B). $K_{leaf}$ was lower for elevated-[$CO_2$] plants than for ambient-[$CO_2$] plants on day 76 ($P = 0.0208$; Figure 5A, B), but [$CO_2$] did not affect $K_{leaf}$ on the other two measurement days. On day 76, the difference in $K_{leaf}$ between [$CO_2$] treatments was primarily driven by the control precipitation plots, which had the highest $K_{leaf}$ of all four treatments at ambient [$CO_2$] and the lowest $K_{leaf}$ of all four treatment combinations at elevated [$CO_2$], whereas there was no difference between $K_{leaf}$ for ambient and elevated [$CO_2$] in drought plants.

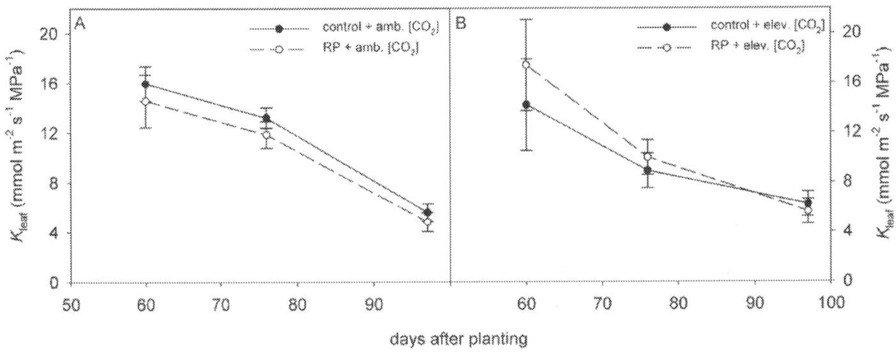


**Figure 5.** $K_{leaf}$ for plants grown in RP plots (closed circles) compared to plants grown in control precipitation plots (open circles). Precipitation treatments were conducted under both ambient [$CO_2$] (385 ppm; A) and elevated [$CO_2$] (585 ppm; B); panels are separated for clarity. $K_{leaf}$ was measured on uppermost, fully expanded leaves sampled before sunrise in the field.

In the growth chamber experiment, by withholding water for 4–5 days during each drought cycle, volumetric soil moisture was decreased by an average of 62% on day 38 and 66% on day 49 in the drought pots compared to control pots. This soil moisture deficit was sufficient to significantly decrease $\Psi_{leaf}$ by 33% on day 38 ($P = 0.0213$) and 50% on day 49 ($P = 0.0546$) (Figure 6C). Stomatal conductance ($g_s$) was 24 and 66% lower on days 38 and 49 in drought than in control plants, a response to soil drying and $\Psi_{leaf}$ (Figure 6B). The $g_s$ decrease was only significant at $\alpha = 0.01$ on day 49 ($P = 0.0661$). However, despite declines in soil moisture, $\Psi_{leaf}$ and $g_s$, $K_{leaf}$ in drought plants was not different from $K_{leaf}$ in control plants on either day ($P = 0.37$ and $P = 0.95$) (Figure 6A), although $K_{leaf}$ for both treatments was higher on day 49 than on day 38.

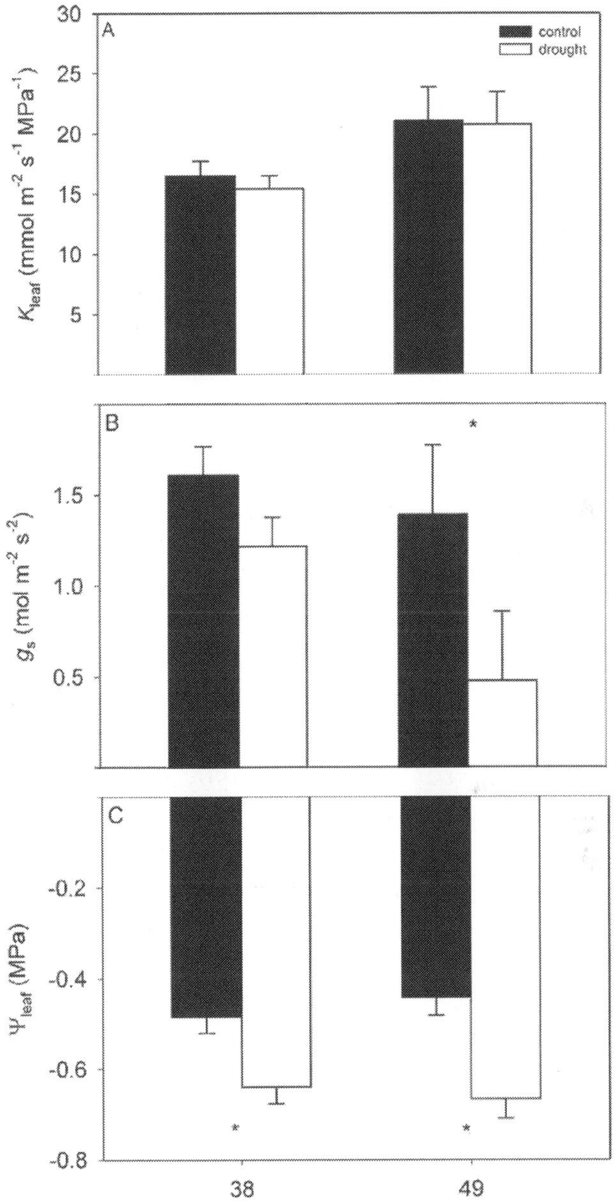


**Figure 6.** Pre-sunrise $K_{leaf}$ (A), midday $g_s$ (B) and midday $\Psi_{leaf}$ (C) for drought and control plants grown in growth chambers. Measurements were taken at the end of a 4–5 day period during which water was withheld from drought treatment plants, while control plants were watered as usual. All measurements were made on uppermost, fully expanded leaves. $K_{leaf}$ was measured with the evapo-

rative flux method, $g_s$ was measured with a porometer and $\Psi_{leaf}$ was measured with thermocouple psychrometers. Asterisks indicate a significant difference between drought and control treatments. Error bars represent standard error.

## DISCUSSION

Field and growth chamber data both showed a trend of decreasing $K_{leaf}$ as soybean leaves age, although the decrease was not always consistent over the course of the plant's lifespan. Similar $K_{leaf}$ behavior observed in well-watered growth chamber plants as compared to field-grown plants and the absence of a $K_{leaf}$ response to deliberately manipulated soil moisture in separate experiments support the conclusion that the observed declines in $K_{leaf}$ in leaf age experiments were linked to leaf aging rather than varying soil water availability in the field. $K_{leaf}$ was as unresponsive to short, sudden drought periods when grown in pots as it was to prolonged drought in the field. The drought treatments imposed in the chamber experiments were substantial enough to decrease both $\Psi_{leaf}$ and $g_s$, supporting the conclusion that $K_{leaf}$ in soybean does not acclimate to protect against cavitation or loss of $g_s$ during drought.

The observed decrease in $K_{leaf}$ as leaves age may be a result of down-regulation or inactivation of aquaporin proteins in living cells or of xylem blockages, such as emboli or tyloses. The lower $\Psi_{leaf}$ in senescing leaves compared with young leaves observed in both field-grown and chamber-grown soybean in this experiment would increase the risk of cavitation in xylem of older leaves (Tyree and Sperry, 1989). There is evidence that repeated cycles of cavitation and refilling over the course of the growing season can weaken xylem pit membranes, making the xylem more vulnerable to cavitation over time (Sperry *et al.*, 1991; Hacke *et al.*, 2001). This vulnerability increase in conjunction with decreasing water potential has also been implicated in the decline of $K_{leaf}$ in *Rhedera trinervis* and *Calycophyllum candidissimum* during leaf senescence (Brodribb and Holbrook, 2003b). By this mechanism, the lower $\Psi_{leaf}$ observed in older leaves could lead to a build up of emboli, which the leaf becomes unable to completely refill overnight as the growing season progresses. An increasing number of emboli may also allow the formation of tyloses, which have been implicated in leaf abscission (Sexton and Roberts, 1982). Careful measurements of aquaporin expression and activity, as well as xylem imaging, could illuminate the mechanism(s) by which $K_{leaf}$ decreases in the long-term. *A* decreased similarly to $K_{leaf}$ as leaves aged. Although lower *A* in older leaves could be attributed simply to lower light intensity within the canopy, reductions in *A* with leaf age are common even at near-saturating irradiance (Vos and Oyarzun, 1987). Light response curves confirmed that photosynthetic capacity was reduced in older soybean leaves (Figure 3). All $K_{leaf}$ measurements were taken at the same near-saturating light intensity, so the observed correlations between $K_{leaf}$ and *A* suggest that the long-term regulation of these parameters is functionally coordinated in soybean. The varying strengths of the correlations between $K_{leaf}$ and *A* as compared to $K_{leaf}$ and $g_s$ indicate that *A* may respond to hydraulic capacity in a manner that is not solely mediated by a stomatal limitation to $CO_2$ intake.

$K_{leaf}$ for the uppermost fully expanded leaves (stage A) was usually lower when the plants were older, as observed in the field-based leaf age experiments and in the DRI-FACE experiment (Figs 2 and 6). However, such consistent patterns were not observed for *A* and $g_s$, leading to variability in the slope of the relationships between $K_{leaf}$ and *A* as well as $K_{leaf}$ and $g_s$ (Figure 3). The differences in *A* for leaves at the same growth stage, but different nodes, appeared to be driven by changes in $g_s$ rather than $K_{leaf}$ (Figure 1D–I). This suggests that *A* is likely not limited by $K_{leaf}$, except possibly during senescence. The 2011 late-season increase in upper-canopy *A* is consistent with reports of whole-plant photosynthesis peaking during the seed filling period, when sink strength is greatest (Wells, 1991), although this pattern was not observed in 2013.

While canopy WUE frequently decreases over multiple growing seasons as tree stands age (Köstner *et al.*, 2002), WUE did not change consistently as leaves age in a single growing season for soybean (data not shown), which is consistent with observations in *Gossypium hirsutum* and *Lepechinia calycina*(Constable and Rawson, 1980; Field and Mooney, 1983). Although $K_{leaf}$ was correlated separately with both *A* and $g_s$ in soybean, decreases in $K_{leaf}$ over the growing season apparently do not function to maintain a balance between water lost and carbon gained. The variability in WUE support the finding that $K_{leaf}$ is sometimes more strongly coordinated with *A* than with $g_s$ in soybean; both suggest that $g_s$ can be regulated in a more transient manner by microenvironment, whereas *A* is more tightly controlled by gradual biochemical acclimation to overall shifts in microenvironment as leaves age.

Variation in the coordination between $K_{leaf}$ and $A$ as compared to $K_{leaf}$ and $g_s$ may further be attributable to the different degrees of leaf uncoupling from the atmosphere experienced by field-grown plants and chamber-grown plants. In the field, older, lower-canopy leaves become greatly uncoupled from the atmosphere after canopy closure, experiencing calmer, moister air in addition to lower light intensity. This greatly decreases transpiration demand in addition to triggering light acclimation of photosynthesis. In growth chambers, however, plants were grown alone in pots, resulting in a much lower effective planting density, so lower canopy leaves were both not as shaded and less uncoupled from the 'atmosphere' than upper leaves.

$\Psi_{\Pi}$ decreased as leaves aged in growth chambers, but not consistently in the field. Although declines in $\Psi_{\Pi}$ over the course of the growing season have been observed in some evergreen tree species and woody understorey species, these decreases were linked to drought conditions (Sobrado, 1986; Ishida *et al.*, 1992). As the chamber-grown plants for which this decline was most pronounced were always well-watered, osmoregulation could be a mechanism for soybean leaves to maintain turgor when $K_{leaf}$ declines in older leaves. This would facilitate continued, if decreased, $A$ in older leaves.

Because $K_{leaf}$ was measured for leaves sampled before sunrise (or before growth chamber lights turned on for the day), any emboli that may have formed in xylem during the previous day had likely refilled overnight (McCully *et al.*, 1998; Yang *et al.*, 2012). Thus, the observed $K_{leaf}$ values represent the maximum $K_{leaf}$ as determined by venation architecture and mesophyll pathways, and any difference in $K_{leaf}$ between treatments in the drought experiments would have been due to acclimation of the leaves to soil moisture conditions rather than transient midday decline in $K_{leaf}$ by refillable embolism. A decrease in maximum $K_{leaf}$ could protect the leaf from daytime $K_{leaf}$ decrease due to embolism (Sadok and Sinclair, 2010b). Because no acclimation was observed, however, soybean likely does not have phenotypic plasticity to respond to soil moisture conditions either by adjusting vein density during leaf development or by aquaporin regulation in mature leaves. This is similar to the lack of $K_{leaf}$ plasticity we have previously observed for soybean in response to growth at elevated [$CO_2$] and temperature (Locke *et al.*, 2013). Because maximum $K_{leaf}$ is the same for plants in both control and RP treatments while soil moisture is decreased, plants in the RP plots are likely more vulnerable to cavitation during transpiration, particularly when vapour pressure deficit is high during the middle of the day. Diurnal cycles of embolism and vessel refilling driven by vapour pressure deficit are thought to occur frequently, and low soil moisture would increase midday tension in the xylem even further, causing more cavitation (Hacke *et al.*, 2001). The inability of $K_{leaf}$ to acclimate to decreasing soil moisture may leave soybean leaves more vulnerable to cavitation during peak midday transpiration demand. This vulnerability could contribute to the observed depression in midday $g_s$ in chamber-grown, water-stressed soybean leaves. The decrease in $g_s$ without maximum $K_{leaf}$ acclimation suggests that stomatal sensitivity to dry soil protects against hydraulic failure in soybean (Brodribb and Holbrook, 2004).

The reduced precipitation treatment left RP plots with rainfall levels equivalent to some of the driest years of the last 60 in the Champaign, IL area. Although the lowest average soil moisture achieved during dry-down periods for chamber-grown soybean was about 30% v/v, which is typically well above the permanent wilting point, there was enough variation in drought treatment pots that some drought treatment plants were already visibly losing leaf turgor. Furthermore, the pots were watered with fertilizer that had a high solute concentration, which likely made root water uptake more difficult for plants even at a soil volumetric water content that would be sufficient in central Illinois soil.

Contrary to previous findings, there was a slight difference in $K_{leaf}$ between ambient- and elevated-[$CO_2$] plants on one measurement day, 76 days after planting (Figure 6), but this effect disappeared when the field data from all three measurement days were analysed as a repeated measures model. At this time in the growing season, there was a slight difference in soil moisture between ambient- and elevated-[$CO_2$] plots that could have contributed to this difference in $K_{leaf}$ (Gray *et al.*, 2013), although this short-lived difference likely had no impact on photosynthesis or water use on timescale of the whole growing season.

The effects of drought on $K_{leaf}$ in a major field-grown crop had not been previously examined, and, taken together, these field and chamber experiments suggest that $K_{leaf}$ in soybean does not acclimate to drought. Because maximum $K_{leaf}$ does not adjust to decreased soil moisture conditions, soybean leaves may be extra vulnerable to cavitation and loss of $K_{leaf}$ during daytime transpiration when grown in drought conditions. Thus, inability of $K_{leaf}$ to acclimate to drought has the potential to limit stomatal conductance and photosynthesis under severe soil moisture deficit.

Studies with other species suggest that hydraulic failure throughout the plant initiates the process of leaf senescence and shedding (Rood *et al.*, 2000; Salleo *et al.*, 2002; Brodribb and Holbrook, 2003b). While it cannot be concluded from these data if $K_{leaf}$ decline in soybean triggers photosynthetic decline and senescence, these results show that hydraulic decline, accompanied by gradual decreases in $A$ and leaf water status, is a part of leaf maturation and senescence in soybean. If $K_{leaf}$ is limiting $A$ in older leaves, then an improvement in hydraulic maintenance could have the potential to increase canopy-level photosynthesis, which is a critical target for crop yield improvement (Zhu *et al.*, 2010).

## SUPPLEMENTARY MATERIAL

**Supplementary Table 1:** Measurement dates and LI-6400 settings for midday gas exchange measurements. Light and temperature for gas exchange measurements were based on ambient weather conditions. $A$, $g_s$ and $\Psi_{leaf}$ were measured at midday on the dates shown, and leaves were sampled before sunrise the following morning for $K_{leaf}$ measurements

| Node | Stage | Date | PAR ($\mu$mol m$^{-2}$ s$^{-1}$) | Block temperature (°C) |
|---|---|---|---|---|
| 3 | A | 8 July 2011 | 1700 | 30.0 |
| 3 | B | 4 August 2011 | 875 | 29.0 |
| 3 | C | 12 September 2011 | 300 | 28.0 |
| 10 | A | 28 July 2011 | 1600 | 31.0 |
| 10 | B | 27 August 2011 | 1500 | 28.0 |
| 10 | C | 14 September 2011 | 300 | 20.0 |
| 3 | A | 7 July 2013 | 2500 | 28.0 |
| 3 | B1 | 25 July 2013 | 500 | 28.0 |
| 3 | B2 | 9 August 2013 | 300 | 30.0 |

| | | | | |
|---|---|---|---|---|
| 3 | C | 4 September 2013 | 300 | 27.0 |
| 10 | A | 31 July 2013 | 1000 | 24.0 |
| 10 | B1 | 9 August 2013 | 500 | 30.0 |
| 10 | B2 | 27 August 2013 | 1000 | 31.0 |
| 10 | C | 13 September 2013 | 1000 | 24.0 |
| 5 | A | 5 February 2013 | 1000 | 25.0 |
| 5 | B | 19 February 2013 | 300 | 25.0 |
| 8 | A | 17 February 2013 | 1000 | 25.0 |
| 8 | B | 2 March 2013 | 1000 | 25.0 |

## ABBREVIATIONS

A: photosynthesis

DRI: Drought by Rain Interception

$g_s$: stomatal conductance

$K_{leaf}$: leaf hydraulic conductance

$\Psi_{leaf}$: leaf water potential

$\Psi_{\Pi}$: leaf osmotic potential

$\Psi_{P}$: leaf hydrostatic potential

PAR: photosynthetically active radiation

RP: reduced precipitation

WUE: sisintrinsic water-use efficiency.

## ACKNOWLEDGEMENTS

We would like to thank Colleen Cahill, Nathan Couch, Pablo Fenton-Ortiz, Sharon Gray, Charlie Mitsdarfer, Kannan Puthuval, David Rosenthal, Becky Slattery and Adrian Zimbelman for assistance with field maintenance and measurements.

## REFERENCES

1. Allen LH Pan D Boote KJ Pickering NB Jones JW . 2003. Carbon dioxide and temperature effects on evapotranspiration and water use efficiency of soybean. Agronomy Journal 95, 1071–1081.
2. Allen LH Valle RR Jones JW Jones PH . 1998. Soybean leaf water potential responses to carbon dioxide and drought. Agronomy Journal 90, 375–383.
3. Aslam M Lowe S Hunt L . 1977. Effect of leaf age on photosynthesis and transpiration of cassava (Manihot esculenta). Canadian Journal of Botany 55, 2288–2295.
4. Barrett DJ Hatton TJ Ash JE Ball MC . 1996. Transpiration by trees from contrasting forest types. Australian Journal of Botany 44, 249–263.
5. Blackman CJ Brodribb TJ Jordan GJ . 2009. Leaf hydraulics and drought stress: response, recovery and survivorship in four woody temperate plant species. Plant, Cell and Environment 32, 1584–1595.
6. Boyer JS . 1982. Plant productivity and environment. Science 218, 443–448.
7. Brodribb TJ Holbrook NM . 2003a. Stomatal closure during leaf dehydration, correlation with other leaf physiological traits. Plant Physiology 132, 2166–2173.
8. Brodribb TJ Holbrook NM . 2003b. Changes in leaf hydraulic conductance during leaf shedding in seasonally dry tropical forest. New Phytologist 158, 295–303.
9. Brodribb TJ Holbrook NM . 2004. Stomatal protection against hydraulic failure: a comparison of coexisting ferns and angiosperms. New Phytologist 162, 663–670.
10. Brodribb TJ Holbrook NM . 2006. Declining hydraulic efficiency as transpiring leaves desiccate: two types of response. Plant, Cell and Environment 29, 2205–2215.
11. Brodribb TJ Holbrook NM Zwieniecki M a Palma B .2005. Leaf hydraulic capacity in ferns, conifers and angiosperms: impacts on photosynthetic maxima. New Phytologist 165, 839–846.
12. Bucci SJ Scholz FG Campanello PI et al . 2012. Hydraulic differences

along the water transport system of South American Nothofagus species: do leaves protect the stem functionality?Tree Physiology 32, 880–893.

13. Bunce JA . 1996. Growth at elevated carbon dioxide concentration reduces hydraulic conductance in alfalfa and soybean. Global Change Biology 2, 155–158.
14. Bunce JA Ziska LH . 1998. Decreased hydraulic conductance in plants at elevated carbon dioxide. Plant, Cell and Environment 21, 121–126.
15. Burke EJ Brown SJ Christidis N . 2006. Modeling the recent evolution of global drought and projections for the twenty-first century with the Hadley Centre climate model. Journal of Hydrometeorology 7, 1113–1125.
16. Burkey KO Wells R . 1991. Response of soybean photosynthesis and chloroplast membrane function to canopy development and mutual shading. Plant Physiology 97, 245–252.
17. Canny M . 1997. Vessel contents during transpiration-embolisms and refilling. American Journal of Botany 84, 1223–1230.
18. Cochard H . 2002. Xylem embolism and drought-induced stomatal closure in maize. Planta 215, 466–471.
19. Cochard H Tyree MT . 1990. Xylem dysfunction in Quercus: vessel sizes, tyloses, cavitation and seasonal changes in embolism. Tree Physiology 6, 393–407.
20. Constable GA Rawson HM . 1980. Effect of leaf position, expansion and age on photosynthesis, transpiration and water use efficiency of cotton. Australian Journal of Plant Physiology 7, 89–100.
21. Domec J-C Palmroth S Ward E Maier CA Thérézien M Oren R . 2009. Acclimation of leaf hydraulic conductance and stomatal conductance of Pinus taeda (loblolly pine) to long-term growth in elevated CO2 (free-air CO2 enrichment) and N-fertilization. Plant, Cell and Environment 32, 1500–1512.
22. Ferrio JP Pou A Florez-Sarasa I Gessler A Kodama N Flexas J Ribas-Carbó M . 2012. The Péclet effect on leaf water enrichment correlates with leaf hydraulic conductance and mesophyll conductance for CO2. Plant, Cell and Environment 35, 611–625.

23. Field C Mooney HA . 1983. Leaf age and seasonal effects on light, water, and nitrogen use efficiency in a California shrub. Oecologia 56, 348–355.

24. Ghorashy SR Colville WL Ashworth DL . 1969. Effects of 2,3,5-triiodobenzoic acid on the morphology and anatomy of Glycine max (L.) Merrill. Crop Science 9, 399–402.

25. Gray SB Strellner RS Puthuval KK Ng C Shulman RE Siebers MH Rogers A Leakey ADB . 2013. Minirhizotron imaging reveals that nodulation of field-grown soybean is enhanced by free-air CO2 enrichment only when combined with drought stress. Functional Plant Biology 40, 137–147.

26. Guyot G Scoffoni C Sack L . 2012. Combined impacts of irradiance and dehydration on leaf hydraulic conductance: insights into vulnerability and stomatal control. Plant, Cell and Environment 35, 857–871.

27. Hacke UG Stiller V Sperry JS Pittermann J McCulloh KA . 2001. Cavitation fatigue. Embolism and refilling cycles can weaken the cavitation resistance of xylem. Plant Physiology 125, 779–786.

28. Hewitt EJ . 1966. Sand and water culture methods used in the study of plant nutrition . London: Commonwealth Agricultural Bureau.

29. Hinckley TM Ritchie GA . 1970. Within-crown patterns of transpiration, water stress, and stomatal activity in Abies amabilis. Forest Science 16, 490–492.

30. Holbrook NM Ahrens ET Burns MJ Zwieniecki MA . 2001. In vivo observation of cavitation and embolism repair using magnetic resonance imaging. Plant Physiology 126, 27–31.

31. Ishida A Yamamura Y Hori Y . 1992. Roles of leaf water potential and soil-to-leaf hydraulic conductance in water use by understorey woody plants. Ecological Research 7, 213–223.

32. Johnson DM McCulloh KA Meinzer FC Woodruff DR Eissenstat DM . 2011. Hydraulic patterns and safety margins, from stem to stomata, in three eastern U.S. tree species. Tree Physiology 31, 659–668.

33. Johnson DM McCulloh KA Woodruff DR Meinzer FC . 2012. Evidence for xylem embolism as a primary factor in dehydration-induced declines in leaf hydraulic conductance. Plant, Cell and

Environment 35, 760–769.

34. Kositsup B Kasemsap P Thanisawanyangkura S Chairungsee N Satakhun D Teerawatanasuk K Ameglio T Thaler P . 2010. Effect of leaf age and position on light-saturated CO2 assimilation rate, photosynthetic capacity, and stomatal conductance in rubber trees. Photosynthetica 48, 67–78.
35. Köstner B Falge E Tenhunen JD . 2002. Age-related effects on leaf area/sapwood area relationships, canopy transpiration and carbon gain of Norway spruce stands (Picea abies) in the Fichtelgebirge, Germany. Tree Physiology 22, 567–574.
36. Kriedmann PE Kliewer WM Harris JM . 1970. Leaf age and photosynthesis in Vitis vinifera L. Vitis 9, 97–104.
37. Leakey ADB Bernacchi CJ Dohleman FG Ort DR Long SP . 2004. Will photosynthesis of maize (Zea mays) in the US Corn Belt increase in future CO2 rich atmospheres? An analysis of diurnal courses of CO2 uptake under free-air concentration enrichment (FACE). Global Change Biology 10, 951–962.
38. Leakey ADB Bernacchi CJ Ort DR Long SP . 2006. Long-term growth of soybean at elevated [CO2] does not cause acclimation of stomatal conductance under fully open-air conditions. Plant, Cell and Environment 29, 1794–1800.
39. Linton MJ Nobel PS . 2001. Hydraulic conductivity, xylem cavitation, and water potential for succulent leaves of Agave deserti and Agave tequilana.International Journal of Plant Sciences 162, 747–754.
40. Locke AM Sack L Bernacchi CJ Ort DR . 2013. Soybean leaf hydraulic conductance does not acclimate to growth at elevated [CO2] or temperature in growth chambers and in the field. Annals of Botany 112, 911–918.
41. Lo Gullo MA Nardini A Trifilo P Salleo S . 2003. Changes in leaf hydraulics and stomatal conductance following drought stress and irrigation in Ceratonia siliqua (Carob tree). Physiologia Plantarum 117, 186–194.
42. Lo Gullo MA Nardini A Trifilò P Salleo S . 2005. Diurnal and seasonal variations in leaf hydraulic conductance in evergreen and deciduous trees. Tree Physiology 25, 505–512.

43. Machado JL Tyree MT . 1994. Patterns of hydraulic architecture and water relations of two tropical canopy trees with contrasting leaf phenologies: Ochroma pyramidale and Pseudobombax septenatum. Tree Physiology 14, 219–240.

44. Maurel C Verdoucq L Luu D-T Santoni V . 2008. Plant aquaporins: membrane channels with multiple integrated functions. Annual Review of Plant Biology 59, 595–624.

45. McCully ME Huang CX Ling LEC . 1998. Daily embolism and refilling of xylem vessels in the roots of field-grown maize. New Phytologist , 327–342.

46. Meehl GA Stocker TF Collins WD et al . 2007. Global climate projections. In: Solomon SQuin DManning MChen ZMarquis MAveryt KBTignor MMiller HL, eds. Climate Change 2007: The physical science basis. contribution of Working Group I to the Fourth Assessment Report of the Intergovernmental Panel on Climate Change . Cambridge: Cambridge University Press.

47. Meinzer FC . 2002. Co-ordination of vapour and liquid phase water transport properties in plants. Plant, Cell and Environment 25, 265–274.

48. Nardini A Salleo S . 2000. Limitation of stomatal conductance by hydraulic traits: sensing or preventing xylem cavitation?Trees 15, 14–24.

49. Nardini A Tyree MT Salleo S . 2001. Xylem cavitation in the leaf of Prunus laurocerasus and its impact on leaf hydraulics. Plant Physiology 125, 1700–1709.

50. Reich PB Schoettle AW Robert G . 1985. Effects of low coecentrations of O3, leaf age and water stress on leaf diffusive conductance and water use efficiency in soybean. Physiologia Plantarum 63, 58–65.

51. Rood SB Patiño S Coombs K Tyree MT . 2000. Branch sacrifice: cavitation-associated drought adaptation of riparian cottonwoods. Trees 14, 0248–0257.

52. Sack L Holbrook NM . 2006. Leaf hydraulics. Annual Review of Plant Biology 57, 361–381.

53. Sack L Melcher PJ Zwieniecki MA Holbrook NM . 2002. The hydraulic conductance of the angiosperm leaf lamina: a comparison of three measurement methods. Journal of Experimental Botany 53,

2177–2184.

54. Sadok W Sinclair TR . 2010a. Genetic variability of transpiration response of soybean [Glycine max (L.) Merr.] shoots to leaf hydraulic conductance inhibitor AgNO3. Crop Science 50, 1423–1430.
55. Sadok W Sinclair TR . 2010b. Transpiration response of "slow-wilting" and commercial soybean (Glycine max (L.) Merr.) genotypes to three aquaporin inhibitors. Journal of Experimental Botany 61, 821–829.
56. Salleo S Nardini A Lo Gullo MA . 2002. Changes in stem and leaf hydraulics preceding leaf shedding in Castanea sativa L. Biologia Plantarum 45, 227–234.
57. Salleo S Nardini A Pitt F Lo Gullo MA . 2000. Xylem cavitation and hydraulic control of stomatal conductance in laurel (Laurus nobilis L.). Plant, Cell and Environment 23, 71–79.
58. Scoffoni C Sack L . 2014. Are leaves "freewheelin'"? Testing for a Wheeler-type effect in leaf xylem hydraulic decline. Plant, Cell and Environment , 1–30.
59. Sexton R Roberts JA . 1982. Cell biology of abscission. Annual Review of Plant Physiology 33, 133–162.
60. Silva FC Shvaleva A Maroco JP Almeida M Chaves MM Pereira JS . 2004. Responses to water stress in two Eucalyptus globulus clones differing in drought tolerance. Tree Physiology 24, 1165–1172.
61. Sobrado MA . 1986. Aspects of tissue water relations and seasonal changes of leaf water potential components of evergreen and deciduous species coexisting in tropical dry forests. Oecologia 68, 413–416.
62. Sobrado MA . 1994. Leaf age effects on photosynthetic rate, transpiration rate and nitrogen content in a tropical dry forest. Physiologia Plantarum 90, 210–215.
63. Sperry JS Perry AH Sullivan JEM . 1991. Pit membrane degradation and air-embolism formation in ageing xylem vessels of Populus tremuloides Michx. Journal of Experimental Botany 42, 1399–1406.
64. Tyree MT Sperry JS . 1989. Vulnerability of xylem to cavitation and embolism. Annual Review of Plant Physiology and Plant Molecular Biology 40, 19–36.

65. Vos J Oyarzun P . 1987. Photosynthesis and stomatal conductance of potato leaves - effects of leaf age, irradiance, and leaf water potential. Photosynthesis Research 11, 253–264.
66. Wells R . 1991. Soybean growth response to plant density: relationships among canopy photosynthesis, leaf area, and light interception. Crop Science 31, 755–761.
67. Wheeler JK Huggett B a Tofte AN Rockwell FE Holbrook NM . 2013. Cutting xylem under tension or supersaturated with gas can generate PLC and the appearance of rapid recovery from embolism. Plant, Cell and Environment 36, 1938–1949.
68. Woodward RG Rawson HM . 1976. Photosynthesis and transpiration in dicotyledonous plants. II. Expanding and senescing leaves. Australian Journal of Plant Physiology 3, 257–267.
69. Yang S Tyree MT . 1993. Hydraulic resistance in Acer saccharum shoots and its influence on leaf water potential and transpiration. Tree Physiology 12, 231–242.
70. Yang S-J Zhang Y-J Sun M Goldstein G Cao K-F . 2012. Recovery of diurnal depression of leaf hydraulic conductance in a subtropical woody bamboo species: embolism refilling by nocturnal root pressure. Tree Physiology 32, 414–422.
71. Zhu X-G Long SP Ort DR . 2010. Improving photosynthetic efficiency for greater yield. Annual Review of Plant Biology 61, 235–261.

# CITATION

## CHAPTER 2

Dong X. How to put plant root uptake into a soil water flow model [version 1; referees: 2 approved, 1 approved with reservations]. *F1000Research* 2016, 5:43 (doi: 10.12688/f1000research.7686.1).

## CHAPTER 3

Basu S, Ramegowda V, Kumar A and Pereira A. Plant adaptation to drought stress [version 1; referees: 3 approved]. *F1000Research* 2016, 5(F1000 Faculty Rev):1554 (doi: 10.12688/f1000research.7678.1).

## CHAPTER 4

Mei T, Fang D, Röll A, Niu F, Hendrayanto and Hölscher D (2016) Water Use Patterns of Four Tropical Bamboo Species Assessed with Sap Flux Measurements. *Front. Plant Sci.* 6:1202. doi: 10.3389/fpls.2015.01202.

## CHAPTER 5

Garrido M, Silva P and Acevedo E (2016) Water Relations and Foliar Isotopic Composition of *Prosopis tamarugo* Phil., an Endemic Tree of the Atacama Desert Growing at Three Levels of Water Table Depth. *Front. Plant Sci.* 7:375. doi: 10.3389/fpls.2016.00375.

## CHAPTER 6

Cocozza C, de Miguel M, Pšidová E, Ditmarová L, Marino S, Maiuro L, Alvino A, Czajkowski T, Bolte A and Tognetti R (2016) Variation in Ecophysiological Traits and Drought Tolerance of Beech (*Fagus sylvatica* L.) Seedlings from Different Populations. *Front. Plant Sci.* 7:886. doi: 10.3389/fpls.2016.00886.

## CHAPTER 7

Singh D and Laxmi A (2015) Transcriptional regulation of drought response: a tortuous network of transcriptional factors. *Front. Plant Sci.* 6:895. doi: 10.3389/fpls.2015.00895.

## CHAPTER 8

Krannich, C.T.; Maletzki, L.; Kurowsky, C.; Horn, R. Network Candidate Genes in Breeding for Drought Tolerant Crops. *Int. J. Mol. Sci.* 2015, *16*, 16378-16400. doi:10.3390/ijms160716378.

## CHAPTER 9

Cieśla A, Mituła F, Misztal L, Fedorowicz-Strońska O, Janicka S, Tajdel-Zielińska M, Marczak M, Janicki M, Ludwików A and Sadowski J (2016) A Role for Barley Calcium-Dependent Protein Kinase CPK2a in the Response to Drought. *Front. Plant Sci.* 7:1550. doi: 10.3389/fpls.2016.01550.

## CHAPTER 10

Juan D. Franco-Navarro, Javier Brumós, Miguel A. Rosales, Paloma Cubero-Font Manuel Talón, José M. Colmenero-Flores, "Chloride regulates leaf cell size and water relations in tobacco plants," *Journal of Experimental Botany*, Volume 67, Issue 3, 1 February 2016, Pages 873–891,https://doi.org/10.1093/jxb/erv502.

## CHAPTER 11

Jarzyniak KM and Jasiński M (2014) Membrane transporters and drought resistance – a complex issue. *Front. Plant Sci.* 5:687. doi: 10.3389/fpls.2014.00687.

## CHAPTER 12

Hepworth C, Turner C, Landim MG, Cameron D, Gray JE (2016) Balancing Water Uptake and Loss through the Coordinated Regulation of Stomatal and Root Development. PLoS ONE 11(6): e0156930. https://doi.org/10.1371/journal.pone.0156930.

## CHAPTER 13

Anna M. Locke and Donald R. Ort, "Leaf hydraulic conductance declines in coordination with photosynthesis, transpiration and leaf water status as soybean leaves age regardless of soil moisture," Journal of Experimental Botany, Volume 65, Issue 22, 1 December 2014, Pages 6617–6627,https://doi.org/10.1093/jxb/eru380.

# INDEX